C. Truesdell

Essays in the History of Mechanics

With 126 Figures

Springer-Verlag
Berlin Heidelberg New York 1968

ISBN 978-3-642-86649-4 ISBN 978-3-642-86647-0 (eBook)
DOI 10.1007/ 978-3-642-86647-0

Softcover reprint of the hardcover 1st edition 1968

CARAE · UXORI

CARLOTTAE

QUAM · SOLAM · INTER · MULIERES · SAECULI · SUI

DEAE · TRES

QUAE · OLIM · NUDAE · IDA · PARIDEM · MALUM · POPOSCERUNT

DOTIBUS · SUIS · OMNIBUS · PRAETER · UNAM

LIBERALITER · ET · AEQUALITER

ORNAVERUNT

HUNC · LIBELLUM · INSTAR · DONI

DEDICAT

C. T.

Foreword

This volume collects my shorter articles on the history of mechanics, some already published in various places, some revised from earlier papers, and some never published before. All of them began as lectures, and here they are printed as such, little changed from the last times I read them out to an audience. While the several articles concern different aspects of mechanics, overlap and even some repetition could not be avoided, since mechanics is one great science, and the same original oftentimes served more than one end in its growth.

My three major historical treatises, which were published in Volumes (II) $\mathbf{11}_2$, **12**, and **13** of *L. Euleri Opera Omnia*, are not included. To simplify the printing I have also mostly omitted detailed reference to sources discussed more fully in those treatises, but of course I have added to the texts of the lectures citations of other sources, some notes in answer to questions a reader might ask, and bibliographical notes at the end of each.

I am grateful to the U.S. National Science Foundation for its support of this work through a grant to The Johns Hopkins University.

If these lectures find any favor with professional historians of science, I shall be humbly thankful for their toleration of a book not intended for them. However eager to tell us how scientists of the seventeenth century used their inheritance from the sixteenth, these scholars seem to regard as irrelevant anything a scientist today might think about any aspect of science, including his own debt to the past or reaction against it. Such historians remind me of those taxonomers, perhaps of only fabulous existence, who cannot recognize a particular plant unless they see a sprig of it dead, dried, and pasted to a sheet of paper. For me, mathematical science is alive today, alive not only in its freshest leaves but also in its branches that reach down to the past. I know young men who have read the words of Gibbs and Kelvin and Stokes and Cauchy, even of Euler and Newton, neither so as to decorate a paper of their own by an early reference nor to write a history, but in search of understanding and method, revealed by the speech of giants untranslated by pygmies. For such men, such scientists of our own day, these lectures were composed and are here printed.

The revolutions of world war and socialism and quantum mechanics, however right and necessary and fruitful, have clouded the massive solidity, the serene confidence of classical mechanics. Classical mechanics has weathered through, standing fast behind the smoky, putrid mists. These lectures tell how some parts of it were founded. If the reader may form here his own explanation of how classical mechanics gained solidity so massive and confidence so serene that it not only survived but even is now flourishing anew, my aim will be reached. I will give him a hint. While "imagination, fancy, and invention" are the soul of mathematical research, in mathematics there has never yet been a revolution.

C.T.

Contents

Essays in the History of Mechanics

I. The Mechanics of Leonardo da Vinci

1. Proem

With the rise of the experimental sciences, the objectives, methods, and results of which are quickly grasped by anyone and easily explained in the newspapers, the rational sciences are pushed aside and forgotten even by the learned, until today in any intellectual community it is common to hear from philosopher and scientist definitions of science and knowledge which, if rigidly applied, would keep mathematics outcast from their honored worlds. Yet the technology which today we

---

Fig. 1. Self-portrait of Leonardo da Vinci (by courtesy of the National Library of Turin)

watch, spellbound, as it grinds to destruction the human race, whether in fine grains or coarse, is not experimental. Neither is it rational. Rather, as grim and inexorable as the law of the beehive, technology multiplies myriadfold the results of experiments by blind application of the results of mathematics. The giant computer of today is only the natural extension of the countless drawing boards and slide rules that built bridges, railroads, and power stations one and two generations ago. Other civilizations have stretched strong, smooth roads to nowhere, have built massive slums, and have contrived machines to smash, puncture, or burn human beings in multitudes, but only ours has done so by mathematics. If by the "experimental method" we mean appeal to knowledge gained from practical trial rather than from authority, we have no cause to boast, for such knowledge has been the common property of artisans and inventors in every age. It is only in its application of mathematics to natural science, its replacement of craft lore by calculation, that the technology of the western culture differs in principle from others'.

The main problem in the history of science and technology is to understand this singular union, to find its beginning and to trace its growth. Now that history of science has become a recognized university discipline, supported by chairs, meetings, journals, societies, boards, committees, and subsidy by the taxpayers — in a word, professionalized — this problem has been more obscured than solved, since most who promote the independence of this new field are unable to solve a calculus problem, let alone follow a paper on hydrodynamics a hundred years old. Science, accordingly, is cut down to the size of its professed historians, and a competent scientist who dares raise his voice on the history of science is dismissed as an intruding amateur.

The scientist who approaches LEONARDO DA VINCI in terms of reason draws upon himself indignation from all sides. He must face a hydra: the common false ideas of what science is, the common puerilities on the history of science pattered *ad infinitum* in textbooks and histories and hence believed as unshakable gospel by scientists, and, worst, the common boundless hero-worship of LEONARDO as a universal genius of superlative though otherwise vague achievements. Even in the moonlit cosmos of art criticism, where precision and reason are not drugs on the market, LEONARDO is the subject of a style of raving of order higher than common if not unique. While the art world is not given to the praising of scientists whom it does not consider to have been primarily artists, it does not hesitate to list among the unparalleled achievements of LEONARDO a number of startling inventions such as the submarine, the helicopter, and contact lenses. The one scientist whom the common man can name, though he may well never have heard of NEWTON and LEIBNIZ, is LEONARDO. An advertisement for a current volume adulating LEONARDO quotes RUBE GOLDBERG, the cartoonist of ridiculous gadgets:

"As an inventor of sorts I am humbled by this great man's accomplishments. He predicted everything that is happening today.... I shall always treasure this new book and refer to it when I need a new gimmick."

Not only the public press but also the précis of the paedagogues has taught us to gulp superlatives as daily narcotics. The professor who forgets to class, rate, and rank his subject as the first and finest something will fail to find it mentioned in the students' terminal examinations. The busy modern calls for culture in predigested quintessence pills, packaged in abridged paperbacks, explained by folksy prefaces in pellet-paragraphs of sugared baby-talk, lullabies to smugness. Suggestion to a general audience that historical facts must precede if not replace historical enthusiasms may expect only oblivion, the palm of dullness.

In today's world of cozy pat phrases, LEONARDO remains a habitual magnet for superlatives. He entered every realm of artistry or intellect. From a few pictures, especially two or three long favorite to the journalists of art, has swelled a legend of what he might have done. The publication of his notebooks in the late Victorian era stirred the world of art and drew toward his science and his philosophy the interest of a large public which to other scientists and other philosophers was and is indifferent. Remarkable as are LEONARDO's sketches and designs of mechanisms, still more remarkable, naturally, must they appear to a scientist who, trained to patronize the pre-quantum man as a babbling baby, has never before seen any mechanism from the fifteenth century, or to an artist who, trained to look upon mechanisms of his own day as instruments of black magic, has never thought to ask what makes the hands of a fifteenth-century clock turn round today at the very rate its maker designed them to turn.

2. Leonardo and Early Engineering

Consider, for example, a page (Fig. 2) showing five water pumps and scoops, machines not only elaborate but practicable. Do we see here the genius of LEONARDO, forecasting the technology of later centuries? Scarcely. First, it is not by

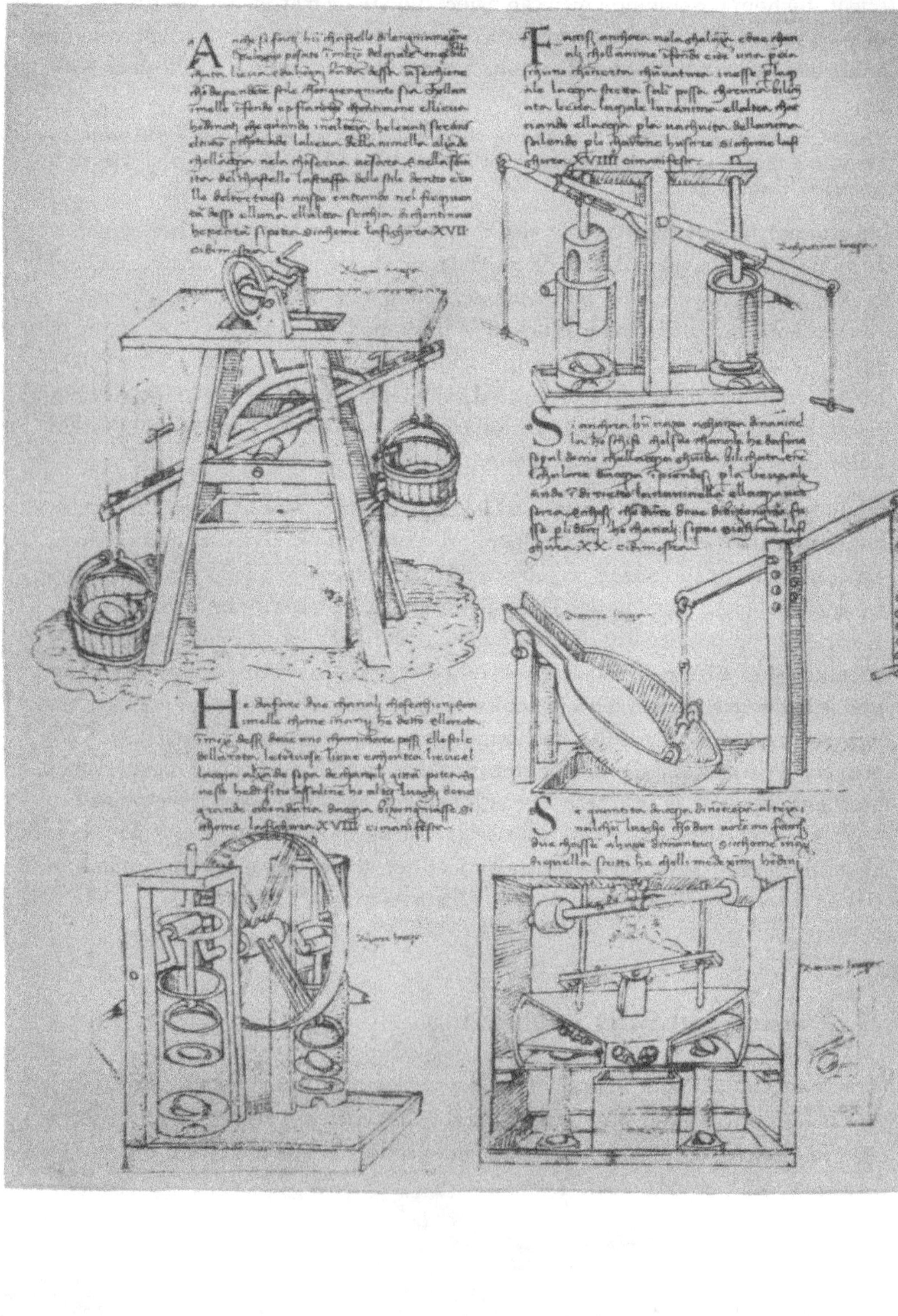

Leonardo. Its author is Francesco di Giorgio Martini (1439—1501). It comes from a work that touches Leonardo in more ways than one, Martini's *Trattato di Architectura*[1], which was written about 1475, when Leonardo was in his early twenties; in 1490 Leonardo and Martini went together to Pavia to advise on the construction of the cathedral; an incomplete copy[2] of Martini's treatise bears marginal notes and sketches in Leonardo's hand; and some of the machines and devices formerly attributed to Leonardo are to be seen in Martini's book.

The existence of *one* earlier collection of this kind is enough to overturn the fable of Leonardo the lonely creator. As the number and variety of machines he drew suggests, Leonardo is now seen as illustrating the common or extraordinary technology of his day and projecting improvements upon it. That little evidence of such technology remains, is not hard to explain. First, the practice of machinery was a mechanic rather than liberal art; we cannot stretch backward the slant of the nineteenth century and expect philosophers and mathematicians of the fifteenth to have thought water pumps and blacksmiths' forges of more inherent interest than, say, the effects of prayer or sidereal configurations upon human fate. All these things belonged to the common experience of man; there is no reason to think the learned of that day were unaware of one any more than of the others, but they had not been taught in school that pumps and forges can be explained by human reason, while religion and astrology lie outside its grasp. Mechanisms belonged to mechanics, and it is in the interest of the mechanic, even if he is literate, to guard his bag of tricks by secrecy. Like Leonardo, Martini was an uncommon mechanic in that he was also a fine painter, sculptor, and architect. A good artist's drawing of an existing machine carries the same forthright veracity as his drawing of an existing man, giving us no reason to presume that he invented either.

---

[1] L. Reti, "Francesco di Giorgio Martini's treatise on engineering and its plagiarists," *Technology and Culture* **4**, 287—298 (1963).

[2] Cod. Med. Laur. 361 in the Laurentian Library, Florence.

Fig. 2. Page from Francesco di Giorgio Martini's *Trattato di Architectura*, c. 1475. Cod. 148 Bibl. Naz. Torino (ex Duca di Genova), *f.* 47 *v.* (by courtesy of the National Library of Turin)

While, like LEONARDO, MARTINI complains that his originality is not enough appreciated, he makes it clear that he does not lay claim to invention of everything he draws:

"Therefore, to conclude my work, in this last part I shall present examples from among the most common, potent, and useful of each of the said devices in spite of having several times decided not to disclose the nature of some of my machines and instruments. Such knowledge of them as I have has been acquired with great toil and at the sacrifice of my means of livelihood, so I am reluctant to show them forth to all, for once an invention is made known, not much of a secret is left. But even this would be a lesser evil if a greater did not follow. The worst is that ignoramuses adorn themselves with the labors of others and usurp the glory of an invention that is not theirs. For this reason the effort of one who has true knowledge is oft retarded. If in all epochs this vice hath abounded, in our own it is more widespread than in any other."

Too little is known about the technology of the late middle ages to justify any positive claim for or against LEONARDO's originality as an inventor[3], except in the case of devices that could not work now and hence could not have been seen working then. We can make only the guarded statement that for several sound machines, LEONARDO's drawing is the earliest now known. We cannot decide whether he was inventing or modifying or merely recording what he saw with the same detail and precision that he recorded the outer aspect or dissected structure of the human body[3a].

---

[3] The old standard reference, now quite out of date, is F. M. FELDHAUS, *Leonardo als Techniker und Erfinder*, Jena, 1922.

Cf. also the idolizing Chapters IV—VI of W. PARSONS, *Engineers and Engineering in the Renaissance*, Baltimore, 1939.

[3a] *Added in proof.* The paragraph above was written about ten years ago. I am reassured to find that the evaluation there stated has been confirmed by the leading modern expert on early mechanical technology, Dr. LADISLAO RETI, who concludes as follows the first competent article to appear on the manuscripts newly rediscovered in Madrid (L. RETI, "The Leonardo da Vinci Codices in the Biblioteca Nacional of Madrid," *Technology and Culture* **8**, 437—445 (1967)):

"This leads to the much discussed and abused subject of LEONARDO's 'inventions.' I think that this problem has not yet received fair and objective treatment. Many early writers accepted all the ingenious mechanical innovations found in LEONARDO's manuscripts as original inventions of LEONARDO. Their emphatic claims often suffer from lack of general information in the field of the history of technology, particularly on the work of the engineers who preceded LEONARDO. But there are others who deny originality to LEONARDO, affirming that almost all of his mechanical ideas were conceived by his predecessors. Very likely, the truth lies somewhere in the middle.

"LEONARDO took note of all interesting mechanical contrivances he saw or heard about during his frequent contacts with fellow artists, scholars, and artisans. Such notes and drawings seldom indicate the source. But LEONARDO did not publish his notes; they were meant for his

3. Did Leonardo Experiment?

LEONARDO's notebooks were orderless, their entries all fragments, while later vicissitudes have brought them nearly into chaos. Not only did LEONARDO write in them his thoughts and projects, but also observations of things he saw, notes from conversations which he wished to record or did not understand, and even Latin conjugations[4], the easiest propositions of EUCLID, exercises in multiplication, and current commonplaces of mechanics and philosophy, which he strove to learn and remember, it seems with little success. He who searches for anything can find it there. The student must beware the mottoes with which LEONARDO was given to exhorting himself. Often quoted to prove his independence from tradition is "If thou art alone, thou belongest all to thyself," but the context indicates only the scholar's personal solitude, and many another lapidary phrase for which no context remains may be equally misconstrued. To LEONARDO's own contradictions and vagueness must be added the errors of the translators, usually from literary or artistic circles and unlikely to recognize what little precision there is. A violent eloquence was LEONARDO's style, but an eloquence blunted by misty generalities fit for a prophet: The concrete, the definite of poet or scientist he rarely achieved in words. He is vague, and his translators usually make him vaguer, but the few phrases where he says something specific they may be trusted to reduce to entire nonsense[5]. Where his words failed, his eye and hand recorded with passionate exactness, so that from the drawings, rather than the

---

own files, and it is now the task of historians to document those derivations. For the time being, we have to recognize an undeniable fact: examining even the incomplete and mutilated legacy of LEONARDO, we are faced with an imposing number of technological ideas for which there are no precedents.

"The Madrid manuscripts, with their rich technological content, will cause a reassessment of the problem. If we are ready to face the facts without passion and prejudice, LEONARDO DA VINCI's drawings and notes may become the most important collection of documents in the study of medieval and Renaissance technology."

[4] The reader who, reared in the popular Leonardian adulation, is shocked by my words, might do well to consult A. MARINONI's essay, "Da Vinci's philology," pp. 215—226 of *Leonardo da Vinci*, New York, Reynal and Co., n.d. [1963], written c. 1938. As MARINONI remarks, "It has not been realized how much harm has been done by the mania for discovering anticipatory divinations in everything LEONARDO wrote... ."

[5] *E.g.* MS I$^{(2)}$ 115 [67] *v.*,
"Sellacqua fussi quantita disscreta comella e chontinua... ."
EDWARD MACCURDY, in *The Notebooks of Leonardo da Vinci*, New York, Reynal and Hitchcock,

words of LEONARDO, we learn his grasp of mechanisms and nature. Thus the usual editions, which give only the text with some few of the more artistic sketches adjoined, furnish a doubly false guide.

LEONARDO dreamt of great things. His pages contain several outlines of books, not the outlines of a creator who will some day write his masterpiece, but mere lists of questions. To most, he gives no answer. Such answers as he does give he usually justifies by "pruovasi", which means "to try" or "to be tried" or "to show" or "to be shown", and often the following text is purely theoretical. Sometimes the reason given is "sperienza", which usually means "experience" but is usually translated "experiment". He tells us he believes in asking questions directly of nature herself, but did she ever answer him? Of the hundreds of things he plans to try, or tells us to try, he never once records having tried a single one; from the hundreds of "experiments" he projects, he never once records a datum obtained by measurement[6].

However, LEONARDO always defies summary: To every conclusion one may make about him, there seems to be at least one exception. In the many hundreds of

---

1939 (several times reprinted), translates this passage as follows (p. 725):
"If the water was a quantity endowed with sense as it is a continuous one... ."
The meaning is,
"If water were a discrete quantity instead of a continuous one... ."

[6] I believe my statement is strictly correct, but it may have unfair connotations. LEONARDO may have known the difference between experience and experiment. In at least one case, he describes a piece of apparatus projected especially for experiment, namely, a canal with one side of glass so that the flow within it may be observed. This passage is quoted on p. 67 below. So far as I know, however, he nowhere records a table of measured numbers.

In addition to the commoner word "sperienza", LEONARDO once in a while uses "sperimentatione" (MS I$^{(2)}$ 115 [67] *r.*) and "isperimentare" (Atl. 126 *v.* a). It is not clear whether he intends a distinction. BATTISTI & ALESSIO in their *Dizionario Etimologico Italiano* cite GUITTONE D'AREZZO and JACOPONE DA TODI for "spermento" and date "esperienza" as only from the fourteenth century; since they do not present supporting quotations, we cannot be sure the divergence of the words is more than literary. The *Grande Dizionario della Lingua Italiana* has not yet reached the letter E.

Both "experientia" and "experimentum" are classical Latin words, the former being the older, but both seem to have meant at first "trial, proof, experiment", while by post-Augustan authors both are sometimes used to mean "experience". Thus the context, not the word, has been essential for the distinction from the very beginning until times much later than LEONARDO's.

Finally, while "experiment" in the sense that word is used in modern science seems to have been extremely rare in classical and mediaeval times, it did exist, so no-one can claim LEONARDO invented it. Certainly some of the passages in HERON's *Mechanics* describe crude experiments.

pages he wrote about mechanics, I have found one statement, and one only, that can scarcely have come from any other source but experiment, namely:

"Every body resists in its friction with a power equal to the fourth of its heaviness if the motion is plane [or slow?] and the surfaces dense and polished [or clean?]." (Atl. 72*v.* b)

In modern terms, LEONARDO here asserts the law usually attributed to AMONTONS (1699), that the frictional resistance of a sliding body is proportional to the normal force, and moreover LEONARDO assigns to the coefficient of friction the value $^1/_4$. While LEONARDO does not describe any tests, on a different page (Arund. 41*r.*) he draws a figure which seems to project one (Fig. 3). The general law and the drawing by themselves are typical, as we shall see, of LEONARDO's studies of mechanics: conjecture of a simple proportion, and project of an apparatus for testing it. The specific number $^1/_4$ — not a bad value for many materials — is not at all typical, nor is it an accident, for LEONARDO repeats it in at least five places (Atl. 81*r.* b, 81*v.* b, 112*r.* b, 197*r.* b, 209*v.* a). He does not tell us where he got it, but such a number can scarcely come from anything but an experiment. While it would be unfair, therefore, to assert that he never measured anything, it would be still more unjustified, on the basis of what he wrote, to claim that measurement was anything but exceptional in his approach to nature. Thus when he tries to take account of the variation of friction according as the surfaces in contact are rough or smooth, he relapses into his usual vague remarks (Atl. 198*v.* a, 209*v.* a) (Fig. 4).

Fig. 3 (pp. 10 and 11). Atl. 198*r.*a and *v.*a, showing some of LEONARDO's researches on sliding friction (by courtesy of the Biblioteca Ambrosiana, Milano)

Fig. 4 (p. 12). Arund. 41*r*, showing further researches on sliding friction (by courtesy of the Trustees of the British Museum)

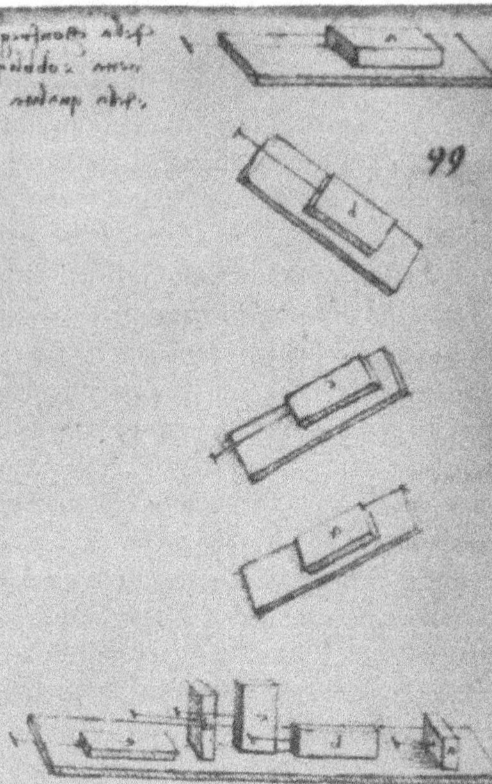

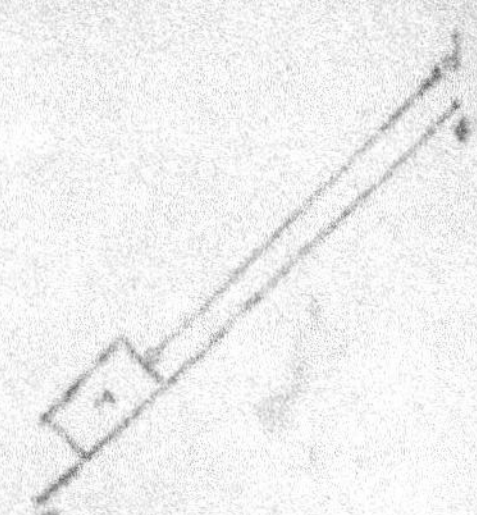

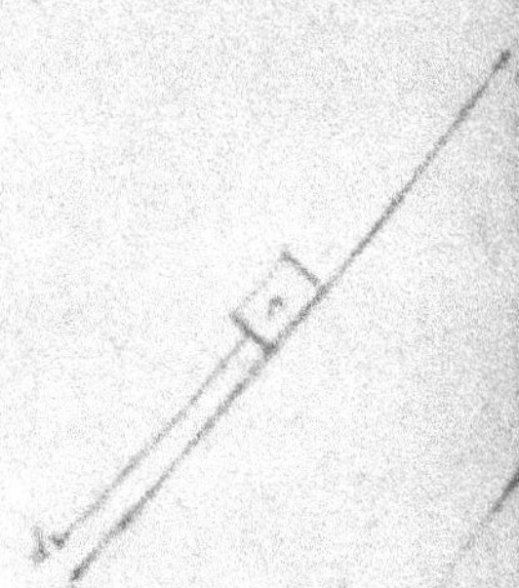

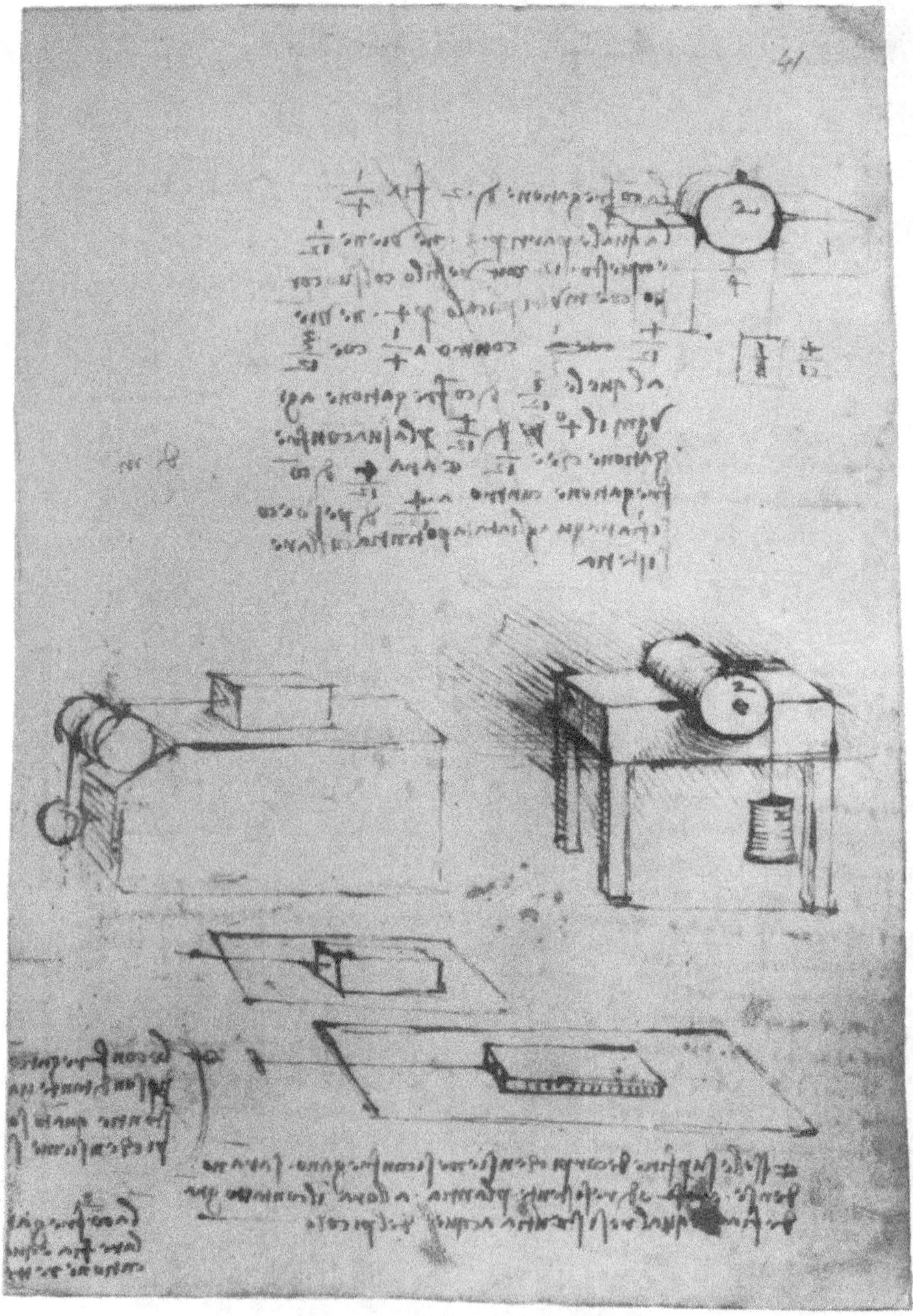

4. Leonardo on Beams

Of the many perplexities which entangle him who would assess rationally LEONARDO's work in mechanics, the most arise from the pages on structural elements: beams, pillars, and arches. Here the drawings are most impressive (*e.g.* Figs. 5—8). Should we encounter such sketches as these in the notebook of a scientist of the baroque period, we could be sure that they illustrated an account of experimental measurements or a presentation of mathematical theory. LEONARDO's texts do not bear out any such expectation. He lists question after question; few are the answers, and they are invariably speculative, adorned with hypothetical round numbers in simple or geometric proportion like 100 and 20 and 4 or $^1/_2$ and $^1/_4$ and $^1/_8$. To the reader of modern "humanistic" origins, the presence of any numbers whatever seems evidence of the dry, "mathematical" exactitude he associates with science and measurement; to the scientist, such sequences as these show positively that, at best, if any experimental data were found, they have been altered to fit some preconjectured rule, and in the absence of records he will simply conclude that no tests were made. The historian's trouble here lies not in discovering what LEONARDO did, but rather in our almost total ignorance of the mediaeval and earlier Renaissance views on materials and members. We cannot be so naive as to think that the builders had *no* rules for the strengths of beams: The men who designed and built Chartres and the Kremlin and St. Peter's cannot have merely copied trade practice, since for them no sufficient precedents existed. LEONARDO saw that understanding of the elements of which structures are composed — beams, pillars, arches, vaults — is the proper basis for the science of construction, but this rather simple thought may well have been common in his day and earlier. Indeed, there is no reason to suppose architecture ever was exempt from the standard human pattern of building up a complex skill by practice gained from mastery of component tasks.

Fig. 5 (p. 14). Atl. 152*r*.b, with hypothetical loadings of a beam which does not deform (by courtesy of the Biblioteca Ambrosiana, Milano)

Fig. 6 (p. 15). Atl. 332*r*.b, showing LEONARDO's studies of the bending of a beam supported at both ends (by courtesy of the Biblioteca Ambrosiana, Milano)

Fig. 7 (p. 16). Atl. 211*r*.b, showing LEONARDO's projects for the bending of beams (by courtesy of the Biblioteca Ambrosiana, Milano)

Fig. 8 (p. 17). MS A 45*v*., with LEONARDO's sketch of a buckled beam (by courtesy of the Institut de France)

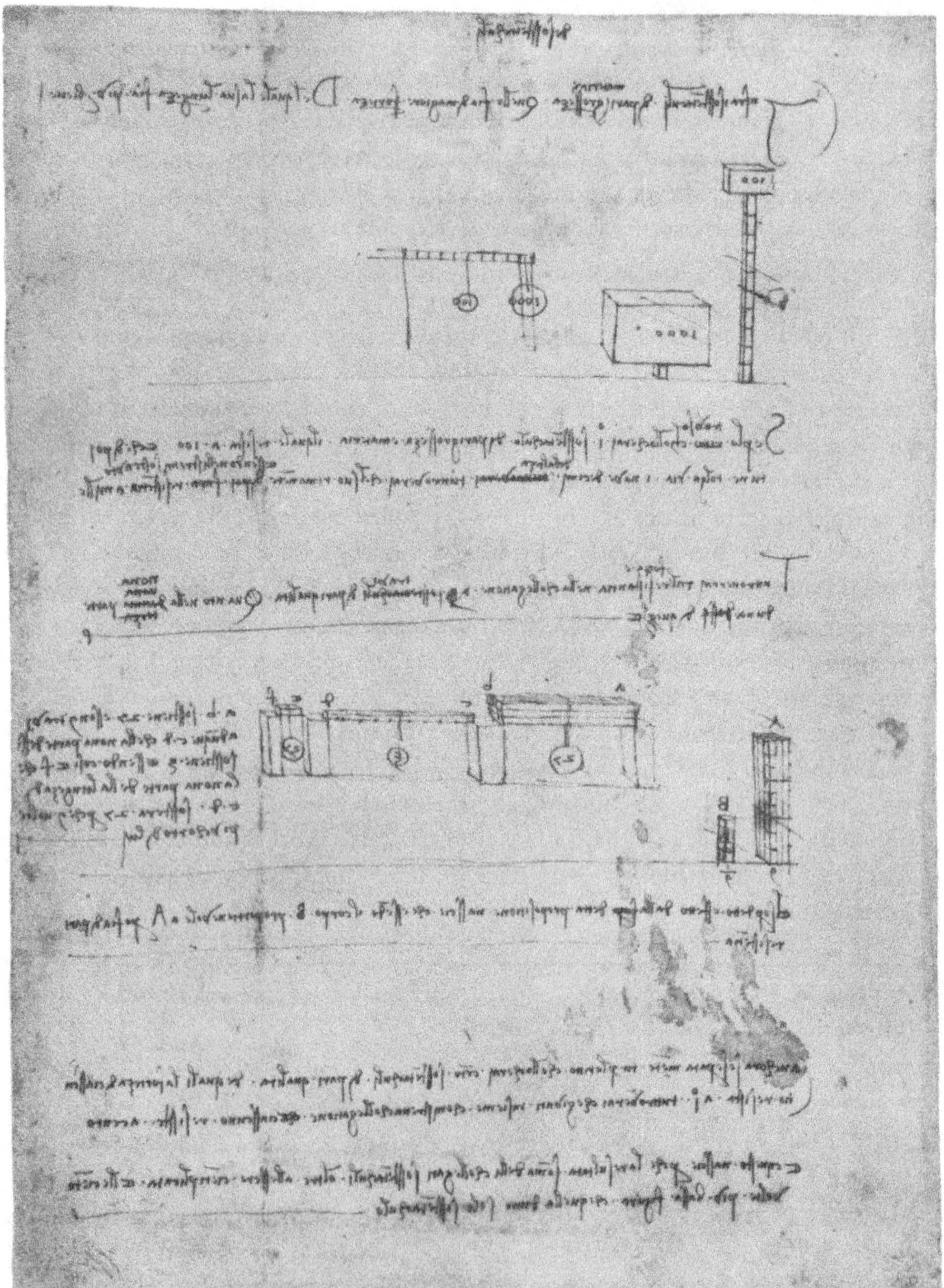

29

[illegible]

[illegible]

[illegible]

[illegible]

[illegible]

[illegible]

[illegible]

[illegible]

[illegible]

[illegible]

[illegible]

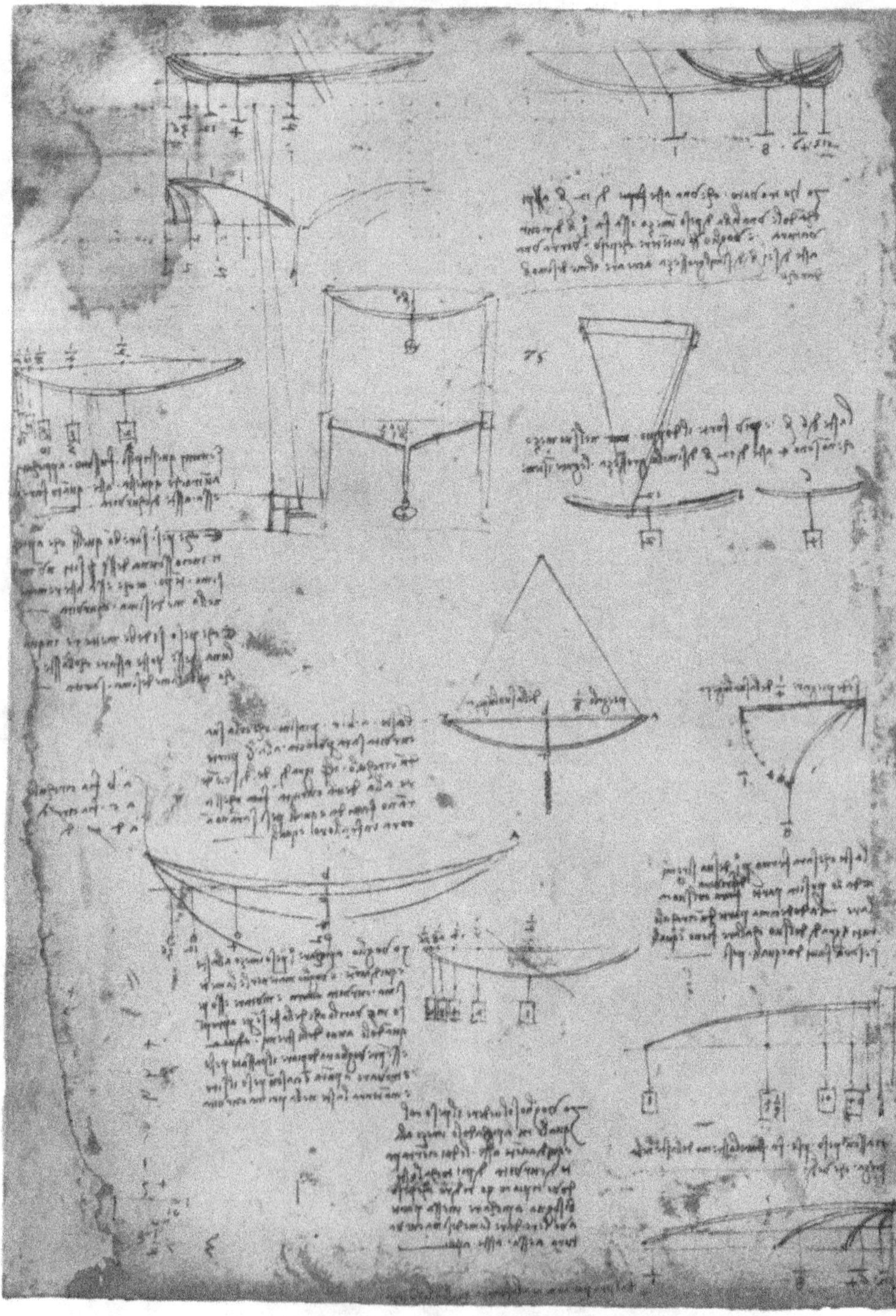

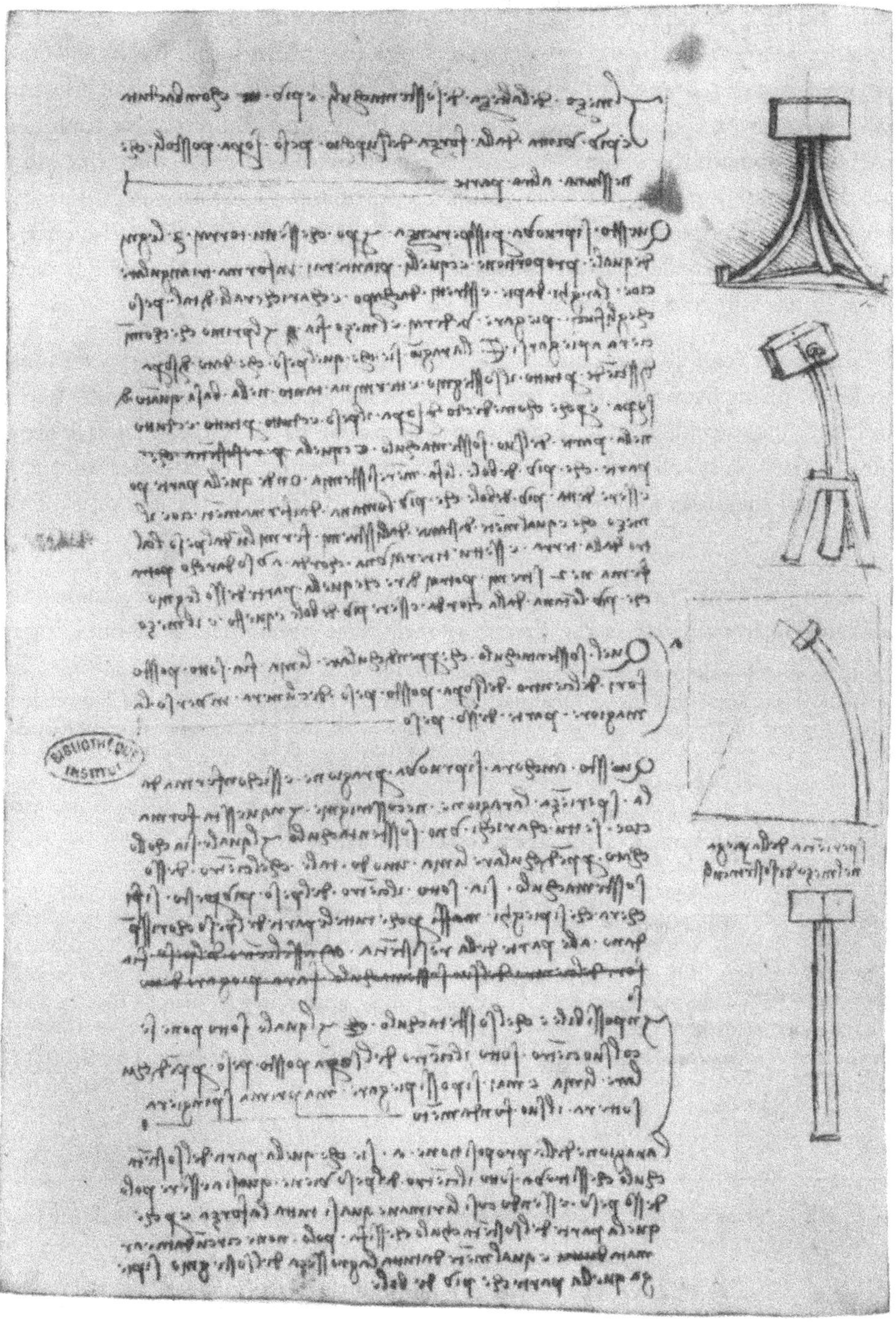
BIBLIOTHEQUE INSTITUT

Clearly, LEONARDO set himself the problem of determining the breaking strength of a beam in terms of its length, cross-section, modes of support, and loads, and the deflection of a beam in terms of its load and support. In view of the difficulty of these problems, which were attacked with little success by GALILEO and were not given any halfway adequate treatment until toward the end of the Enlightenment, we cannot be surprised that LEONARDO did not solve them. His notebooks contain the earliest statements we know of any of the main questions, and the earliest guesses, not always consistent, of answers to some of them, but to call LEONARDO on no better ground the founder of the rational science of construction is as naive as to attribute the discovery of houses to the builder of the oldest house still standing today.

The best we can do is merely to state what LEONARDO accomplished. So far as I can learn, of his proposed rules the only one that survives as valid now is rather obvious: The strength of a beam is proportional to its cross-sectional area[7]. I notice that every carpenter today seems to know this rule, whether or not he has studied mechanics, and I suspect it was well known to the builders of the Parthenon.

When we turn from numerical law to pictures of things seen, LEONARDO comes into his own. He is the first to describe the phenomenon of buckling:

"... if the superimposed weight should be more to one side of the support than to the other, the support will bend [on the side] across from which it is pressed more by the greater load, and will break in the middle of the opposite side, that is, the part farthest from the ends." (MS A 3*v*.)

He is the first to draw the form of the catenary curve (Fig. 9) and to propose to study it by means of a discrete model. He is the first, it seems, to use a

[7] Practically all the passages relevant to the strength of structural members have been collected and compared by A. UCCELLI in an article written about 1938 and published as "The Science of Structures," pp. 261—274 of *Leonardo da Vinci*, N.Y., Reynal and Co., n.d. (copyright 1956). UCCELLI's description and analysis are just, but the unwary reader may be tripped by use of some prochronistic terms like "stress" and "strain". *Cf.* also my remarks on pp. 19—22 of *The Rational Mechanics of Flexible or Elastic Bodies*, 1638—1788, L. Euleri Opera Omnia, (II) $\mathbf{11}_2$ Zürich, Füssli, 1960.

Fig. 9. MS A 48*r*., showing LEONARDO's sketch of the catenary curve (by courtesy of the Institut de France)

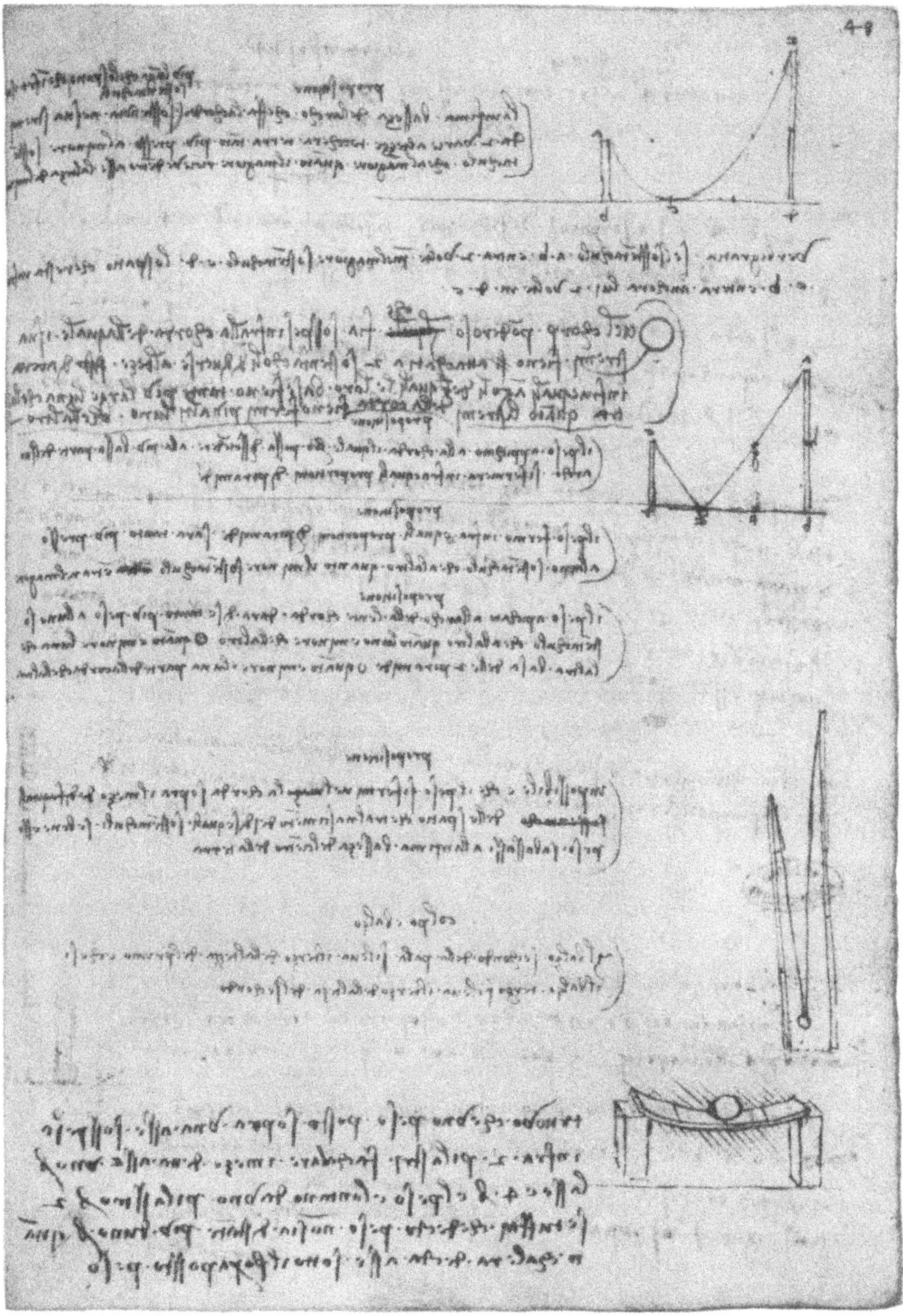

light rider so as to demonstrate the well known phenomenon of sympathetic vibration:

"The blow given in the bell makes another, like bell answer and move a little, and the sounded string of a lute makes another, like string, of like voice, in another lute answer and move a little, and this you will perceive by placing a straw upon the string like to the one sounded." (MS A 22*v*.)

He is the first to record observation of a fact which HUYGENS, two centuries later, was to use to infer the existence of nodes in a vibrating bar:

"If you beat a flat board, you will see the dust on it come together in little heaps." (MS A 32*v*.)

Over 400 years later, CHLADNI was to develop this simple observation so as to form the basis of a brilliant experimental technique, but LEONARDO, as usual, stops at the plain record of what happens and goes no deeper.

In all this work, such mathematical rules as are stated seem to be hypothetical; theoretical concepts are wanting; such experiments as are projected seem never to have been performed; but scattered among the tantalizing lists of good questions are jewels of recorded experience.

5. Leonardo's Method of Inquiry, I

Now experience in general rather than experiment in particular may be the better basis for science, not the worse. In our day, the mania for planned, controlled, contrived, subsidized experiment, which, like war, appeals to the masses because of its exorbitant cost and its unanimous call for many democratic hands, is a syndrome of man unhinged in more ways than one. That LEONARDO's main effort on natural questions reflects experience and theory first and experiment last and little, is no reproach to him. Perhaps, in the end, it is praise, but in such an approach he has no priority, for the sages of Greece and the Oxford schoolmen had followed it long before him and had gotten much more by it than he ever did.

LEONARDO's curiosity was boundless, like none else could he see, could he draw, but order and hence theory were not given to him. If the disorder of the notebooks, only in part intended for the public and never prepared in final form, is not fair evidence, we may read in VASARI's *Lives of the Painters*,

"It is said that when the Pope commissioned him to make a painting, he straightaway began to distil oils and herbs to make the varnish, because of which Pope LEO said: 'Alas, this man will never do anything, for he begins to think of the end of the work before the beginning.' "

When LEONARDO was almost sixty, he began some book or other, one of the many he intended to write for publication but never did, with the words

"And let this be a collection without order, drawn from many sheets I have copied here, hoping to set them later in order at their places, according to the subjects they treat..." (Arund. 1*r*.)

He goes on to tell the reader to expect repetitions, for which he excuses himself:

"And if I did not wish to fall into such an error, it would be necessary that for every item I wished to copy down, so as not to repeat it, I should have always to read all that done before..."

Thus LEONARDO admits that he does not perceive any natural ordering. The best he can do is to sort his remarks by title, and even so he confesses being unable to remember what he has said already. The essence of science is the *order* it discovers in or imposes on phenomena; nowhere is this property of science clearer than in the writings of EUCLID and ARCHIMEDES, who lived long before LEONARDO; ordering was cherished by the mediaeval tradition; and it flourished anew in the hands of great later figures like STEVIN. Science has *never* been a heap of observations and guesses, even true observations and good guesses. Thus no great competence in science and its history is required to dismiss nine-tenths or more of what LEONARDO wrote as contributing nothing whatever to science. What a man gives to the ages, however, is not the algebraic sum of his endeavors, but the best of them. Accordingly, we must search LEONARDO's bulky dross for such nuggets as may lie hidden in it.

6. Leonardo on Force and Motion

Seldom did LEONARDO write more than a paragraph at a time on any subject; often, only a line. When his statements on one topic are brought alongside each other, they appear to be outbursts of sporadic repetitions and contradictions.

"Force is generated from famine and plenty; she is the daughter of material motion, and nephew of spiritual motion, and mother and origin of weight; and this weight is finite in the element of water and earth, and this force is infinite, because infinite worlds could be moved with it, if one could make instruments with which this force could be generated." (Arund. 151*r*.)

"All violent motions grow feebler the more they depart from their causes." (MS $H^{(2)}$ 77 [29] *v*.)

"By so much as the natural motion departs from its cause, by that much it makes itself swifter." (MS $H^{(2)}$ 78 [30] *r*.)

"Force I say is a spiritual virtue, an invisible power which, by an accidental exterior violence caused by motion, is located and infused in bodies which are drawn aside and wrested from their natural [movement], giving to them an active life of marvellous power, constrains all created things to a change of form and place, runs with fury to its desired death, goes on diversifying itself by means of the causes. Slowness makes it great, and quickness makes it weak. It is born by violence and dies by freedom. And by so much greater it is, so much faster it consumes itself. It drives away with fury whatever stands in the way of its destruction, it desires to conquer, to kill its cause, its opposition, and in conquering kills itself. It makes itself more powerful where it finds greater opposition. Everything gladly flees its own death. Every constrained thing constrains. Nothing moves without it. The body where it is born does not increase in weight or in form. No motion made by it is lasting. It grows in exertions and falls through rest. The body where it is constrained is beyond liberty. And often by means of motion it generates new forces." *Etc.* (MS A 34*v.*)

"Heaviness, force, along with percussion not only may be called exchangeably mother and daughters the one of the other and all sisters together, since they are generated by motion, but also as generatrices and daughters of that motion; since without them among us motion cannot be created, nor without motion can such powers show themselves." (Forster II 117*r.*, *cf.* Arund. 37*r.* & *v.*)

This type of utterance draws a certain class of readers, as do *Also sprach Zarathustra* and the prophecies of NOSTRADAMUS. While some enthusiasts have high regard for LEONARDO's views on force, a rational assessor after reading the above quotations is less likely to regard as a precursor of "NEWTON's second law of motion" the following[8]:

"Motion is an accident born from the inequality of weight or force" (Arund. 205*r.*);

and

"Heaviness, force, as well as percussion, are to be called as much generators of motion as generated by it" (Arund. 184*v.*);

and

"Force is the cause of motion, motion is the cause of force." (Arund. 34*v.*)

Pronouncements of this kind have given rise to proceedings toward sanctification of LEONARDO as a precursor of GALILEO and NEWTON by naive enthusiasts who select a sentence in which such words as "gravity" and "force" can be interpreted in more or less their modern meanings and enshrine it as a prophecy of modern science. It is not that the statements made are wrong. They are not right, either.

[8] This sentence is written in the margin of a sheet of pronouncements on force and motion, including "Motion is created by hot and cold" and "lightness and heaviness at the same time are mother and children of each other" and "No element has in itself heaviness or lightness unless it moves."

Had NEWTON's three laws, word for word, appeared in one of LEONARDO's notebooks, they would not be right or wrong. The truest of scientific statements loses all scientific meaning if separated from definitions of the terms and from mathematical application to cases. For example, at one point LEONARDO writes,

"Every action done by nature is done in the shortest way" (Arund. 85*v.*, *cf.* Q. Anat. IV 16*r.*),

and his enthusiasts have interpreted this pronouncement as implying FERMAT's principle of least time in optics and MAUPERTUIS' principle of least action in analytical dynamics. Since LEONARDO does not tell us how either "action" or "way" is to be interpreted, I can see here nothing but a vague speculation which happens to be borne out when the words are given modern senses unknown to LEONARDO or to anyone else in his day[9]. The poetical quality of LEONARDO's writings on force has often been praised. For some they may be poetry, but they are science for no one.

While denial that any motion may be perpetual has often been listed as one of LEONARDO's achievements, I have found only two passages that can be wrested toward the subject. The first:

"No thing without life can push or pull without going along with the pushed thing. And these pushers cannot be anything else than force or weight, and if the weight pushes or pulls, it makes only this motion in the thing because it desires to stand fast, and when no thing moved by its falling motion will be sufficient ever to rise back to its first height, and then the movement is over.

"And if the thing that moves the other thing becomes [makes?] force, then that force accompanies the thing moved by it, and by so much as it moves it, it consumes itself. Being consumed, no thing moved by it becomes able to cause it again. Then no thing moved can have a long operation, since when the causes are wanting, so are the effects." (MS A 21*v.*)

---

[9] In fairness I must note here that LEONARDO wrote the sentence about "the shortest way" just after he asserted that the angle of incidence equals the angle of reflection in a mirror, and just before he wrote,

"As is the power of percussion, so is the bounce."

Thus he sensed that the three ideas were in some way connected, as indeed far later work has shown they are, but *what* the connection may be, he does not even hint. I am reminded of the way GALILEO set the vibration of a sounding string alongside the oscillation of a pendulum. While GALILEO is second only to LEONARDO in inspiring ridiculous special pleading from his modern devotees, even they do not suggest that he really discovered the principle of small oscillations.

MACH[10] quoted the beginning of this passage, just a little modernized and simplified, and pronounced it to be of "high clarity". He did not quote the end, which, in effect, denies the principle of inertia *in vacuo*, because only causes can produce effects. Neither did he quote the very next sentence, which states that motion on a perfect plane is perpetual in some cases and gives as an example sliding on ice. The second passage:

"No insensate thing moves of itself; hence, when in motion it will be moved by an unequal power, that is, of unequal time and movement, or of unequal weight, and when the desire of the first mover is stopped, the second will stop right away." (MS A 22*v*.)

To the extent I can make anything out of this sentence at all, it seems a mixture of misunderstood ARISTOTLE with misunderstood mediaeval physics. More impressive than these passages, which may seem to imply that perpetual motion is impossible in theory, are LEONARDO's attempts to design a hydraulic machine so as to move perpetually in practice. That he was wrong on both counts is not a reproach; he was unusual in being able to think independently on both problems, no small ability, and not unusual in being unclear in his approach to them; but I cannot see that he equalled his predecessors in what he said on either.

Science does not consist in mottoes, even true ones; a scientist is not a mystic seer, prophet of the future, healer of body or spirit, or material benefactor of the masses; nor is a scientific work holy writ to be interpreted by priests, professors of philosophy, or art critics.

The notebooks reward anyone who approaches them with preconception. For FREUD, who was intrigued by LEONARDO's indifference to the opposite sex except on canvas, a fanciful reading of certain drawings, borne out by one entry among thousands, gave rise to a book proving that LEONARDO was in love with his mother. In other passages VALÉRY found a "method" for writing poetry. To imagine history is allowed a poet by license; for some psychologists it is a professional function; but for a historian it seems a somewhat uncommendable activity. Unfortunately, the scientists who comment on the history of their subject, from MACH to the teaching textiscribes of today, have usually chosen the approach of the poets and psychologists.

---

[10] E. MACH, *Die Mechanik in ihrer Entwicklung, Historisch-Kritisch dargestellt,* Leipzig, Brockhaus, 9th ed. (1933), Kap. I, § 5, ¶ 9.

7. Birth of the History of Mechanics

By an accident of ignorance, LEONARDO has given us a gift which few know of even today, namely, the history of Western science in the Middle Ages. Apart from scattered remarks in his own times, LEONARDO's scientific work was first described in print by GIAMBATTISTA VENTURI in an enthusiastic and inaccurate essay in 1797. Historians of letters had meanwhile created the myth of the "Renaissance". According to this myth, in the Middle Ages man hibernated beneath a pall of scholastic repetition, borrowed from ARISTOTLE and enforced by the Church; the Renaissance, casting all this aside, opened its eyes and discovered man and the world by personal sensation. While for the arts this theory may be tenable as an expression of taste, in science other than anatomy it is not: The "early" Renaissance, 1450—1500, stands in the front rank of competition for the most sterile half-century of Western mathematics and physics, and the only exact science from the "late" or "high" Renaissance, 1500—1550, is algebra, which grew, not from open-eyed wonder at the world, but rather from bookish study of Arabian mathematics. Where is the empirical science which should have crowned the "rebirth" of knowledge? GALILEO, whom some physicists enthrone as founder of the empiricism they claim as their own, came one hundred years later, in the late Mannerist and early Baroque periods. Some historians of the last century found in VENTURI's essay the guide to the missing link: the notebooks of the singular LEONARDO, who at least showed what the Renaissance *could* have done, even though his work remained incomplete, unpublished, and unknown for centuries. The notebooks were then published in facsimile in the years 1881 to 1936.

Among the readers of the first volumes was the great man who founded the modern history of science: PIERRE DUHEM (1861—1916). A scientist, as a sound historian of science must be, he has left permanent achievements in physical chemistry, thermodynamics, hydrodynamics, and elasticity. Reading the notebooks of LEONARDO drew DUHEM into research on the history of science. From his discoveries developed slowly a second, almost opposite view of LEONARDO's work, a view that is held today by most of the inframinority that has first-hand knowledge of early science. While the critics gushed, such historians of science as then there were — a paltry scatter of semi-philosophers and science teachers — found several of GALILEO's ideas, more or less, in LEONARDO's notes. Previously historians had believed that GALILEO thought these things out of "genius"

applied to thin air. In 1788 LAGRANGE[11] had written of ARCHIMEDES and GALILEO, "the interval separating these two great geniuses disappears in the history of mechanics," and this simple view, as satisfying as the account of creation given in Genesis, was good enough for MACH a century later[12]. "Discovery" of LEONARDO transferred the point of application of this same theory a century backward. He, too, had the same materials to work with: "genius" and thin air, and the remaining problem for this group of historians was only to see how LEONARDO's ideas got to GALILEO, thus making the latter a true grandson, if not son, of the Renaissance.

But DUHEM considered the enormous bulk and range of LEONARDO's writings. Today we may see collected and organized into a single volume his fragments on the general principles of mechanics; these alone, excluding most of what he wrote on mathematics, mechanisms, flexible solids, and fluids, fill 550 large pages[13]. Another volume of 400 pages is required to reprint only a part of his

[11] In preparing this essay for the press, I have been unable to trace the quotation. A somewhat more specific statement is found at the beginning of Part II of J.-L. LAGRANGE, *Méchanique Analitique*, Paris, Desaint, 1788: "Dynamics is due entirely to the moderns, and Galileo is the one who laid its first foundations."

[12] Thus MACH in the first edition of his *Die Mechanik in ihrer Entwicklung*, 1883, mentioned no author living after ARCHIMEDES and before LEONARDO.

[13] LEONARDO DA VINCI, *I Libri di Meccanica nella ricostruzione ordinata da Arturo Uccelli, preceduti da un'introduzione critica e da un'esame delle fonti*, Milano, Hoepli, 1940.

The most extensive study of LEONARDO's mechanics is that of R. MARCOLONGO, "La meccanica di Leonardo da Vinci," *Società Reale di Napoli, Atti della Reale Accademia delle Scienze Fisiche e Matematiche* (2) **19**, No. 2, 148 pp. (1932). The chapters are as follows:

Part I. Statics
1. Mediaeval statics and LEONARDO's sources
2. Straight and bent levers and balance
3. Concept of moment and composition of forces
4. Equilibrium on the inclined plane. Stability of the balance. Polygon of support
5. Centers of gravity
6. Pulley, tackle, and the problem of constrained reactions
7. Researches on the resistance of materials, on the theory of the arch, and on friction

Part II. Dynamics
1. Greek and mediaeval dynamics
2. Force, percussion, impetus, weight
3. The laws of motion
4. Free motion of a heavy body. Motion on an inclined plane. Motion of projectiles and shock

Fig. 10. PIERRE DUHEM (1861—1916)

LES SOURCES DES THÉORIES PHYSIQUES.

LES

ORIGINES DE LA STATIQUE

PAR

P. DUHEM

Correspondant de l'Institut de France,
Professeur à la Faculté des Sciences de Bordeaux

TOME PREMIER

PARIS
LIBRAIRIE SCIENTIFIQUE A. HERMANN
Libraire de S. M. le Roi de Suède et de Norvège
6, RUE DE LA SORBONNE, 6

1905

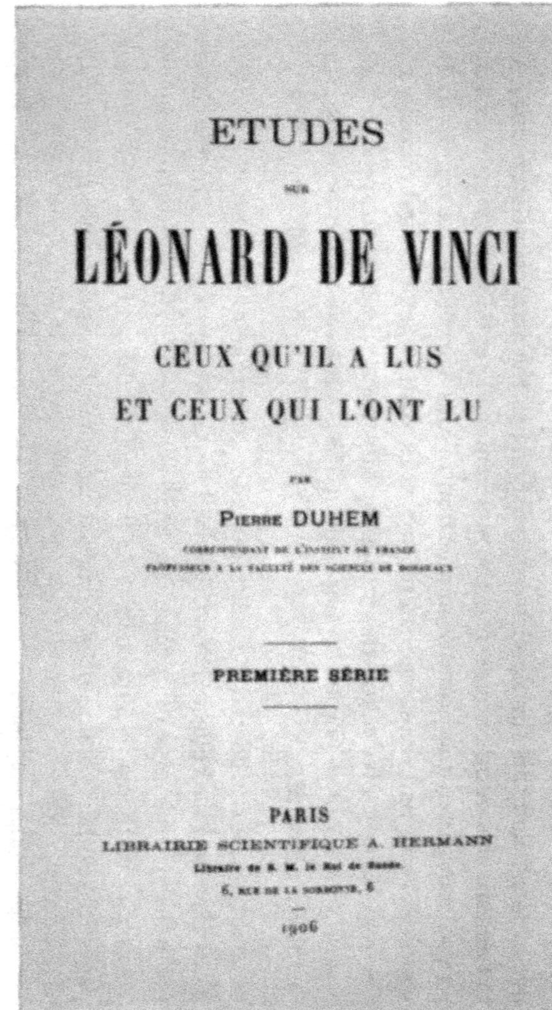

ETUDES

SUR

LÉONARD DE VINCI

CEUX QU'IL A LUS
ET CEUX QUI L'ONT LU

PAR

PIERRE DUHEM

CORRESPONDANT DE L'INSTITUT DE FRANCE
PROFESSEUR A LA FACULTÉ DES SCIENCES DE BORDEAUX

PREMIÈRE SÉRIE

PARIS
LIBRAIRIE SCIENTIFIQUE A. HERMANN
Libraire de S. M. le Roi de Suède
6, RUE DE LA SORBONNE, 6

1906

writings on fluid flow[14]. Whatever a litterato may consider a reasonable activity for a "genius", few scientists will believe that a thousand pages on mechanics, right or wrong, have *ever* sprung from the application of genius to thin air. The question DUHEM asked himself was, where did LEONARDO learn all this, and what fraction of it did he modify or add from his own thought or experience? In the attempt to answer these questions, DUHEM became the first person since the Renaissance actually to read what the mediaeval schoolmen wrote on mechanics and physics. In a real sense, he discovered the science of the Middle Ages, and he added, attached to concrete and definite discoveries and theorems, a long list of names never before encountered in histories of science: JORDANUS DE NEMORE, JEAN BURIDAN, NICOLE ORESME, RICHARD SWINESHEAD, WILLIAM HEYTESBURY, and others. In the midst of his immortal historical discoveries DUHEM had already reached the conclusion that [15] "In the mechanical work of LEONARDO DA VINCI there is no essential idea that does not come from the geometers of the Middle Ages."

According to my custom with secondary sources, I did not consult this one until my own studies were complete. MARCOLONGO's essay contains many of the best quotations from LEONARDO's works, but I cannot participate in his generous extrapolations of his great compatriot's researches. The text of UCCELLI is more complete, his interpretations more measured. Neither of these essays takes up LEONARDO's work on fluids.

I have never been able to see MARCOLONGO's *Studi Vinciani*, Napoli, 1936. In Chapters VIII and IX of his general biography, *Leonardo da Vinci, Artista-Scienziato*, Milano, Hoepli, 1939, MARCOLONGO gives a brief and generous survey of LEONARDO's mathematics, mechanics, physics, and engineering.

There is a book in English on the subject: I. B. HART, *The Mechanical Investigations of Leonardo da Vinci*, Chicago, Open Court, 1925. Its level may be judged from these two quotations:
"For the Schoolmen, however, the mere pronouncements of the Peripatetic were what mattered. Initiative and independence were non-existent in the field of physics." (p. 13)
"Now Leonardo, more than one hundred years before Galileo, was most definitely an experimentalist. If the great Tuscan philosopher was indeed the 'Father of Experimental Philosophy,' then da Vinci was its grandfather.... The appeal to experiment pervades all his writings on scientific topics." (p. 78)

[14] LEONARDO DA VINCI, *Del Moto e Misura dell'Acqua, libri nove ordinati da F. Luigi Maria Arconati editi sul codice archetipo Barberiano a cura di E. Carusi ed A. Favaro*, Bologna, Zanichelli [1923].

[15] P. 192 of Vol. 1 of *Les Origines de la Statique*, Paris, Hermann, 1905—1906. DUHEM's main further publications on this subject are included in his *Études sur Léonard de Vinci*, 3 vols.,

---

Fig. 11. Title pages of DUHEM's first books on the history of mediaeval science

8. Mediaeval Kinematics and Dynamics

Later studies by a school of painstaking historians, along with the publication of the texts DUHEM used and of others unknown to him, have revised much of the detail but little of the general aspect of the picture of mediaeval achievement drawn by DUHEM. The now published sources[16] prove to us, beyond contention, that the main kinematical properties of uniformly accelerated motions, still attributed to GALILEO by the physics texts, were discovered and proved by scholars of Merton College — WILLIAM HEYTESBURY, RICHARD SWINESHEAD, and JOHN OF DUMBLETON — between 1328 and 1350. Their work distinguished *kinematics*, the geometry of motion, from *dynamics*, the theory of the causes of motion. Their approach was mathematical. They succeeded in formulating a fairly clear concept of instantaneous speed, which means that they foreshadowed the concepts of function and derivative, and they proved that the space traversed by a uniformly accelerated motion in a given time is the same as that traversed by a uniform motion whose speed is the mean of the greatest and least speeds in the accelerated motion. In principle, the qualities of Greek physics were replaced, at least for motions, by the numerical quantities that have ruled Western science ever since. This work was quickly diffused into France, Italy, and other parts of Europe. Almost immediately, GIOVANNI DA CASALE and NICOLE ORESME found how to represent the results by geometrical graphs, introducing the connection between geometry and the physical world that became a second characteristic habit of Western thought — a habit so deep-seated that it is known to every carpenter and passes unremarked only in certain highly specialized professions.

Mediaeval dynamics is closer to Aristotelian views than to modern physical thinking in its emphasis on resistance rather than what we call today inertia. Nevertheless, unlike Greek science, it is quantitative, seeking to describe numeri-

---

Paris, 1906—1913, and *Le Système du Monde*, Paris, vols. 1—5, 1913—1916, and vols. 6—10, 1954—1959.

[16] A. E. MOODY & M. CLAGETT, *The Mediaeval Science of Weights*, Madison, University of Wisconsin Press, 1952.

M. CLAGETT, *The Science of Mechanics in the Middle Ages*, University of Wisconsin Press, 1960.

E. LAMAR CROSBY, Jr., *Thomas of Bradwardine. His Tractatus de Proportionibus*, Madison, University of Wisconsin Press, 1961.

The book by CLAGETT contains the main texts, translations of them, and a scholarly and penetrating commentary and analysis. It is the basic source for modern study of mediaeval mechanics. My quotations from JOHN THE GRAMMARIAN and BURIDAN, below, are taken from this work.

cal velocities in terms of other numerical quantities. A precursor of the later ideas of inertia, momentum, and energy may be found in the theory of "impetus" put forward by JEAN BURIDAN at Paris during the period when kinematical concepts were developing at Oxford. Rejecting ARISTOTLE's views on the motion of projectiles, BURIDAN adopted an idea mentioned by JOHN THE GRAMMARIAN some 800 years earlier, namely, that "some incorporeal motive force is imparted by the projector to the projectile." BURIDAN replaced this assertion of quality by a statement of quantity: "... by the amount more there is of matter, by that amount can the body receive more of the impetus and more intensely," and "... by the amount the mover moves that moving body more swiftly, by the same amount it will impress a stronger impetus." Thus BURIDAN comes close to defining as "impetus" what we now call "momentum". That a quantity of this kind measures the tendency of a body to persevere in motion, BURIDAN infers from what he calls "experiences" — observations of simple phenomena concerning tops, lances, and ships.

Unlike the riddles of LEONARDO, the results of the schoolmen are explained in detail, developed by means of proved mathematical theorems in plodding, consecutive treatises, and illustrated by examples drawn from everyday life. In these treatises is not a single proposition asserted to be true by authority, be it of ARISTOTLE or of churchman.

CLAGETT[17] has cited much evidence to show that these ideas, which originated in England and France in the early fourteenth century, were discussed back and forth in periods of varying activity and inactivity in France, the Empire, and Italy in the latter half of the same century and were taught in Italian universities in the next one, at the end of which a flood of printed books opened the subject to everyone — everyone who could understand Latin and mathematics.

The books themselves, not only the main texts of the greatest of the mediaeval natural philosophers but also many minor works, were published in Venice, among other places, at the very time when LEONARDO was working. (That such

[17] See the important Chapter 11 in his book, cited on p. 30. Also RANDALL, in the essay cited below, p. 39, writes,
"Having myself done some reading in the anticlerical Italian writers on natural philosophy of the Quattrocento,... I have found that all the presumably 'novel' ideas associated with LEONARDO before DUHEM's discoveries were known and taught in the fifteenth-century Italian universities — in Padua, Bologna, Pavia, and the rest."

dusty scholastic works could sell in 1500, despite the contempt of the literary humanists, incidentally overturns the still popular image of the Renaissance as the "rebirth of knowledge".)

While the schoolmen of the Middle Ages esteemed ARISTOTLE with a veneration now difficult to share, they did not think him infallible, and they regarded themselves able, upon occasion, to go beyond him, far beyond him, even so far as to deny some of what he said. In a word, they looked upon him much as we today look upon NEWTON. The humanists of the literary Renaissance introduced a different attitude toward the ancients, a rigid reverence for authority that the Middle Ages had expected in the deeds of peasants and nobles but not in the thoughts of scholars. For the humanists, man and nature were to be found in the pure classics of Greece, cleared of monkish gloss and corruption. The Renaissance of humanism was also the rebirth of intellectual servility. Condemning the mediaeval Aristotelians *a fortiori* since they had not been able to drink from the pure sources, the then lost Greek texts, the humanists regressed in time and in science to PLATO. In revolution, which is only a special case of war, slogan replaces concept and method, and destruction is king. PLATO in his turn was a pupil of that great revolutionary, SOCRATES, who diverted Greek thought from the hard nuts of natural philosophy which his forerunners, now preserved only through fragments ironically called "pre-Socratic", had brilliantly debated. Instead, he taught a philosophy which everyone could sense, a philosophy which disposes of science and mathematics by serene conventions and revolves endlessly the savory wheel of morals and politics. Such a philosophy was easily revived and refreshed by the humanists, who dismissed the arguments of the schoolmen in scornful phrases about the number of angels that could dance on the point of a pin. The humanists, who stood in rebellion against the doctrines accepted in the bodies which commonly fostered the life of the mind, had to make a living apart from church and university; to strike out as intellectual entrepreneurs, they had to advertise themselves: Journalism is a creation of the Renaissance. A philosophy to be sold to princes and merchants cannot stink of cloister or lectern; it must be simple, fit for polite conversation with ladies in winter evenings at Urbino. Such is Platonism and neo-Platonism, but such is not science.

9. Leonardo and the Tradition

This was LEONARDO's world. Uneducated and nearly illiterate[18], he avoided and distrusted those with book learning. He worshipped ARCHIMEDES by reputation, but the only demonstrated connection between the two is a loose page by ARCHIMEDES found in one of LEONARDO's notebooks. From his youth a companion of artists and artisans, LEONARDO wrote,

> "I know well that, since I am not a man of letters, some presumptuous persons will think they may with reason censure me by alleging that I am totally unlettered. Fools! Do they not know that I might retort as Marius did to the Roman patricians by saying, 'They who deck themselves out in the labors of others will not allow me my own.' They will say that I, since I am not lettered, cannot properly speak of that which I wish to treat. But they do not know that my subjects are to be treated by experience rather than by the words of others, [experience] who was the mistress of whoever wrote well, and thus I take her as mistress, and her will I adduce in all cases." (Atl. 119*v*. a)
>
> "Though I may not, like them, be able to cite the authors, it is a much greater and worthier thing to cite by citing experience, mistress to their masters. They go blown up and pompous, dressed and ornamented not by their own labors, but by others'; and they do not allow me my own; and if they scorn me, an inventor, how much more will they, not inventors but trumpeters and reciters of others' works, be blamed." (Atl. 117*r*. b)

Indeed, LEONARDO could not quote Greek, like a humanist, but neither had he the mathematics or the Latin sufficient for reading the books by the great scientists from the two centuries preceding his. While by our fathers' generation ignorance of Latin was praised as a sign of independent intellect, and nowadays that ignorance is so diffused as to be taken without special pride, in LEONARDO's age an unlatined man could learn science as easily as today a person who understands only Eskimo can learn calculus. Lack of Latin not only shielded LEONARDO from the Platonic blight but also kept him from first-hand knowledge of the mathematics and mechanics of the schoolmen. While in the middle period of Vincian studies it was generally asserted that LEONARDO knew the work of JORDANUS DE NEMORE, he nowhere cites him[19], and he nowhere gives evidence of mathematical compe-

---

[18] The editions for the general reader, even those in Italian, smooth out LEONARDO's spelling and grammar. Too much has been made, in praise or blame according to the writer's temperament, of LEONARDO's literary skill or lack of it. While his Italian is not only dialectal but often illiterate, his manuscripts are mainly (though not entirely) jotted notes, not drafts of essays, and it is only our own day that has laid upon literature the curse of writers who are no more than tubes for casual speech.

[19] In Q. Anat. II, 22*v*. he mentions "Giordano de ponderibus", but it is not clear which book he means or whether he ever saw it. The great and original work by JORDANUS DE NEMORE is the *Liber Jordani de ratione ponderis*. According to MOODY & CLAGETT, the work called *Liber*

tence sufficient to follow one of the real mediaeval sources on physical science. Although LEONARDO's remarks on the principles of motion are closer to Western science than are the views of the literary humanists, they seem to be mere garblings of university commonplaces, perhaps picked up in conversation[20]. Indeed, they remind me of the views of modern poets on quantum theory and nuclear fission. Advance in basic mechanics had to wait for such scholars as CARDANO and GALILEO, who studied systematically and exhaustively before asserting independence.

Finally, even he who takes LEONARDO's mottoes seriously need not choose the ever-quoted paeans to empiricism we have quoted again above. Instead he may carve upon his architrave

"No effect occurs in nature without a reason. Understand the reason, and you need no experience." (Atl. 147*v.* a.)

Perhaps it was in his moods of following this motto that LEONARDO studied Peripatetic and mediaeval physics, but he never mastered them.

10. Uniformly Accelerated Motion in the Void and on the Inclined Plane

Of all the discoveries in the principles of mechanics formerly attributed to LEONARDO, today there is only one not shown to belong to some earlier author. While the mediaeval scholars invented and proved the kinematical theorems for uniformly accelerated motion, they did not tell how to produce it [21]. Thus in a famous passage, JOHN THE GRAMMARIAN, who wrote over 3000 pages of commentary on ARISTOTLE in the fifth or sixth centuries A. D., had stated that

---

Jordani de ponderibus shows "not the slightest trace of any influence of the proofs which JORDANUS [NEMORARIUS] gave..."

[20] *E.g.*, BURIDAN's "impetus" seems to be misunderstood in the passages in MS E 22*r*., 40*v*. Likewise, the seven assertions about a "potenza" or "virtù" and the "body" (*i.e.* body of given weight) it can move in a certain time LEONARDO records in MS F 51*v*.—52*r*. (also 26*r*.) are not, as some have claimed, a definition of mechanical power but rather merely a reflection of Peripatetic views on dynamics. *Cf.* UCCELLI's note on pp. 306—307 of LEONARDO DA VINCI, *I Libri di Meccanica*, Milano, Hoepli, 1940.

[21] THOMAS BRADWARDINE had inferred from his hypotheses that "all mixed bodies of similar composition will move at equal speeds in a vacuum." This famous tenet of "equality of velocities in the void", also commonly attributed to GALILEO, according to later views is either untrue or else far less specific than the statement that the earth's gravity uniformly accelerates a body in a vacuum.

Aristotle's view of free fall

"is completely erroneous, and our own view may be corroborated by actual observation more effectively than by any sort of verbal argument. For if you let fall from the same height two weights, of which one is many times as heavy as the other, you will see that the ratio of the times required for the motion does not depend on the ratio of the weights, but the difference in time is a very small one."

However, he states clearly that this experimental fact refers to motion in a medium; he thinks that in a void the velocity of fall would be proportional to the weight. Leonardo wrote much on falling bodies, never telling whether he considered air resistance or not. When he projects an experiment of dropping two lumps of sugar from a campanile, one weighing two ounces, the other, one, he says to keep your eye on the lighter one and mark the brick it is passing when you hear the heavier one hit the ground first (MS A 30*v.*). He does not say whether he ever did this experiment. Again,

"if two unequal spherical bodies, heavy and of the same material and descent, fall from high to low, the one will fall as much more swiftly than the other as the diameter of the one goes into the other." (MS A 22*v.*)

Two other passages contradict:

"Therefore, in every doubled quantity of time, [a falling body] doubles the length of its descent and the speed of its motion." (MS M 44*v.*, *cf.* 45*r.*)

Here he has asserted that *both* the distance and the speed increase in arithmetical progression[22]; if he had understood the Merton kinematical theorems, he would have seen that this statement is self-contradictory. Finally,

"The thing that moves in natural motion acquires in every degree of motion degrees of velocity, which degrees will have the same proportion, the last to the next-to-last, as has the second with the first." (Forst. II² 152*v.*)

There it is, the earliest[23] statement we have today of the law of free fall.

But this is not all. Leonardo studied also the problem of motion on an inclined plane. Indeed, he seems to be the first to have done so. While his diagram indicates that he was trying to take account of the size of the ball, which certainly he could not successfully have done, the accompanying words say:

"the thing will go down by the line *bc* that much slower than by the line *bd* as that line *cb* is longer than *bd*, the bodies being of equal heaviness and roundness and of equal motion, which thing is impossible." (MS M 42*r.*) (Fig. 13)

---

[22] For the same at greater length, with a diagram in Oresme's style, see MS M 44*r.* & *v.*, 45*r.*, 47*r.* & *v.*, 48*r.*, 49*r.*

[23] On pp. 553—556 of *The Science of Mechanics in the Middle Ages*, Clagett quotes an unclear statement of Oresme that may have the same meaning, and he attributes the first publication to Domingo de Soto, 1555.

In two other passages (MS A 33*r*., Atl. 210*v*. a) he states even more clearly the same result, which we have all learned in high school in different terms: The times of descent on inclined planes vary inversely as the sines of the angles of inclination. Here is an instance for the believers in "genius". How did LEONARDO obtain the result? By experiment? Theory? Plausible guess? For the first, there is no evidence; for the second, no then known starting point. From many other examples we know that LEONARDO often proposed rules of simple proportion, and that usually they came out wrong. Here is one that is right. Did LEONARDO know it? Was not this, for him, just one of the hundred linear rules he proposed and never got around to trying out, the only difference being that this one concerns a problem later seen to have central importance, and this answer just turned out to be right?

I fear so. We remember that in regard to free fall, LEONARDO proposed several linear laws, mutually contradictory, and in the single one that turned out to be correct, he may simply have forgotten to repeat his other, incorrect, statements.

The facts before us are simple:

1. In physics, some relations are linear and some are not.
2. LEONARDO *never* proposed any relation other than a linear one[24].
3. LEONARDO *did* propose dozens of linear relations.
4. From 1) and 3), some of LEONARDO's rules may be expected to come out right.

Counting only the cases when he was right and disregarding those when he was wrong may give a somewhat distorted estimate of his capacity as a natural philosopher. We cannot ascertain the order in which LEONARDO wrote his various entries. When, as often is the case, he writes one time an oracular pronouncement that is now considered wrong, and another time its opposite, as for example regarding perpetual motion, his enthusiasts describe the former as an early question and the latter as his final "discovery". So far as I know, however, it is not a law of nature that those who speculate always get righter as they grow older.

---

[24] Of course, his knowledge of mathematics was insufficient for him to have considered quantitatively anything else.

Fig. 12. Atl. 338*r*.b, with LEONARDO's diagram of a ball on an inclined plane (by courtesy of the Biblioteca Ambrosiana, Milano)

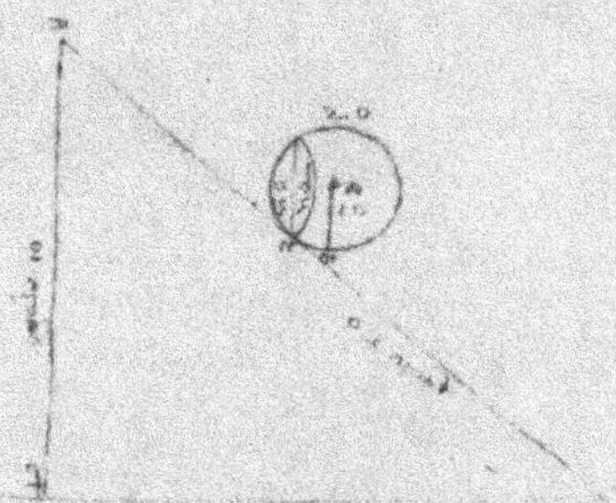

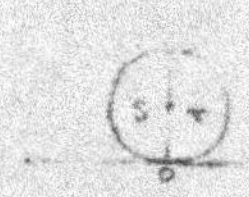

28

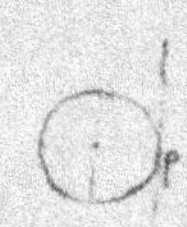

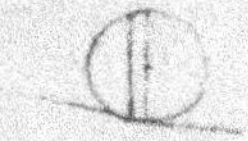

[illegible]

[illegible]

[illegible]

[illegible]

42.

11. Leonardo's Method of Inquiry, II

We can now summarize the method LEONARDO seems to have used, doubtless unconsciously, in scientific questions:

1. Observe the phenomenon and list the quantities having numerical magnitude that seem to influence it.
2. Set up such linear relations among pairs of these quantities as are not obviously contradicted by experience.
3. Propose these "rules of three" for trial by experiment.

Beyond one case, mentioned above, he has left us no evidence that he ever added what seems now the natural fourth step, namely, to carry out the proposed trials.

Thus a recent historian[25] has written

"... for all his great gifts as a careful observer, LEONARDO, strictly speaking, has no place whatever in [the] process... by which modern science came into being...

"1. LEONARDO was not himself a scientist, in the sense in which he and his contemporaries understood science — or in any sense in which anybody else has ever understood science...

"2. There is not discoverable, in LEONARDO's Codici, a single theoretical scientific idea that is essentially new, or that was unknown in the organized scientific schools of Italy in his day. The notion that not unnaturally arose when his Codici were first printed after 1881, that they were full of startlingly original insights, arose from the sheer ignorance prevailing at that time of what was familiar to LEONARDO's scientific contemporaries.

"3. Even if LEONARDO had had original ideas in scientific theory, he would still have no 'place' in, and would have exerted no 'influence' on, the emergence of modern science. For these ideas of his remained quite unknown until the Paris Codici were published, 1881—1891, and the Codice Atlantico in 1894."

This panegyric was sung on the 500th anniversary of LEONARDO's birth. (On the 600th, perhaps art will have progressed so far that an art historian can inform the world that LEONARDO was not an artist.)

It is impossible, I know, to say anything about LEONARDO that will not outrage one or another segment of the audience. For peaceful co-existence with the standard view, shared with the general semiliterate mass by ninety-nine percent

---

[25] J. H. RANDALL, Jr., "The place of Leonardo da Vinci in the emergence of modern science," pp. 207—218 of *Roots of Scientific Thought*, N.Y., Basic Books, 1957.

Fig. 13. MS M 42*r*., showing LEONARDO's statement of the correct rule of motion on an inclined plane (by courtesy of the Institut de France)

of university professors, he must be labelled a "genius" straight off and described by no adjective of less than superlative degree. For the historical initiate, he must be dismissed as a curious artist who learned a bit of science. That we may bow respectfully to the standard view, regarding it as an aspect of devotion to a religion which need not be ours, does not prove the historians' alternative to be the only one. I do not think it important to decide whether or not LEONARDO was a "scientist" according to one or another definition of that ugly word, but it does seem obvious that *anyone* in the fifteenth century who wrote as much about mechanics as LEONARDO did, whether or not he was able also to paint pictures and to design forts, deserves the careful attention today of him who searches the origins of mechanics. If LEONARDO is not the towering genius, the all-prophetic wizard of his popular image, this is no reproach. The fault lies not in him but in the eager credulity of the unscientific, who insist upon regarding science as magic, revealed through mystic inspiration to seers who are then enshrined as objects of minutial veneration like that paid by professors of literature to SHAKESPEARE and BOSWELL. While such wizard geniuses may exist in religion, politics, and folklore, they have never existed in science. To clear away the preposterous popular image of LEONARDO is not a mere exercise in debunking but rather a preliminary, unfortunately a necessary one, to a sober study of what he really did in natural philosophy. Granted that the historians have shown us conclusively that LEONARDO did not invent the mechanics of GALILEO any more than GALILEO did, those who have progressed past the freshman year in mechanics are less likely to think that subject begins and ends with changes rung on the formula $s = \frac{1}{2} g t^2$. Granted that buckets of blather have been and doubtless forever will be poured over LEONARDO by the litterateurs, LEONARDO's own words and drawings remain undirtied by them. Granted that, contrary to Vincian folklore, LEONARDO did not discover *everything*, we have seen already that, contrary to the sweeping denials just quoted[26], he did discover *something*, and soon we shall find more of his solid if limited achievements.

---

[26] Moreover, RANDALL is unfair in concluding that only those ignorant of mediaeval mechanics can find originality in LEONARDO's science. ROBERTO MARCOLONGO, a leading expert on the modern mechanics of his day, was a careful and understanding reader of the mediaeval sources as quoted by DUHEM, but his knowledge did not prevent him from granting to LEONARDO, in a burst of enthusiasm for his great countryman, priority in the statement of NEWTON's first and third laws!

12. Science and Art

We hear too often that only an artist deserves to speak of LEONARDO, because LEONARDO was first and foremost an artist. A man who in a life of seventy years painted only ten pictures, half of them abandoned unfinished, perhaps can be called primarily an artist if he did nothing else. LEONARDO, however, spent only a small part of his time in work now regarded as appropriate for an artist. He seems not to have had a primary interest in any one domain: Rather, he attempted all. To know all and to do all, to be a pantamath and a pantapoet, "l'uomo universale", was an ideal of the Renaissance. To be a pantamath LEONARDO struggled blindly and in vain: Lacking both formal study and an ordering mind, he could never be learned. But as a pantapoet, one man who can do all that any other does, he was a noted success. Judgment of such a man is often said to require a staff of experts, one for each of his fields of work. This may be less a credit to LEONARDO's genius than a reproach to today's ignorance, generally called "specialization". Perhaps also it reflects the common antithesis of "science" and "humanities": An artist is to seek the artistic behind LEONARDO's inventions, a scientist the scientific basis of his pictures. The antithesis itself, born of the foul orgies of romantic unreason which burst like pus from the French Revolution to NAPOLEON to WAGNER to PICASSO to HITLER to nuclear bombs, reflects a more than commonly tedious vulgar blunder and is nothing but a puerile excuse for professionalized crime against the human mind. Each of us is ignorant if not also stupid, but there is no reason to codify darkness as being just a new kind of light. However out of fashion for him to admit it, I must believe a good painter able to reason[27] and to learn the laws of nature; I am sure that there is no mathematician who does not seek beauty in mathematics. While it may be that some of those who have made intuition a profession are too lazy to acquire the discipline of science, I am certain that a scientist who lacks intuition is no scientist at all. Mathematics is the paradise of disciplined intuition; not only in mathematics but also in painting, crude impulse or "self-expression", contemptuous of traditional measures of excellence, yields no more than the breathless rush of ugly fads in which our museums and critics and art merchants rub our reluctant noses today. It is only a phoney who claims for his racket exclusive standards to defend his own mediocrity against detection by

[27] *Cf.* the famous gloss GOYA wrote to No. 43 of the *Caprichos*, which he had originally intended to be No. 1:
"Imagination deserted by reason creates impossible, useless thoughts. United with reason, imagination is the mother of all art and the source of all its beauty."

those he contends unqualified to judge. LEONARDO in method and expression was much the same as a painter, as a mathematician, and as a physicist: In each domain, his eye and hand leave far behind his brain and spirit.

13. Leonardo as a Mathematician

For a second look at LEONARDO's mechanics, let us begin with mathematics, for what LEONARDO could do and could not do in mathematics illuminates his work in all other fields. He wrote,

"... mechanics is the paradise of the mathematical sciences, because with it one comes to the fruits of mathematics" (MS E 8*v*.),

and

"There is no certainty where one can neither apply any of the mathematical sciences nor any of those which are based upon the mathematical sciences" (MS G 96*v*.),

and

"Therefore, O students, study mathematics, and do not build without foundations" (Quad. Anat. I 13*v*.),

and

"Let him not read me who is not a mathematician of my principles." (Quad. Anat. IV 14*v*.)[28]

However, these are no more than mottoes over the door, such as PLATO's "Let none not a geometer enter here," which might well be taken in warning that if one is ever to learn geometry he must do so before entering, since in PLATO's Academy he is sure never to learn any. So it is with LEONARDO. Some of his mathematical passages are mere lessons from the beginnings of EUCLID, poorly remembered from popular translations. Of the deeper problems of Greek geometry, such as the theory of conics, the construction of irrationals, or the method of exhaustion, he has no ken. Heedless of the restriction to ruler and compasses, for him a practical or approximate solution is enough, as when he says,

"... ARCHIMEDES never found the square of any figure with curved sides, but I have obtained the square of a circle minus the smallest possible portion that the intellect can conceive, that is, the smallest point visible." (Windsor Drawings 12280*v*.)

[28] The meaning of this motto is not clear: "No mj leggha chi non e matematicho nelli mja pricipj." Italian enthusiasts who quote it are not in agreement in their tacit conversions of this statement into grammatical Italian, and English translations are various, *e.g.* "Let no one read me who is not a mathematician in my beginnings" (MACCURDY), and "Let no man who is no mathematician read the elements of my work" (GOLDSCHEIDER).

This manner of thinking is not mathematical, and LEONARDO nowhere shows any sense of need for mathematical definition and proof. Also, however much LEONARDO admired ARCHIMEDES, at least when he wrote the passage just quoted he did not know ARCHIMEDES' work on the quadrature of curved figures.

LEONARDO's was the day and his the country of the great Italian school of algebraists: SCIPIONE DEL FERRO (1465—1526), ANTONIO FIOR (c. 1535), TARTAGLIA (1499?—1557), CARDANO (1501—1576), and FERRARI (1522—1565). To these men, solving a quadratic was easy. They discovered the general solutions of the equations of third and fourth degrees and struggled with higher cases. The work of this school began in LEONARDO's lifetime, but he took no part in it or notice of it; he never solved a quadratic correctly, and even with equations of first degree he was clumsy[29]. Since he could add fractions only by guesswork, often wrong, it is not surprising that he was equally unaware of the numerical problems treated by the mediaeval Italians such as LEONARDO FIBONACCI of Pisa (1170 to 1250). While LEONARDO adored mathematics as a goddess, he made his suit to her with no more constancy than to his other muses, and she seems to have granted him few favors[30].

Although a man is not altogether responsible for his friends, it is not insignificant that LEONARDO chose his from among artisans or the fringes of the learned. That he was a friend and collaborator of the mathematician LUCA PACIOLI is often noted, but less often that PACIOLI was rather a mystic than a thinker. While LEONARDO may have learned some simple geometry from PACIOLI, it is to his credit that he drew his opinions on artistic proportion from experience, from his own eye rather than from PACIOLI's baseless number games of "divine proportion".

---

[29] Even MARCOLONGO must agree here. On p. 156 of his *Leonardo da Vinci, Artista-Scienziato*, Milano, Hoepli, 1939, he cites two vague references to algebra by LEONARDO and continues, "But beyond these various and brief remarks, he never took up algebra. It is certainly true that some simple mechanical questions... lead to easy linear equations, but LEONARDO never solves them, even in the easiest cases, by means of algebra, and in cases of any complexity the results are not right."

[30] After finishing my study of LEONARDO's manuscripts, I came upon the article by the respected mathematician U. CISOTTI, "The mathematics of Leonardo," pp. 201—203 of *Leonardo da Vinci*, N.Y., Reynal & Co., n.d. [1963], written c. 1938, where a fair summary of the positive side of LEONARDO's mathematical studies is presented. With the elegant tact characteristic of Italian assessments of colleagues, dead or alive, CISOTTI concludes, "it adds nothing to the splendor of the many-sided figure of Leonardo to isolate him as a mathematician..."

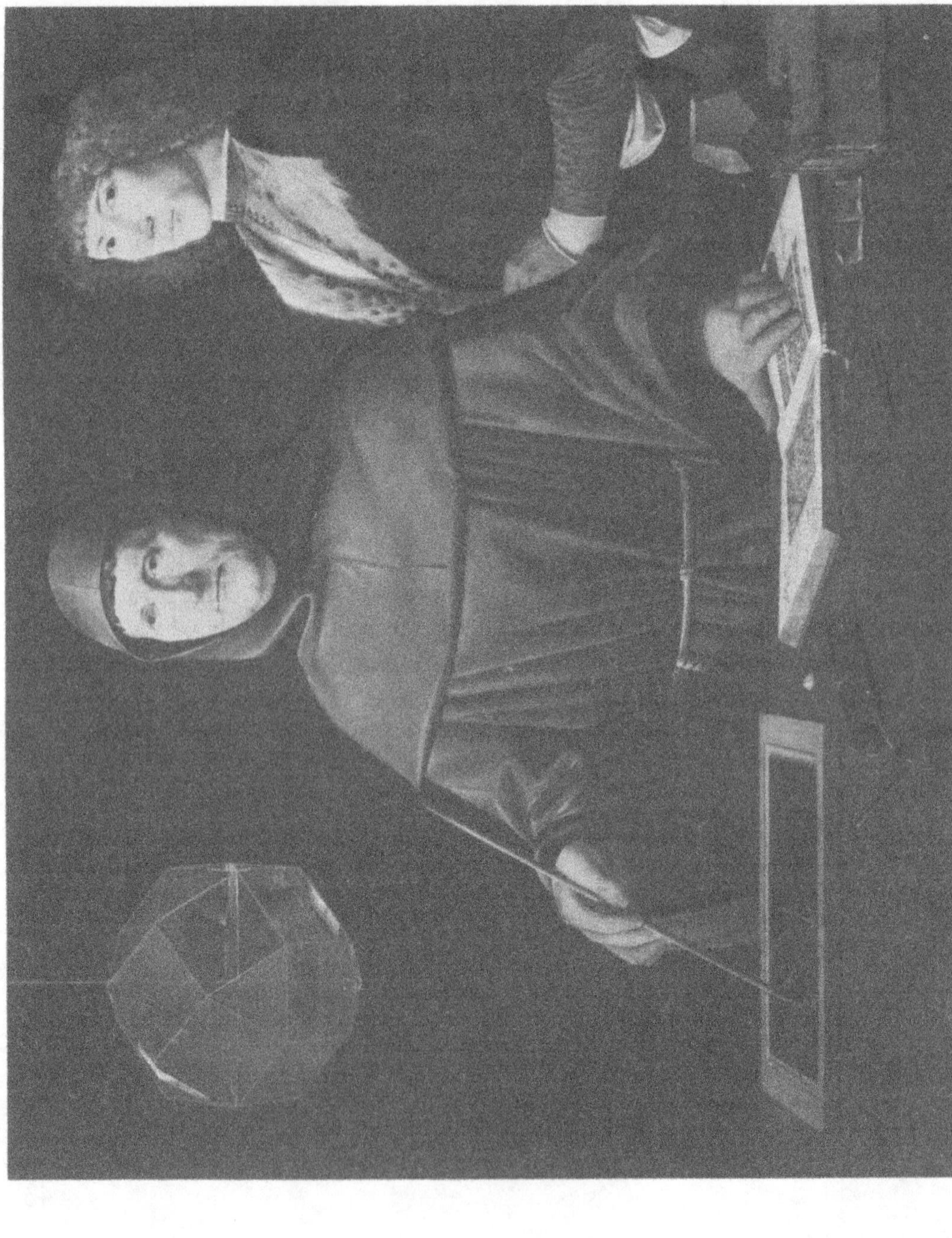

It was LEONARDO's eye that led him to the one original discovery in mathematics that it now appears he did in fact make. Known from antiquity was the position of the center of gravity of a triangle at the intersection of the medians. On one page (Atl. 194*r. v.*) (Fig. 15) we may see sixteen attempts by LEONARDO to prove this result, all of them crossed out. While none constitutes a real proof, they make clear LEONARDO's approach. One reads,

"Every diameter divides the triangle in two parts of equal weight; therefore that diameter contains in itself the center of gravity of all the triangle... "

Thus LEONARDO was thinking in terms of motion: The triangle if set on a knife edge along the median shows equal tendency to move to either side and hence does not move at all. On several different pages[31] LEONARDO attacks the problem of finding the center of gravity of a pyramid. Some passages give uncertain or wrong answers, but most of them assert that the altitude of the center is one quarter the altitude of the pyramid; there is no proof, but in one attempt LEONARDO constructs a diametral plane for the pyramid. I think we can best summarize his method by saying that he finally *saw* the center by balancing the pyramid in his mind and drawings (Figs. 16 and 17).

Another example is furnished by LEONARDO's approach to a problem proposed by IBN-AL-HAITHAM (965—1020), known in the west as ALHAZEN. A candle is seen in a circular mirror (Fig. 18): Which of its rays strikes the eye? The law of reflection requires a ray to turn back from the mirror at the angle with which it came. The problem, being purely geometrical, may be visualized just as well by a mechanical system: On a circular billiard table, how shall a ball at one place be aimed so as to strike a given second ball on the first return? Undoubtedly

[31] Arund. 65*r.*, 123*v.*, 124*r.*, 193*v.*, 218*v.*; Atl. 73*r.* b., 100*v.* c, 146*v.* c; MS K 89 [9]*r.*, MS F 51*r.*; and others.

Fig. 14 (p. 44). Portrait of LUCA PACIOLI and a student, painted by an unknown artist (photograph by courtesy of the Soprintendenza Gallerie Napoli)

Fig. 15 (pp. 46 and 47). Atl. 194*r.*a and Arund. 72*v.*, showing LEONARDO's sketches for the centroid of a triangle (by courtesy of the Biblioteca Ambrosiana, Milano, and of the Trustees of the British Museum)

Fig. 16 (p. 48). Arund. 123*v.*, showing LEONARDO's insight into the center of gravity of a tetrahedron (by courtesy of the Trustees of the British Museum)

Fig. 17 (p. 49).. Arund. 193*v.*, with LEONARDO's studies of the centroids of pyramids (by courtesy of the Trustees of the British Museum)

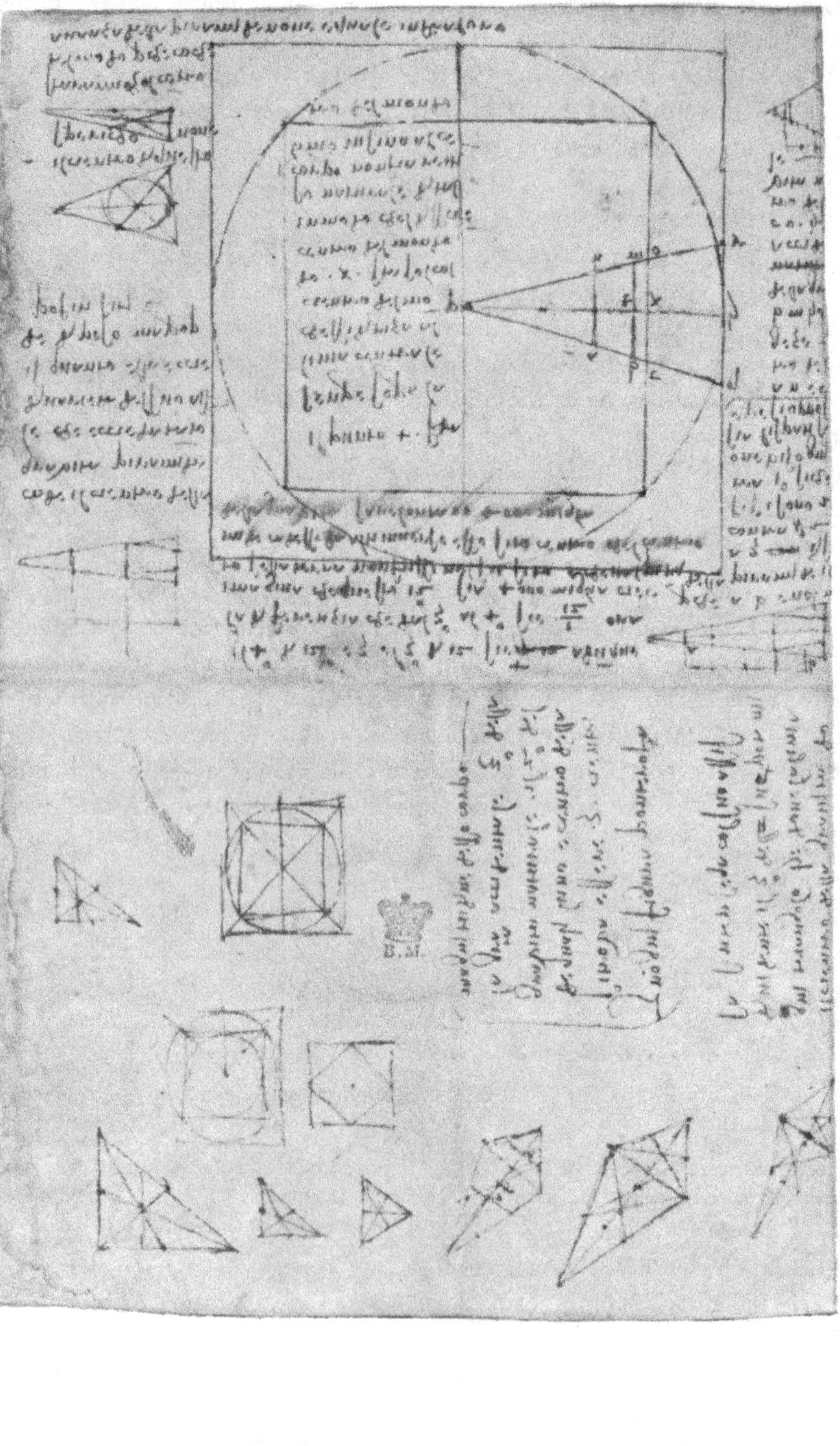

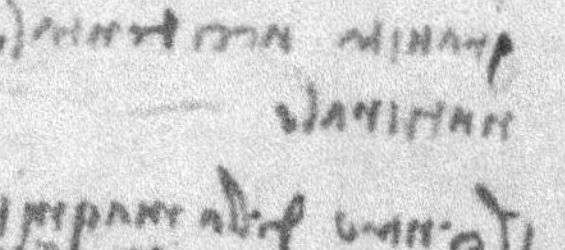

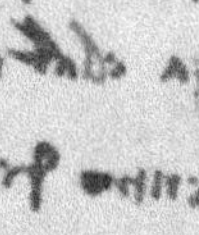

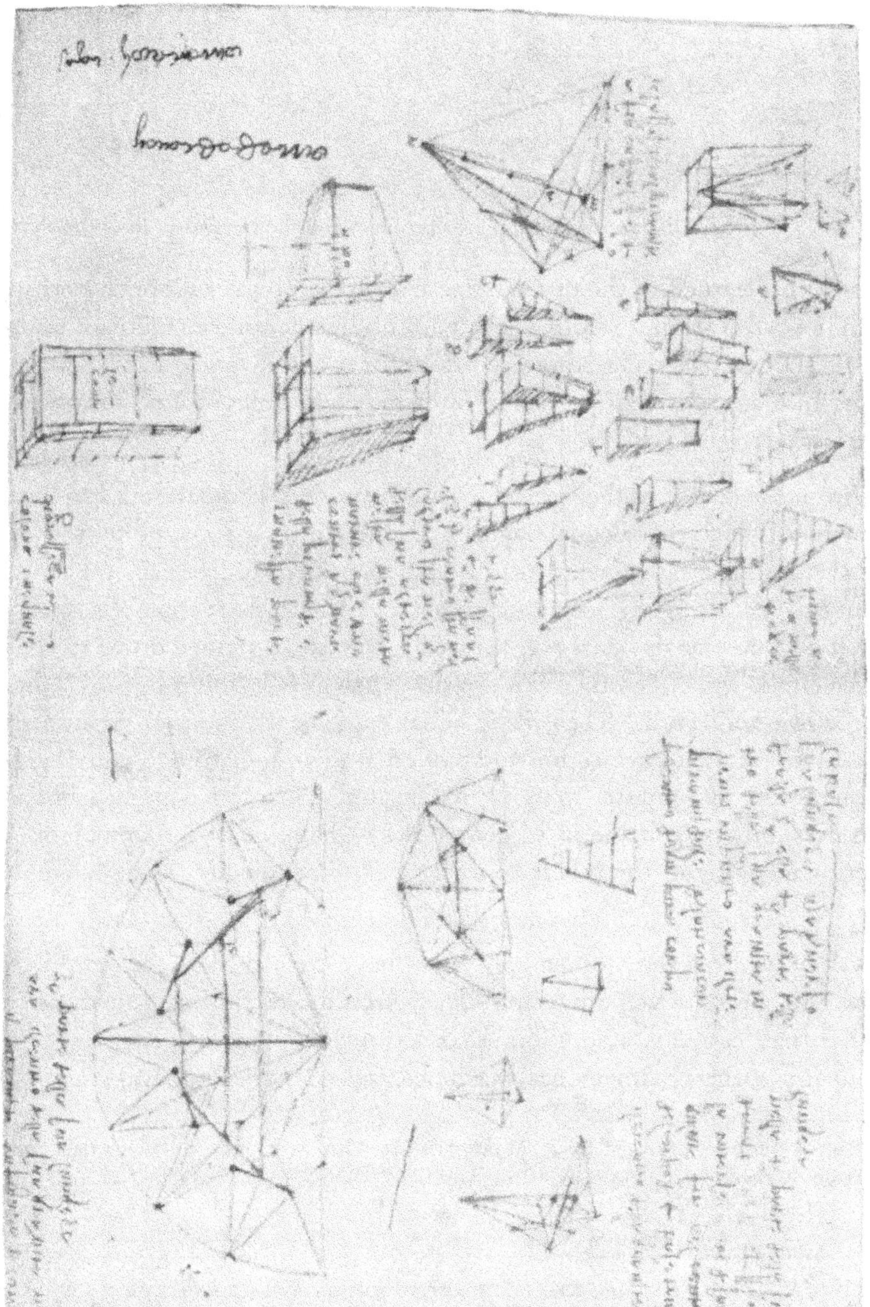

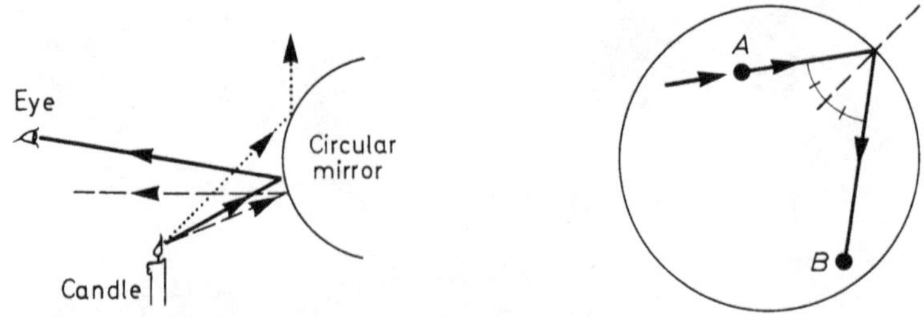

Fig. 18. Diagrams of ALHAZEN's problem and the circular billiard problem

LEONARDO learned of the problem from an Italian book on optics, perhaps even from IBN-AL-HAITHAM's book itself; with no less doubt, he could not understand IBN-AL-HAITHAM's complicated solution, for it employs the theory of conic sections. Instead, he invented a solution of his own, a mechanical one based on the simplest geometry (Atl. 191 *r*. a)[32].

From the center of the mirror, extend any radius. From the eye and the candle, draw lines which make equal angles with the radius at some point, say *P*. Now, preserving this arrangement, turn the radius until *P* touches the mirror at some point *C*. Then *C* is the solution. To achieve this construction, LEONARDO projected a linkage of five bars (Figs. 19, 20). One bar is pinned down to the center of the circle, while the other two are slotted so as to admit pins for the positions of the eye and candle. They make equal angles with the first because they are pinned to it at their intersection and pinned at their ends to two equal bars joined to each other by a pin free to move in a slot in the first bar. This elegant construction uses only two theorems of geometry: A diagonal of a rhombus bisects the interior angles, and the radius of a circle is perpendicular to the tangent at the point of tangency.

All of us, even the "humanities" folk, have learned these theorems in school. How few of us, even mathematicians, would see this elegant construction! Rather than calculate a solution, LEONARDO draws one by designing a device, one of the earliest mathematical machines. Indeed, LEONARDO was among the first

[32] This passage was analyzed by R. MARCOLONGO, "Lo strumento inventato da Leonardo da Vinci per la risoluzione del problema di Alhazen," *Rend. Accad. Sci. Fis. Nat. Napoli* (3a) **34**, 22—24 (1928).

Fig. 19. Atl. 181 *r*.a, including LEONARDO's treatment of ALHAZEN's problem (by courtesy of the Biblioteca Ambrosiana, Milano)

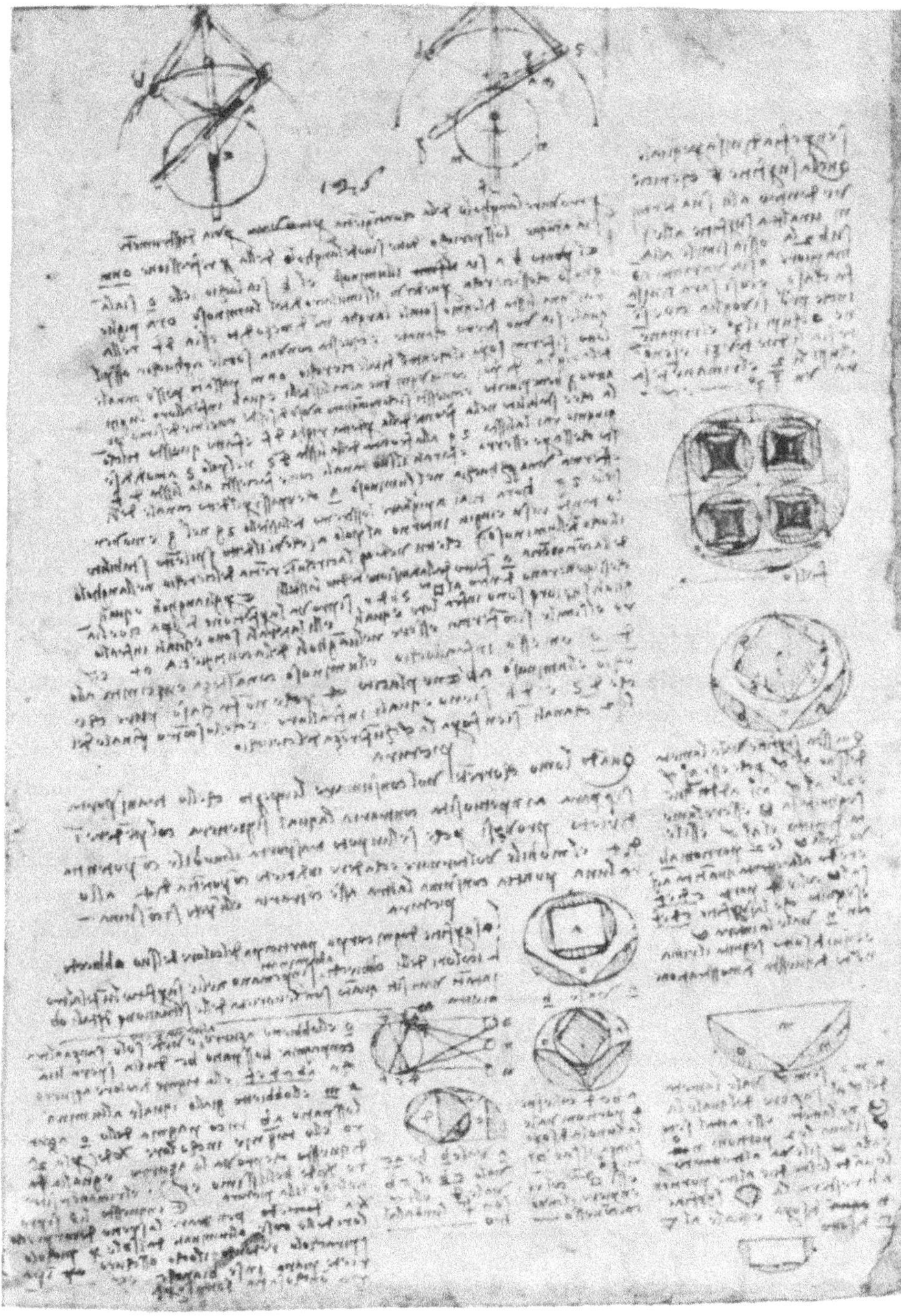

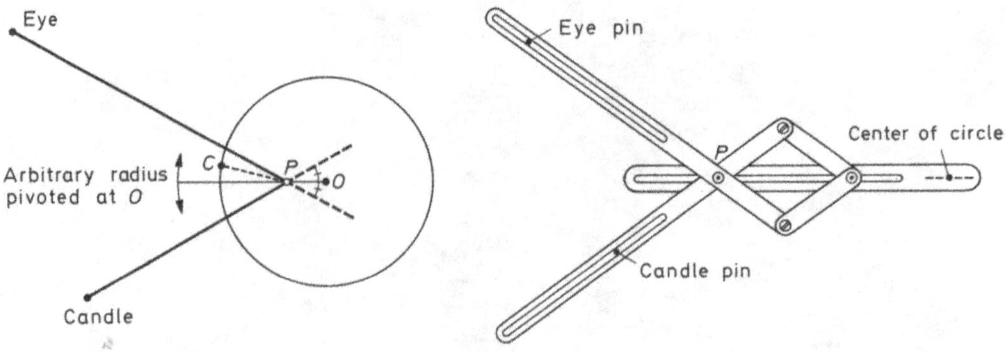

Fig. 20. LEONARDO's solution and linkage, as interpreted and realized by MARCOLONGO

to sketch linkages jointed so as to interconvert linear, reciprocal, and rotating motion. They help us to grasp some of his remarks on geometry, for he says,

"... point is not part of line" (Tr. 34*r.*),

but

"line is a length made by the motion of the point..." (Arund. 160*r.*),

and

"the line is the transit of the point, and points are boundaries of the line" (Arund. 176*r.*),

and he goes on in this style regarding surfaces and bodies. Although these imprecise words did not lead him to definite results, they suggest a visualization of space-time for which I know no earlier occurrence. One historian[33] has written that they foreshadow "a world envisaged in terms of the calculus — like the world of LEIBNIZ —." They offer us a key to understanding LEONARDO's successes and failures. For the theorist ALBERTI, a line was the totality of its points, to be obtained by projection one by one. For LEONARDO, a line is what the moving pencil draws. Of such lines, always produced in motion, are built his paintings, his machines, and his world.

While the word "forza" is ever dropping from LEONARDO's pen, it is a weapon of incoherent eloquence rather than a magnitude. The kinematical theories of the mediaeval schoolmen he lacked the mathematics to understand, but he could see motions. Although the forces that produce motions remained for him dark titans of the shadows, the motions themselves were the elements of his world.

[33] RANDALL, *op. cit.* p. 39.

The link between kinematics and dynamics, motions and forces, could not be seen: It had to be thought, and he was not the man to think it. But motions he could see, arrange, represent, and relate.

14. Perspective and Motion in Leonardo's Painting

We might expect that with his interest in visualized mathematics, LEONARDO would have had a part in the development of perspective. While just after his own time he was considered to have been one of the inventors of the system of perspective practised in the high Renaissance in Italy, it is difficult to learn what his specific contribution was[34]. VASARI's description of what LEONARDO gave to painting is as follows:

"LEONARDO... began the third style, which I will call the modern, notable for boldness of design, the subtlest imitation of nature in trifling details, good rule, better order, correct proportion, perfect design and divine grace, prolific and diving to the depths of art, endowing his figures with motion and breath."

This is one painter's appreciation of another. Writers on art often sneer at VASARI, perhaps because he is straightforward. VASARI has only admiration for LEONARDO's painting, but it is admiration of the kind one scientist may express for another today: no soul-rending ecstasy, no cosmic cataclysm, but a simple statement that LEONARDO was good at nearly everything a painter does. Although in saying a word about art I run the risk of seeming to question the principle of art for the art historians' sake, I suggest that those who reproach VASARI with noticing only the technical side of LEONARDO's art overlook the possibility that only this side may represent the LEONARDO of fact rather than myth, the LEONARDO of Renaissance flesh rather than the mystic idol of aesthetes. VASARI speaks of the *Mona Lisa* as "an exact copy of nature". The fact that VASARI himself can never have seen the original of the *Mona Lisa* rather strengthens than weakens the point, since it shows that he is reporting to us here an opinion general among artists of his own day. Discovery of infinities of psychic depths in the picture, while it is a tribute to the exorbitant sensibilities of recent critics of art, may well have nothing to do with LEONARDO himself as an object of historical study. A suitably

---

[34] For a discussion of LEONARDO's perspective, see the latter part of the article by JOHN WHITE, "Developments in renaissance perspective — 1," *J. Warburg Inst.* **12**, 58—79 (1949). WHITE seems to feel that LEONARDO's ideas on perspective "are... not of the sort to have immediate influence on practising artists...," but his article concerns only perspective in the strictly geometrical sense, not the "perspective of color" sometimes attributed to LEONARDO.

aesthetic person may fall into ecstasies over a clump of trees, a pile of rocks, or even, nowadays, a heap of old rags, but even he does not suggest that necessarily some great precursor of modern art dug the holes, unloaded the truck, or turned out the laundry bag. To insist on seeing in the *Mona Lisa* more than an extraordinarily fine portrait, of the same quality as the best by GIOVANNI BELLINI, ANTONELLO DA MESSINA, RAPHAEL, and TITIAN, is perhaps to imagine in LEONARDO a psyche precociously cursed with the ism fads of our own century. Thus one of the few pictures LEONARDO himself ever completed, the portrait of Ginevra de' Benci, despite its good documentation was until as late as 1939 generally attributed to a student, LORENZO DI CREDI. In relegating this great painting to a minor artist the critics of the last century and the first quarter of ours necessarily denied to it the unique, superlative mysticism they read into other works by LEONARDO — the qualities, they claimed, LEONARDO alone possessed. Thus the recent sale of the *Ginevra* at a record price proves that the older image of LEONARDO, the one still presented in works for laymen and usually accepted and multiplied by them with enthusiasm, is regarded by art historians today as being, if not false, at least insufficient to cover all his few known works. While VASARI called the *Ginevra* only "a beautiful portrait", it deserves also his praise of the *Mona Lisa*, "an exact copy of nature", where the words "copy" and "nature" are to be taken in the broad senses of their root meanings. From VASARI we learn that it was LEONARDO's technical innovations, the skills of eye and hand, that drew the universal admiration of rivals and connoisseurs in his day and made him, with his ten paintings half unfinished, an acknowledged master, a master admired more for key examples of technique than for his total production as an artist[35].

[35] According to a journalist writing in a cultural magazine issued while this essay was in press, "Until Leonardo, it has been said, Western art bears a family resemblance to the art of the rest of the world. After Leonardo, Western art took a course uniquely its own." The revulsion this typical utterance of capsular clichés from our summit of cosy enlightenment for surveying the world in precis on a small, bleary screen calls up in a thinking man should not prevent him from acknowledging that LEONARDO has had an *influence* on Western painters second to few if any. For example, echoes of LEONARDO's poses, faces, colors, and compositions resound through hundreds if not thousands of splendid, imaginative, intellectual paintings from what is now often called "the Mannerist period" (1520—1620), although these paintings rarely if ever probe the psychic depths critics, at least until quite recently, regarded as the most important content of LEONARDO's own products.

Fig. 21. GIOTTO's *Lamentation Over Christ* (by courtesy of the Gabinetto Fotografico del Museo Civico di Padova)

Fig. 22. Ercole di Roberti's *Pietà* (by courtesy of the Walker Art Gallery, Liverpool)

Fig. 23. Andrea dal Castagno's *Youthful David* [by courtesy of the National Gallery of Art, Washington, D. C. (Widener Collection)]

Leaving to others the field of imagining what it was that LEONARDO invented in art, let us return to VASARI's summary and note the virtue he did *not* attribute to LEONARDO. He says not one word about force. He spoke of GIOTTO, who to a Mannerist three centuries later might well have seemed archaic, as depicting "the warmest love", "the emotion of weeping", "tenderness and devotion"; of ERCOLE DI ROBERTI as "most moving", and as portraying "almost inconceivable grief"; of ANDREA DAL CASTAGNO, although he detested that "brutal man", as showing "the suffering of the flesh, and the divinity hidden in the body preserving a certain splendor of nobility", and the display of "hatred and fury" in faces; of MASACCIO as having painted a head "with such vigor that it seems to lack only the power of speech" and as showing "an invincible and divine spirit"; and there are many more examples to prove that VASARI did see force of expression, when it was there.

While VASARI's long biography of LEONARDO is full of praise of his skill as an artist, vigor and force are never mentioned. The summary refers to figures endowed with "motion and breath", and this quality VASARI mentions again and again. People flocked from far and near to see the cartoon of the *Virgin and St. Anne*; what they admired was "the simplicity and beauty", the "modesty and humility of a Virgin contentedly happy, ... while St. Anne smiles in her great joy."

It is motion, irrespective of the forces needed to produce it, that LEONARDO presents in the final painting. The arrangement, showing the Virgin sitting upon the lap of her mother, is traditional and symbolic. Perhaps in an attempt to render the scene less improbable as a glimpse from common life, he replaces the solid balance of weight on weight by motion on an inclined plane. The sitter, about to slide off, stretches out her arms to the Child. The arms have just reached their full extension so as to pick up the little boy as the woman slips to the ground. No mighty gesture of joy or grief or menace or anger, but, rather, an ordinary, unposed human motion is recorded. Here is the virtuosity that excited the wonder of artist and connoisseur in LEONARDO's own day. In an age which

---

Fig. 24 (p. 58). MASACCIO's *Tribute Money*

Fig. 25 (p. 60). LEONARDO's Cartoon for *The Virgin and Child with Saints Anne and John the Baptist* (by courtesy of The National Gallery, London)

Fig. 26 (p. 61). LEONARDO's painting of *The Virgin and Child with Saint Anne* (by courtesy of the Musée du Louvre)

has turned its back on skill, which triumphantly proclaims the artist's right to depict nothing at all, on a basis of no technique whatever, as an "experiment" of no interest to anyone, our art critics are compelled to seek and find something different in LEONARDO than the master of the craft of "photographic" likeness. True, the mark of a masterpiece is a basic veracity, which succeeding ages can reinterpret in the light of their own experiences and tastes, but the Victorian appreciators verge on arrogance in claiming themselves able to see as the most obvious and important in LEONARDO's paintings an aspect to which such giants of art as GIAMBOLOGNA, the CARRACCI, and GENTILESCHI were, so far as we can tell, stark blind. Rather, I suggest, the critics of the past hundred years have deluded themselves and us into seeing something that was not there, something as foreign to the sixteenth century as would be democracy or Freudism or strategic bombing or public welfare. If we could look upon LEONARDO's paintings dispassionately, without feeling the social duty to spout superlatives over them between each gasp of wonder, we might apply to them a simple term from the science of mechanics: LEONARDO's image is kinematic, not dynamic. He shows us motions, with little or no grasp of the forces that give rise to them.

15. Leonardo on General Aspects of Fluid Motion

Fearful of being haled before the bar for practice of art criticism without a license, I hop back into the frying pan of unprofessional history of science. LEONARDO's style of painting, added to what we have seen of his mathematics, helps us to infer requirements hedging possible original work in natural science by him:

1. It could not involve any mathematics beyond visual geometry.
2. It could not require knowledge of previous science.
3. It could not require experiment or measurement.
4. It would necessarily consist in the observation of configurations or motions, irrespective of the forces or other causes.

This may seem to leave only a narrow field, but it is the kind of field in which a human being, rather than a "genius", can work. The motion of fluids, as that subject stood in 1500, fits these requirements, and here LEONARDO's results are real and, to the best of my knowledge, original.

Here, too, we find mainly mere questions or statements of what ought to be investigated or muddy, silly theory, as when he compares the ocean to the heart, the "pool of blood" which "fills the body of the earth with an infinite number of veins of water," or "water, which is the vital humor of the terrestrial machine, moves by its own natural heat." All this we brush aside in favor of what LEONARDO reports from observation.

In LEONARDO's notes and drawings we find what seems to be the earliest statement of the law of communicating vessels:

"... the surfaces of all liquids at rest, which are joined together below, are always of equal height" (Atl. 219*v.* a),

no matter what be the sizes and shapes of the vessels. He observed also the rise of fluid at the wall of a spinning vessel and of a stationary capillary tube (Atl. 11*r.* b, 35*r.* a, MS G 44*v.*, 68*r.*).

LEONARDO's work on flying machines has been inflated by his devotees [36]. In 1500, naturally, no-one understood fluid mechanics sufficiently to do more than toy with such devices, and it is no reproach to LEONARDO that he merely imitated birds and fishes. While DAEDALUS had been content to affix his wings with wax, LEONARDO designed a harness and a flapping mechanism. LEONARDO's cultists, undauntable, again experimenting with machines sketched by him, do him injustice, since not every drawing should be interpreted as a production design. If a mathematician today should project something so patently absurd as a man-powered helicopter (Fig. 27), he would be met by the patronizing smiles the practical bestow upon eggheads. Even LEONARDO in 1490 cannot have thought such a device could work. Of course, the human figures were added by a modern enthusiast, one of those who by their unreasoning abjectness have been and are in fact detractors of him they worship. In LEONARDO's own drawing, reproduced below the model, the capstan should be interpreted merely as a schema to show that the screw should turn. When we see in a scientific work of the seventeenth century a diagram showing cannon balls lifted by hands unsupported except by elegant lace cuffs, we do not suspect the scientists of that day of having thought a hand severed from the body could lift, nor do we form clubs to experiment with the loose hands of cadavers in hope that HUYGENS or VARIGNON had somehow dreamt a feasible system of sky-hooks. Here, at least, LEONARDO is not showing us a silly gadget. What he illustrates is the Archimedean screw lift for water, except that, since the fluid is to remain and the machine is to rise, the sides are left open. This idea is simple, but it is based on a fact which is not obvious. That fact is that *air and water are essentially alike in respect to their motions.* Time after time in LEONARDO's

[36] The standard reference is R. GIACOMELLI, *Gli Scritti di Leonardo da Vinci sul Volo*, Roma, Bardi, 1936.

HELICOPTER
Scale model of a design for an aerial screw, heralding the helicopter of the present day and also credited by many authorities as being the forerunner of the modern propeller.

manuscripts we read,

"... in all cases of motion water has great conformity with air" (MS A 61*r.*),

and

"... the motion of water within water proceeds like that of air within air[37]",

and

"The wind bears likeness with a movement of water" (Atl. 361*v.* a),

and many similar statements (*e.g.*, Atl. 373*r.* b, MS A 9*v.*, 60*r.*).

Such a conclusion can come only from observation. While simple, it is a necessary preliminary to rational aerodynamics, and LEONARDO was the first to assert it clearly and strongly. It is not so small a thing as the unscientific might think: NEWTON, a great experimenter, almost two centuries later failed to see this unity and hence constructed a largely irrelevant beginning for aerodynamics, a beginning whose influence hangs on today in the sophomoric notices to be seen in many a fixit shop, praising the bee for disregarding the theorists' dictum that he cannot fly. LEONARDO did not invent the airscrew, since whirligig toys existed long before he was born[38]. Rather, we may guess, this application joined other data from which he inferred the first vague concept — so far as I know, he was the first to do so — of a *fluid* as supersuming both water and air.

Here, too, as in everything we try to conclude from LEONARDO's notebooks, we must face negative evidence along with positive. LEONARDO knew that air and

[37] MACCURDY in his English translation of the notebooks, p. 645, gives this passage as being in Atl. 108*v.* a, where it is not to be found. I transcribe it above because I am sure it occurs somewhere, and it is the clearest of LEONARDO's statements to this effect. Of the passages quoted from LEONARDO in this essay, this is the only one for which I have not revised the translation.

[38] There is considerable literature regarding LEONARDO's famous drawing; references to it are included in the latest item, L. RETI, "Helicopters and whirligigs," *Raccolta Vinciana* **20**, 331—338 (1964). RETI supports the thesis of C. G. GIBBS-SMITH:

1. The idea was not original with LEONARDO, since whirligig toys existed at least as early as 1440 and are represented occasionally in fifteenth-century art. RETI has found a diagram of one in a manuscript on military engineering.
2. LEONARDO's great contribution was to conceive a helicopter carrying its own power unit on board and designed to sustain human weight.

I do not see evidence for the second claim, which seems to me to confuse "design" with "dream".

Fig. 27. LEONARDO's project of an airscrew (MS B 83*v.*), and a modern model for it (from the collection of the Fine Arts Department, International Business Machines Corporation)

water differ in one essential respect; indeed, he states that air is infinitely compressible, while water is incompressible (Atl. 108*r*. a, Tr. 42*r*.). In another passage he writes,

"Air does not resist unless it grows denser ...
"Air does not grow denser unless it is moved." (Atl. 180*r*. a.)

If the analogy between water and air were precise, then, water could not resist at all, but LEONARDO nowhere draws this logical result of comparing his several statements about resistance, air, and water.

There are a number of passages in which LEONARDO considers the resistance of fluids in general terms.

"A thing makes as much resistance against the air as the air against the thing." (Atl. 381*v*. a.)
"In regard to the motion of water, it is as hard to move the oar against the stationary water as it is to move the water against the stationary oar." (Atl. 175*r*. c.)
"The operation of the motion of water against the motionless air is equal to that of the air against the motionless water." (Arund. 135*r*.)
"Every body flees in a direction opposite to the place from which it is driven by the thing that strikes it.... The body is to strike the object by so much as that object is to strike the said body." (Arund. 85*r*.)[39]

While such assertions as these are useful today in the calculation of resistances, they are hedged by restrictions. In fact, two different ideas are involved: (1) in any mechanical action, linear momentum is unchanged unless forces are applied, and (2) a system of reactions is unchanged by translation of the frame of reference at constant velocity. As LEONARDO nowhere illustrates the use of his own statements, we cannot know what he really meant, and it is not just to take them in modern senses. Since dynamical laws sufficient to justify such principles were not known by others in LEONARDO's time or stated by him, it is more likely that here he is merely extrapolating, without sufficient reason, the ancient principle of equilibrium of static forces.

How can one determine water flow precisely? Says LEONARDO,

"... by the leaves of the trees which are borne in considerable quantities from the surface to the depth of the flowing and transparent water." (MS F 44*r*.)

But one need not wait for leaves:

"Here is the exit of the water near its surface, and one asks what part of the surface of water will take on the swifter or slower motion at such an exit. To make a rule, I should put particles

[39] The translation of the latter phrase is not certain. LEONARDO wrote, "tanto e al corpo a percotere ne lo bietto quanto aesso obietto a percotere in detto corpo."

of things that float, and which are equal [in size], as are some small seeds of grasses, and put them in a circle, equidistant from the exit..., and take note of the first that comes to the mouth, stop the water, and look at the circle." (Atl. 126*v*. a.)

To determine the currents in a boiling cauldron,

"... with slow motion you may drop a few grains of millet, because by means of the motion of these grains you can quickly know the motion of the water that carries them with it. And with this experiment you can study many beautiful motions which happen from one element's having penetrated another." (MS F 34*v*.)

"If you throw sawdust onto a running stream, you will be able to observe where the water, turned upside down after striking against the banks, throws the aforementioned sawdust back toward the center of the stream, and also the revolutions of the water and where other water either joins it or separates from it; and many other things." (MS K 1*r*.)

To observe the efflux from a vessel with glass sides,

"scatter little seeds..., such as to swim throughout the depth of the water.... The motion of the said seeds will let you notice which of the water goes most swiftly to the aperture, and from what location it moves." (Atl. 81*r*. a.)

This same technique he proposes to employ in the earliest project of experimental apparatus in hydraulics (MS I$^{(2)}$ 115 [67] r.)[40].

It was two centuries before these techniques were introduced again, and they are still in use today. One of LEONARDO's drawings is shown as Fig. 28. Other artists have drawn waterfalls as beautiful, but few as accurate as this, which, like LEONARDO's anatomical studies, must be the result of many observations. The waterfall obliges by holding its pose; a few grains are dropped in at one point, and their paths are sketched; then a few at another; then at another, and so on for hours until the whole waterfall is drawn.

Here we should pause a moment to remark how this drawing differs in concept from any even so little as fifty years older. In saying that it is more "accurate" than a late mediaeval drawing of the same object could have been, we recognize a convention of optical uniformity that was adopted, it seems, by most of the Renaissance draughtsmen, and among them LEONARDO surely was the most influential.

---

[40] "Experiment of jumps upon a uniform canal. Let one side of a canal be made of glass, and the rest of wood, and let the water which strikes it be mixed with meal or a pasta of horsetail (pasta di palperi), so that the course of the water is better seen through their motions. And once you have the experience of these jumps, fill the bottom with sand mixed with fine gravel, then smooth out the bottom, and make the water jump onto it, and see where it lifts it or puts it down. Then make the bank on the wooden side of fine sand, and look through the glass at its effects, and make it in flowing water." It is not clear what this experiment was intended to show.

Such a convention is necessary if the object itself, not merely an impression of it or a feeling of its place in an artistic scene, is to be reconstructed. The drawings of machines by the thirteenth century architect VILLARD D'HONNECOURT, which were not intended to be "artistic", we find today so "vague" that we sometimes cannot know exactly how he meant the parts to fit together. A fine Renaissance drawing of a machine, even if it were intended for nothing but background or detail in a painting, is so faithful as to leave no doubt about the size of every bevel or bolt shown. Without such a rigid optical convention[41], no such thing as scientific illustration would be possible, and here we see a point where graphic art indeed bears close upon the rise of science.

16. Leonardo on Waves and Vortices

Returning to the study of water flows, we find LEONARDO the first to state the principle of superposition: A small disturbance of the surface of the water pursues its course, unaffected by other small disturbances.

"Although the voices which penetrate this air separate from one another with circular motions from their causes, nonetheless, the circles moved from different beginnings meet together without any impediment and penetrate and pass through one another, keeping always for center their causes.

"Since, in all cases of motion, water has great conformity with air, I will cite this [fact] as an example for the proposition stated above. I say, if you cast at the same time two little stones some distance apart into a motionless area of water, you will see to be caused two separate quantities of circles around the two said percussions, which quantities, growing, come to encounter, one circle intersecting another, always maintaining for centers the places struck by the stones. And the reason is, although there appear some proof of movement, the water does not leave its location, because the openings made in it by the stones close up again at once, and this motion

[41] Writers on art seem to assume rather naively that the Renaissance optical convention is the *only* one that can serve for scientific illustration. Of course this is not true, as may be seen from the assembly of Euclidean projections that form the main if not the only design drawings for a piece of equipment today. In many cases it suffices to give two or three or a few more orthographic projections, which were first mentioned, according to the *Oxford Dictionary,* in the *Philosophical Transactions* for 1668, and which became universal in nineteenth-century engineering by the influence of the ever wider use of rectangular Cartesian co-ordinates in mathematical geometry from the mid-eighteenth century onward.

Fig. 28. LEONARDO's sketch of a waterfall (Windsor Drawings 12660*v*., reproduced by the gracious permission of Her Britannic Majesty)

[illegible]

[illegible]

[illegible]

[illegible]

made by the sudden opening and closing of the water makes in it a certain shaking which may be called trembling rather than motion. And to make this that I say plainer, take heed of those straws which by their lightness stand on the water; notwithstanding the wave made under them by the coming of the circles, they do not leave their first locations. Since this sort of response of the water is a trembling rather than a movement, [the circles] cannot break one another in meeting. Since all the parts of the water are of one kind, it is necessary that those parts pass on that trembling, one to another, without changing their place, since the water, remaining in its location, easily takes that trembling from the neighboring parts and transmits it to other neighboring parts, the power always lessening until the end." (MS A 61*r.* (Fig. 29), *cf.* also MS H 31*v.*, Forster III 76*r.*)

"The impetus is much swifter than the water, for many are the times the wave flees the place of its creation, and the water does not move from its place, in likeness with the waves made in May in the stalks by the drive of the wind, when the waves are seen to run over the fields, and the stalks do not change their places." (MS F 87*v.*, *cf.* also Leic. 14*v.*)

Thus by observation LEONARDO is the first to learn the true nature of waves: It is motion, not matter, that they transport. He seems to have seen also that water waves are effected through vertical rise and fall, for he writes,

"The valley interposed between the waves is lower than the common surface of the water, as is seen in the water which comes back to fill up again the places struck by the fallings of the waters" (MS E 71*v.*),

and

"By so much higher are the waves of the sea than the ordinary height of the surface of its water, by that much lower are the bottoms of the valleys interposed between these waves." (MS F 25*v.*)

Many passages in his notebooks describe the reflection of waves from the banks of channels, the distortion of wave patterns by the current, and the production of composite wave forms by superposition of simpler ones.

LEONARDO's careful and varied observations of the flow patterns in rivers and canals caused him to give much attention to vortices, which indeed he seems to have been altogether the first to describe ("l'onde colunnali", "l'onda semicolunnale", "retrosi").

In distinguishing the motions of the small eddies from those of the large, LEONARDO gave the first description of turbulence:

"Of the things carried by water, that will be of greater revolution which is of lesser form. This happens because the great revolutions of eddies are rare in the courses of rivers, and the small

---

Fig. 29. MS A 61*r.*, showing LEONARDO's remarks on the superposition of surface waves (by courtesy of the Institut de France)

eddies are almost numberless, and large things are rotated only by large eddies and not by small ones, and small things are turned by both small eddies and large." (MS F 3*r.*)

LEONARDO stated also that

"the water at the bottom makes its eddies in a movement contrary to those above." (MS A 61*r.*)[42]

While the reason given is purely theoretical, and unsound, the fact itself doubtless was observed and may be the first recorded example of secondary vortices carried in a great one whose plane is perpendicular to theirs. LEONARDO observed and drew the vortices formed at a corner:

"If the canal experiences a sudden widening on each side, it generates vortices on each side. . . ." [MS F 91*v.* (Fig. 30)],

and he noticed the damage caused by these vortices in eroding banks. He drew also the vortex wakes behind obstacles[43] (Fig. 31). Such vortices were not to attract much attention until they became an object of intense study in our own century.

LEONARDO saw also that in a river, the water flows more slowly near the sides and the bottom:

"Water of equal depth, breadth, and inclination, the swiftest is that nearest the surface." (Atl. 124*r.* a.)

"In the middle of straight canals the water runs faster than by the sides." (MS H$^{(2)}$ 36*v.*, *cf.* also MS H$^{(2)}$ 45*r.*, I$^{(2)}$ 58*r.*, 105 [57] *v.*, Atl. 175*r.* c.)

---

[42] ARCONATI in Chapter 53 of Book IV of the work cited in the next footnote corrected this to read "eddies which turn in movement contrary to that above", which may be presumed to give the sense LEONARDO intended.

[43] It is safer to qualify this statement. Several drawings are given in Book IV, Chapters 29—34 of *Del Moto e Misura dell'Acqua*, ed. E. CARUSI & A. FAVARO, Bologna [1925], from the manuscript compiled in 1643 by F. LUIGI MARIA ARCONATI on the basis of manuscripts by LEONARDO. Many passages in this book are quoted more or less exactly from entries in the still existing notebooks of LEONARDO. For others, including those about vortex wakes, I have found no Leonardian original. While ARCONATI made some improvements and additions of his own, there is no reason to suspect the descriptions of vortex wakes, since they are quite similar to what LEONARDO did write on vortices at corners, and LEONARDO's drawings show that vortex wakes attracted his attention.

I note here that the passages from LEONARDO's own notes that CARUSI & FAVARO append as footnotes in their edition are neither exhaustive nor always the best or the closest to ARCONATI's text.

Fig. 30. MS F 91*v.*, containing LEONARDO's sketch of a vortex generated by the sudden widening of a canal (by courtesy of the Institut de France)

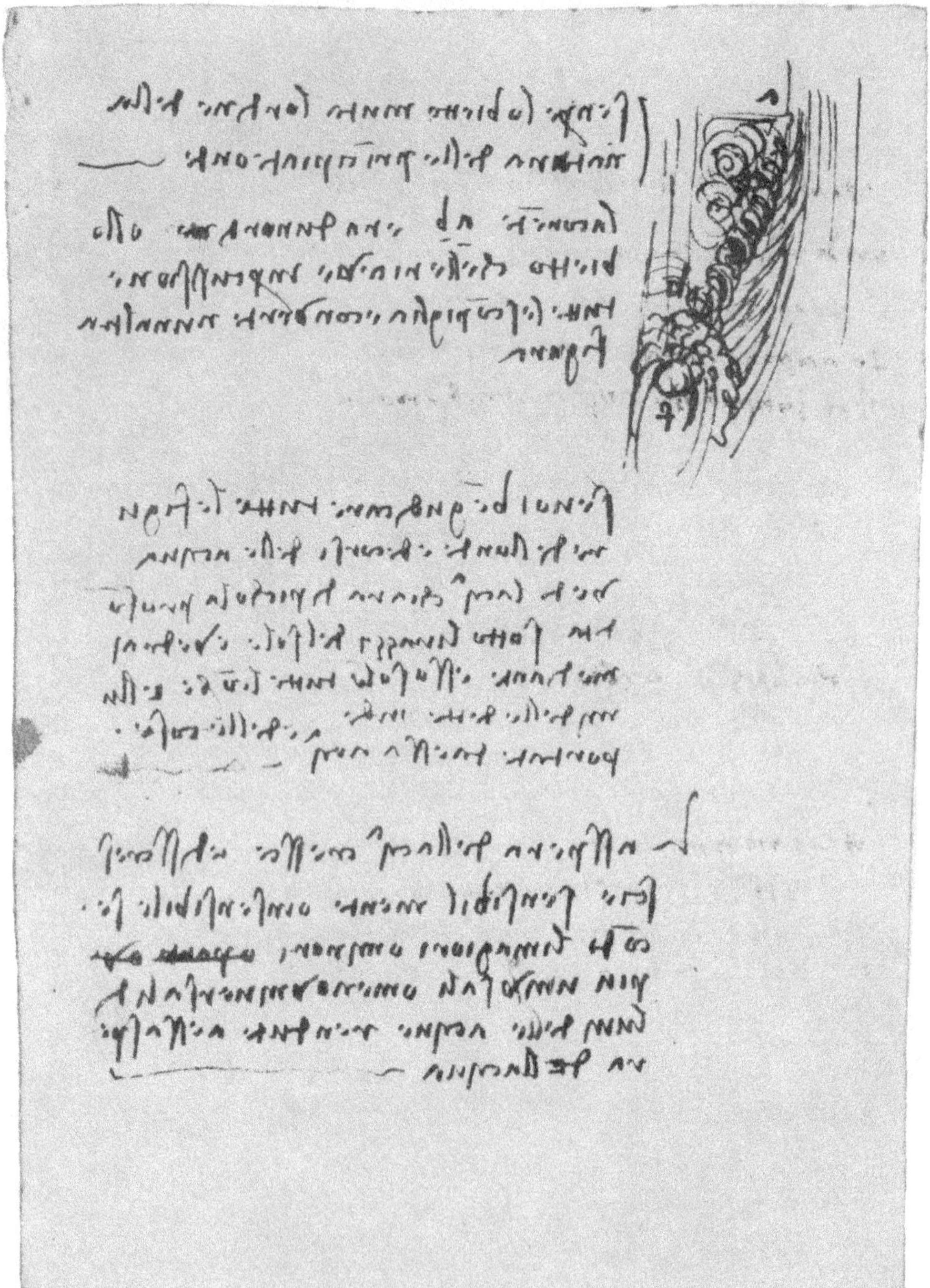

He remarked also the friction of denser or rarer bodies, perhaps fluids, upon water (Arund. 136*r.*). He was the first to propose the problem of finding the cross-section of given area leading to the greatest discharge, and he asserted the solution to be a half circle (MS K$^{(3)}$ 25*r.*). He noticed the friction of the walls of a tube upon the water within and suggested it be minimized by making the tube as big as possible (MS E 12*v.*). He made many observations of the flow through orifices of various shapes (*e.g.* MS F 9*v.*). Finally, he seems to have been the first to remark that the discharge is proportional to the speed of flow for a given orifice (MS F 16*r.*). He attempted to observe the forms of jets and the effect of increasing the inclination of a canal, but no one should be surprised that his results on problems so far beyond his day in theory and so difficult to study by mere observation are not right (*e.g.* MS I$^{(2)}$ 25*r.*, 85*v.*, MS C 5*v.*, Atl. 251*v.* d, MS B 4*v.*).

These examples show LEONARDO watching mechanical situations in much the way that ARISTOTLE had observed and described the growth of a chicken's embryo and the mating of sharks. Such acute observation indeed witnesses to a special talent, but it is no more science than are the activities of the bibliographer or textual critic. Of course I am using the word "science", not as a synonym for knowledge or custom, but in the precise sense common from the eighteenth century onward and defined as follows in the *Oxford Dictionary* (1910)[44]: "A branch of study which is concerned either with a connected body of demonstrated

[44] The *Dictionary* notes, in fact, a later usage, supported from 1867 onward, "sometimes with implied exclusion of pure mathematics"; it is to this sense, more or less, that the opening sentence of this lecture refers.

There is a still newer sense, now in process of overwhelming all the older ones; this newer sense could be defined as follows: "A branch of study which is concerned with customs and mores, methods of manufacture or processing, and beliefs inferred from common experience or repeated trials, usually with the aid of statistics, and whose cultivators have formed professional groups seeking the intellectual recognition formerly accorded to experts in the mathematical, physical, and biological sciences." A short random search of the catalogues of Graduate Schools delivered, in addition to the ubiquitous "Social Science", "Political Science", and "Computer Science", some special examples: "Meat and Animal Science" (Univ. Wisconsin), "Administrative Sciences" (Yale), "Speech Sciences" (Purdue), "Library Science" (Indiana Univ.), "Forest Science" (Harvard), "Dairy Science" (Univ. Illinois), "Mortuary Science" (Univ. Minnesota).

Fig. 31. MS F 65*v.*, containing LEONARDO's drawing of the trail of vortices behind an obstacle (by courtesy of the Institut de France)

truths or with observed facts systematically classified and more or less colligated by being brought under general laws, and which includes trustworthy methods for the discovery of new truths within its own domain." Science, real science, has no monopoly on intelligence or curiosity or logical reasoning or patience or industry, qualities which may be shown just as well by an animal breeder or a stock-broker or a painter. At the beginning of this lecture I stated that the characterizing essence of Western natural science is its being *mathematical*, and that the task of the historian of science is to trace the origin and growth of the mathematical way of thinking about the world. Now obviously "mathematics" does not here mean "pure" mathematics divorced from nature, nor do we expect LEONARDO or anyone else to integrate the differential equations of dynamics before they had been discovered. By "mathematics" I mean the attachment of mathematical quantities, usually numbers or geometrical figures or operations delimited by formal rules, to natural phenomena and the subsequent use of strict mathematical reasoning to replace the qualities, tendencies, experimental measurements, and rules of thumb of the sciences of other cultures and the crafts of all cultures.

17. Leonardo's Principles of Continuity and Circulation

There are in LEONARDO's writing on water flows two[45] striking assertions about nature that are both true and mathematical, namely, the earliest special cases of the *principle of continuity* and the *principle of equal circulation*. Both are obscured by the vagueness and complication we can always expect as companions to the conception of a major idea.

According to the principle of continuity, as we know it today, the speed of steady flow varies inversely as the cross-sectional area of the channel (Fig. 32). LEONARDO seems to have approached this principle through several special cases, subject to unnecessary restrictions[46].

---

[45] A third, "PASCAL's principle" of the equality of the pressures of the air at all points of the wall of a vessel which it fills, has been attributed to him, but I do not find it in the passage cited, *viz* Atl. 351 *r*. b.

[46] *Cf.* F. ARREDI, "Gli studi di Leonardo da Vinci sul moto delle acque," *Annali dei Lavori Pubblici* 1939, No. 4, 357—363.

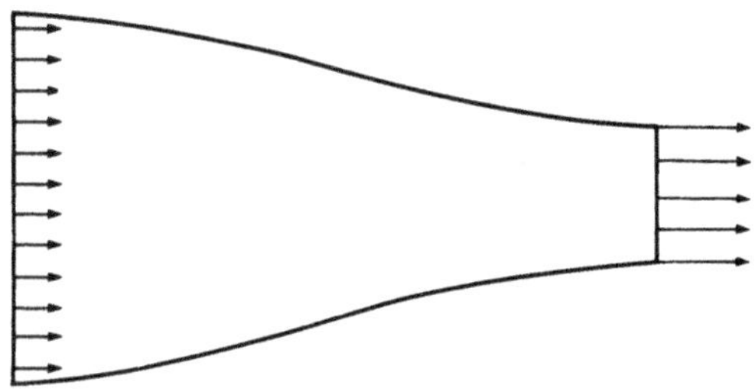

Fig. 32. Principle of continuity for steady flow. In a pipe or channel, the water flows faster as the cross-section decreases: speed $\times$ area is the same at each section.

"Water which runs between banks of equal breadth, that will have less depth which possesses a swifter course. Water which runs upon an equal obliquity of bottom, that will have less depth which is of greater breadth." (MS $H^{(2)}$ 54 [6] *v.*)

"Every movement of water of equal breadth and surface will run that much faster in one place than in another, as [the water] may be less deep in the former place than in the latter" (MS A 57 *v.*),

and

"[for] waters running over the bottoms of rivers of uniform inclination, the proportions of the velocities of their motions is that of their depths." (Atl. 81 *v.* a.)

"The universal movement of water, being of equal bottom, will have so much the greater passage in one place as in another, as its ocean or river has narrower banks." (Atl. 361 *v.* a.)

"By so much as you will increase the river in breadth, by so much you will diminish the speed of its course." (Atl. 80 *r.* b.)

"By so much as you will increase the river in breadth or in depth, by so much you will diminish the quality of its movement." (Atl. 80 *r.* b, *cf.* Leic. 24 *r.*)

Finally,

"In rivers of whatever variety of breadth and depth, it necessarily happens that in every degree of its length an equal quantity of water passes in whatever [given] time. Necessity demands this, since in that degree where less water had passed than in the others the water would be seen to dry up there, since the degree which came below this one could not give more water to its successor than its predecessor had given to it; and thus the degrees where more water had passed, not having anything to clear it away ahead of them, would be seen in time to rise all the way up to the sky." (Leic. 6 *v.*)

Moreover, Leonardo saw that this principle amounted to one of steady discharge:

"If the water is not added to or taken away from the river, it will pass with equal quantities in every degree of its breadth [length?], with diverse speeds and slownesses, through the various straitnesses and breadths of its length." (Atl. 287 *r.* b.)

The principle of equal circulation, when applied to a vortex, asserts that the product of speed by length is the same on each circle of flow (Fig. 33). The earliest statement of this latter fact is to be found in the manuscripts of LEONARDO:

"The helical or rather rotary motion of every liquid is so much the swifter as it is nearer to the center of its revolution. This that we set forth is a case worthy of admiration; for the motion of the circular wheel is so much the slower as it is nearer to the center of the rotating thing. But in this case [*i.e.*, of water] we have the same motion, through speed and length, in each whole revolution of the water, just the same in the circumference of the greatest circle as in the least...; and thus the water is of equal motion in all its circular motion...." (Atl. 296*v*. b.)

For this astonishing fact LEONARDO adduces a purely theoretical reason, and of course an unsound one.

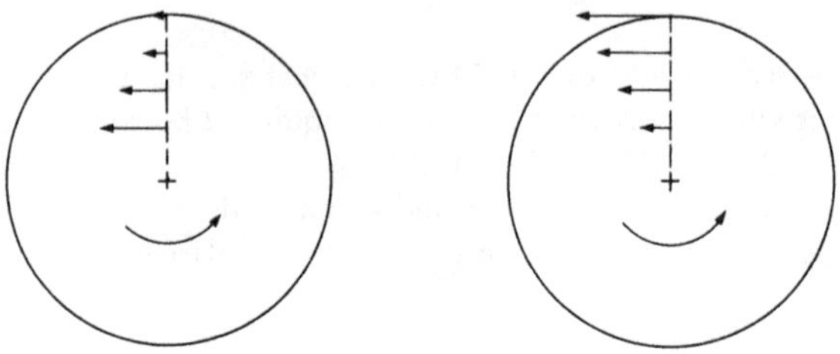

Fig. 33. Vortex and wheel.

Rotation of a vortex: the speed on the rim is the least, the speed at the center is very great. (Law: velocity $\times$ distance $=$ const.)

Rotation of a wheel: the speed on the rim is the greatest, the speed at the hub is zero. (Law: velocity/distance $=$ const.)

It is a rare experimenter who discovers two such laws as these, two fundamental propositions in fluid mechanics. For a single, much lesser, discovery many a man not otherwise famous, let alone a great painter, has earned a permanent name in the history of western science. If LEONARDO discovered these two principles from observation, he stands among the founders of western mechanics[47].

But did he discover them? We have no reason to think so, every reason to doubt it. Recall his method as we have inferred it from simpler cases; recall that the only functional relations he *ever* proposed were linear ones, that he *always* proposed such a relation if it were not obviously false, and sometimes even if it was. In

[47] In a rather enthusiastic essay C. ZAMMATTIO ascribes to LEONARDO the first observation of the *vena contracta*, but without a reference, and I have found nothing to support his claim. See pp. 479—480 of "Hydraulic and nautical engineering," pp. 467—482 of *Leonardo da Vinci*, N.Y., Reynal & Co., n.d. [1963], written c. 1938.

LEONARDO's writing, it seems to me, a statement that we translate as "A is proportional to B" must be interpreted as his way of expressing the idea we should now state as "A increases if B increases" [48].

A scientist, even a very ancient scientist, cannot regard these two statements as generally the same. In the case of the law of continuity and the law of the vortex, luck was on his side, for the simple proportions are right. If LEONARDO had written down a single table of measured data, a single list of discharges for pipes of various cross-sections, we could grant him the rest, we could believe him on faith, but not only has he never given us this, he has never once recorded *any* measured result concerned with flows of water. He always tells *us* what to measure, but he never measures it himself[49].

Thus we read:

"I find that a weight placed on a plank supported by two pilasters makes the middle of the said plank descend one braccio; the plank is four bracci, and the weight is two bracci from one pilaster. If you make the weight to be but one braccio distant, how far will the said plank descend beneath the weight so placed?" (MS A 48*r.*)

This is one of the two earliest statements of the problem of the *elastica*, but it is a question, not an answer. Here is another, more revealing of his method:

"Reminder to do a trial. Do a trial. If a thin piece of wood held crossways by two supports at its ends holds up ten pounds, how much will a beam of the same proportion hold? And see if the rule of three helps you, because trial makes a good rule." (MS A 33*r.*)

18. Leonardo's Fame

Yet we are missing the forest for the trees. After all that has been spouted over LEONARDO for nearly a century, no one seems yet to have remarked the obvious. First, there is the enormous *bulk* of what LEONARDO wrote on mechanics, scientific or unscientific, right or wrong. No one before him had ever written *so much* on mechanics, and few afterward have done so. If he had been employed 500 years later as junior instructor at an American university, and if he had published all

[48] "Tanto..., quanto..." is an expression in common speech as well as in mathematics, and LEONARDO sometimes uses it in describing a natural configuration where precision would be illusory. *E.g.*, Atl. 185*r.* b: "Tanta son magiorj le tortuosita de fiumi quanto esse son piu vicini al lentroito del mjnor fiume nelfiume maggiore."

[49] *E.g.*, MS A 47*r.*

this, he would have risen rapidly to the rank of distinguished senior colleague. Second, though there are no answers, there is *question after question* that is put in terms of quantity, not quality. It is always, or nearly always, "how many bracci?" "how much faster?" "how many pounds?" not "longer or shorter?" or "swifter or slower?" or "heavier or lighter?" There is also LEONARDO's sense of problem. He is the first to consider the problems of motion on an inclined plane and breaking of a beam, problems which were to afford the two main centers of discourse in GALILEO's *Two New Sciences*, a century later. In our day we all know colleagues renowned for their profound insight and brilliant conjectures, and we do not accuse these annoying persons of being unscientific or even useless. The right question to ask is sometimes better than the right answer to the wrong question. The standard histories of science adulate BACON for his having been the first to recommend a now sanctified method of scientific inquiry, though he himself discovered nothing of importance either with it or without it. With equal justice LEONARDO could be enthroned as the father of scientific questions.

We are ready to end our survey of LEONARDO's success and failure in mechanics. His greatest achievement, the record of his observations of fluid motion, lay unfinished and unpublished in his disorderly notes and remained unknown. What shall be the cozy pat phrase with which we put him in his pigeonhole for paperback authors? I prefer to let the facts stand as their own summary, without attaching any other label than the convenient "genius", so satisfying to Victorian thinkers and to modern educators and housewives.

Of the puzzles and contradictions in LEONARDO's writing, painting, and life, one remains unmentioned. While LEONARDO's fame as an artist is supported not only by the few masterpieces he left but also by his clear influence on many of his contemporary artists and on the whole subsequent course of art, how can he, "... the master of the incomplete and the unfinished", of whom VASARI wrote that "he possessed nothing and worked little," and that "he would have made great profit in learning had he not been so capricious and fickle, for he began to learn many things and then gave them up," nevertheless have gained in his own day the fame also of a great "philosopher"? To learn the source, we recall the method of the Renaissance: Self-advertising. Here is a part of his letter applying for a position:

"(1) I have a sort of extremely light and strong bridges: ... others, secure and indestructible by fire and battle, easy and convenient to lift and place. Also methods of burning and destroying those of the enemy. (2) I know how, when a place is besieged, to take the water out of the trenches... (4) Again, I have kinds of mortars, most convenient and easy to carry... (7) Item. In case of need I will make big guns...,"

and on he goes in this style, listing all the things he can do. We have records of what he saw, what he showed, what he told, always of what he could or would do. VASARI speaks only of models and designs, not of achievements. Moreover, "he could so clearly express his ideas in discourse that he was able to confound the boldest opponents," and once in urging a fantastic scheme upon the elders of Florence "he argued with so much eloquence that it was not until after his departure that they recognized the impossibility of such a feat." Thus "the renown of his name so spread abroad that not only was he valued in his own day, but his fame has greatly increased since his death."

In his skill of speech and his self-promotion he was a true son of the Renaissance. Like the humanists, with much adroitness but little solid achievement he blew himself into renown for all time. According to CELLINI, King FRANCIS I said that "he did not believe that any other man had come into the world who had attained so great a knowledge as LEONARDO, and that not only as sculptor, painter and architect, for beyond that he was a profound philosopher." In those days as in ours, qualities of ingratiation with princes and politicians were little meet for a true philosopher, scientist, or artist.

While, as we have seen, LEONARDO's notebooks contain the earliest attempts to solve some of the central problems of Western physics, LEONARDO himself did not give to the passages concerning them any more emphasis than to those on dozens of other matters that now seem to be mere mud. It is *we*, today, who see that LEONARDO touched the keys of some major locks; *he* showed no sign of seeing that they were locks at all. For him, motion on an inclined plane and the circulation of a vortex, like the aim of a bombard and the rig of a sail, are items in a long, monotonous list of mottoes, questions, and recipes. In a historical essay like mine here, search for the positive sets aside any total estimate. This is right, since for a scientist, the history of science must trace the origin and growth of what is *science* today, not the universe of man's concern with nature. If, however, I were to turn aside to cast a rough sum of LEONARDO's writings, I should describe them as notes toward an encyclopedia such as those by PLINY and ISIDORE OF SEVILLE. Like them, LEONARDO did not always master what he recorded from the work of the creators who went before him; like theirs, his collections are in many cases our only source today for the discoveries of those creators, most of whom are lost in name as well as word in the darkness of history; unlike them, he did not edit his books in a form fit for publication.

VASARI'S life of LEONARDO has been pronounced superficial. Perhaps this is only because the critics have thought a man who projected such brilliance as LEONARDO'S *must* also have accomplished; perhaps, though, VASARI told all there was to tell: the notes, the plans, the schemes, the abandoned projects, and the tricks. VASARI writes, "Even though he talked much more about his works than he actually achieved, his name and fame will never be extinguished." Throughout LEONARDO'S long life he continued to give specimens of what he could do, what he might do, what some day he would do. Of the LEONARDO DA VINCI who was gifted as few ever have been and who lived seventy years to bequeath us ten pictures, half unfinished, and a disorderly heap of papers, VASARI tells us that on his deathbed he "showed how greatly he had offended God and man in not having worked in his art as he ought".

Abbreviations used in citing the manuscripts, each of which is available in at least one facsimile publication:
Atl. = Codice Atlantico, Biblioteca Ambrosiana, Milano.
Tr. = Codice Trivulziano, *ibid.*
MS A, B, ..., I, K, L, M = Manuscrits A, ..., Bibliothèque de l'Institut de France.
Arund. = Arundel MSS, No. 263, British Museum.
Forster = Forster Bequest MSS, Victoria and Albert Museum.
Leic. = MS. Leicester, Pierpont Morgan Library.
Q. Anat. = Quaderni d'Anatomia, Royal Library, Windsor.
Windsor Drawings = Drawings in the Royal Library, Windsor.

Acknowledgment

In 1954 a travelling exhibit of models embodying LEONARDO DA VINCI'S designs reached Indiana University. Arrangements for it were left to the Junior Publicity Officer, who invited a professor of art history to give a public lecture explaining LEONARDO'S ideas on mechanics. I was naive enough to be surprised and shocked enough to volunteer to find out something about the subject in the next fortnight and present what I had learned. My lecture was delivered on October 14, 1954.

From that time on I found myself a victim to the recurrent fits of fascinated exasperation that, I suspect, are the only common bond among those who study LEONARDO'S manuscripts. I could not release for publication anything I was not sure of; I could not be sure of anything about LEONARDO. As revisions proceeded, I delivered them in further lectures for the Washington Philosophical Society (1958), the University of Minnesota (1962), the Universities of California at Berkeley and at Los Angeles (1963), as Walker-Ames Professor at the University of Washington (1964), at the Johns Hopkins University (1965), as Gibson Memorial Lecturer at the University of Glasgow (1965), and as Centennial Lecturer at the University of Kentucky (1965). The foregoing text is based on the last of these lectures but has been revised throughout. A much shortened version has appeared in the *Johns Hopkins Magazine*, Spring 1967.

I delayed publication until I felt reasonably sure I had read, weighed, and reread all passages in LEONARDO's notebooks relevant to fundamental principles of mechanics in the broad sense, including the motions of solid and fluid bodies. Were I to wait until sure of understanding them all, I then saw, publication would be never. With misgivings, I now lay my work aside and offer it to the criticism of others — a few days after this phrase was written, the rediscovery of hundreds of further pages by LEONARDO was announced.

As this essay goes to press, I subscribe to the gratitude every student of LEONARDO's mechanics must feel to the three men who made the *texts* available, more or less, in a form so organized as to be usable without loss of more time than the matter is worth:

Frate LUIGI MARIA ARCONATI (c. 1600—after 1643), who compiled the orderly treatise *Del Moto e Misura dell'Acqua*, consisting almost entirely in quotations from or paraphrases of LEONARDO's notes on fluid motions.

PIERRE DUHEM (1861—1916), who first published or described enough of LEONARDO's notes on general statics and dynamics to give a rough picture of his activity and his antecedents.

ARTURO UCCELLI (1889—), who, guided in part by the essays of ROBERTO MARCOLONGO, put into the hands of students an organized and nearly complete collection of LEONARDO's notes and drawings on general mechanics (excluding mechanisms and deformable bodies).

I wish to express gratitude also for the unforgotten guidance of my friend and teacher, PAUL FELIX NEMÉNYI (1895—1952), who was the first, it seems, to cast the lights of reason and knowledge conjoined upon LEONARDO's studies of fluid motion. In 1948—1950 he had told me about his own reading; his manuscript lay unfinished at his death; I used it with profit for my own studies, which time has been granted me to make far more inclusive than his could be; and I was privileged to publish it in 1962 in the *Archive for History of Exact Sciences*, Volume 2, No. 1. Finally, I thank Professors M. CLAGETT, E. GOMBRICH, and J. WHITE for helpful criticism, and Dr. L. RETI for assistance in regard to MARTINI.

Isaacus Newton Eq. Aur.

II. A Program toward Rediscovering the Rational Mechanics of the Age of Reason

NEWTON's epitaph by POPE,

"Nature and nature's laws lay hid in night. God said, let NEWTON be, and all was light,"

is a neat measure of the impact of the *Principia*, published in 1687. This impact, still felt today, was described not much differently by MACH:

Fig. 34. ISAAC NEWTON (1643—1727), after an engraving of 1712 by JOHN SMITH, following a painting by Sir GODFREY KNELLER

"NEWTON discovered universal gravitation and completed the formal enunciation of the mechanical principles now generally accepted. Since his time no essentially new principle has been stated. All that has been accomplished in mechanics since his day has been a deductive, formal, and mathematical development of mechanics on the basis of NEWTON's laws."

Thus MACH asserts that

1. NEWTON discovered a major part of mechanics.
2. NEWTON's system of mechanics is complete.

(In fact, MACH writes,

"The principles of NEWTON suffice by themselves, without the introduction of any new laws, to explore thoroughly every mechanical phenomenon practically occurring.")

3. Nothing essential has been done in mechanics since NEWTON's time.

To find a poet in agreement with a philosopher or scientist is not unusual. Poets have often expressed, and expressed well, the philosophy of the generation before, or the science of the century before. What is unusual is that here the poet seems to predict the views of a physico-philosopher who followed 150 years later.

In 1788, just a century after the *Principia*, was published LAGRANGE's *Méchanique Analitique*, little less celebrated. Every popular history glows in generality over it and quotes HAMILTON's judgment that it is "a kind of scientific poem". In two ways this book has cut us off from earlier researches. First, the working scientist has accepted the *Méchanique Analitique*, if indirectly, as the final repository of all the mechanics that went before it and has not felt need to look behind it to its sources or around it to discover what it left out. Second, LAGRANGE included in the *Méchanique Analitique* short histories of statics, dynamics, and fluid mechanics, and these, too, have been accepted as final in outline if not in detail. Nearly all of MACH's pages[1] about work done before 1800 seem to have been filled either by use of the sources cited by LAGRANGE or by plausible conjecture how discoveries of that kind must surely have been made, or at least would be made by a right-thinking person.

The new school of historians of science, beginning with DUHEM and daily waxing, has destroyed the first of MACH's major tenets, namely, that classical mechanics

[1] I refer to the first edition of *Die Mechanik in ihrer Entwicklung, historisch-kritisch dargestellt*, Leipzig, 1883. In later editions MACH appended passages concerning what he had learned at second hand from WOHLWILL and DUHEM, but he did not modify noticeably his earlier pronouncements.

was created by GALILEO and NEWTON. They have found not only that NEWTON's contemporaries were less far from the light than had been thought, but also that throughout the Middle Ages and the Renaissance the concepts of Western mechanics were developed intensively and at length, though not always steadily. By historians of science MACH's historical picture as such is no longer taken seriously enough to need refuting, yet the historians have let MACH's view of the *nature* of mechanics itself, and in particular of the completeness of NEWTON's enunciation, stand by default. Judged by their works, the historians seem to think mechanics stopped with NEWTON, except for the formal developments which are to be found in textbooks and are there associated with later names. No great historical effort has been spent upon the growth of mechanics since NEWTON's time. The scientists, in so far as they take any note of history at all, not only have shared the historians' neglect of the later mathematical development of mechanics but also, in the main, have ignored what the historians have learned about the earlier periods and have rested content with MACH's whole view or a rudimentary abstract of it. Between LAGRANGE and MACH, between the historians and the scientists, the Age of Reason is left the Dark Ages of the history of mechanics.

Yet it is neither the primitive mechanics of NEWTON nor the physical mechanics of MACH that we are taught as the most successful, the most thoroughly proved and understood, and the most perfect of the sciences of nature—the prototype and paradigm of a mathematical theory for physical phenomena. Rather, it is the easier parts of the rational mechanics of the BERNOULLIS, EULER, and their successors. Whence did it come? Why was it sought? How was it made? Where did it succeed, where fail? What does it owe to mathematics, to experience, to experiment, and what did it give to them in return?

To answer these questions, I call for a program of rediscovery of the Rational Mechanics of the Age of Reason. This note is an outline of such a program. Parts have been completed and published as my three essays listed at the end; parts remain scarcely charted. Rather than let the skeleton of the whole wait to be discerned under the flesh built upon it when, years hence, the work is finished, I have thought others might be led to share the pleasure of discovery for themselves if now I give out this short survey of the riches waiting for anyone willing take down the dusty volumes and read them, as they have not been read for two centuries, with consequence of the mathematics.

1. Newton's Principia (1687)

Since the *Principia* is one of those works everyone talks of but no one reads, anything said about it other than the usual honey-sauced eulogy must stand up against righteous indignation from all sides. But it is a work of science, not a bible. It should be studied and weighed—admired, indeed, but not sworn upon. It has its novelties and its repetitions, its elegant perfections and its errors, its lightning abbreviations and its needless detours, its extraordinary standards of rigor and its logical gaps, its elimination of stated hypotheses and its introduction of unstated ones.

I doubt there be a man alive today who has worked through the *Principia*; only such a one would be in case to issue justly the pronouncements which may be read about it in any popular essay on science or its history. Here I shall put down a hesitant and tentative summary based on detailed study of some parts and an attempt to survey the rest.

The *Principia* consists in three books, each of a different character. Its standard reputation grows mainly from the first half of Book I, concerning the motion of one or two bodies in a vacuum. MACH knew that this material was not entirely original, and subsequent historical research has shown that most of its contents, both in physical principle and in conclusions, may be found in earlier writings. It is small oversimplification to say that on this subject the *Principia* is a retrospective work which selects, marshals, and formalizes the achievements of the century before it.

Although NEWTON's contemporaries were not unaware of this quality in Book I, nevertheless their admiration for it, whether they were of NEWTON's view or not, was great. In the first place, what NEWTON writes is correct, clear, and short; in earlier works the brilliant diamonds of discovery lie concealed in an opaque matrix of wordy special cases, laborious details, metaphysics, confusion, and error, while NEWTON follows a vein of pure gold.

But it was a second and more important quality that struck readers of the *Principia*. At the head of Book I stand the famous *Axioms, or Laws of Motion:*

"I. Every body continues in its state of rest, or of uniform motion straight ahead, unless it be compelled to change that state by forces impressed upon it.

"II. The change of motion is proportional to the motive force impressed, and it takes place along the right line in which that force is impressed.

"III. To an action there is always a contrary and equal reaction; or, the mutual actions of two bodies upon each other are always equal and directed to contrary parts."

Book I starts out as the first general treatise on mechanics, to be organized about and derived from fundamental laws. For readers of that day, it was this deductive, mathematical aspect that was the great achievement. From a few simple axioms, all the major properties of the motions of bodies had been proved.

The diversity of the phenomena successfully ordered within the *Principia* makes its shortness the more remarkable. While the ancients had been ready to make precise mathematical statements, right or wrong, about the doings of the planets, their approach to motions on the earth had been largely qualitative, using the language of cause, effect, and tendency as in biology and medicine. This separation of heavenly geometry from earthly mechanics had been maintained by GALILEO, who, while he had done much to publicize the kinematical properties of uniformly accelerated motion, derived correctly by the schoolmen 300 years before, had widened the gap between geometrical mechanics and practical mechanical phenomena, first by refusal to connect celestial and terrestrial motions in any way, and second by setting up as governing presumably near and familiar motions laws valid only in an ideal or vacuous medium, the unfortunate practiser who has to work on the real earth being put off with effects recognized by name as being due to resistance and friction but explained only as qualities or tendencies in the style of the Aristotelian physics so contemned by GALILEO. KEPLER had sought a physics valid alike in the heavens and on the earth, but while his brilliant success with the solar system brought him to the brink of the law of gravitation, he did nothing with terrestrial mechanics. DESCARTES had categorically asserted that there is one mechanics, which rules alike above and below, but, lacking patience with details, he never pushed his inquiries far enough to learn that his own principles were partly incorrect as well as insufficient. Great and solid gains in dynamics were made by HUYGENS; his work, more than anything done before, showed the pregnancy of treatment by number and figure rather than attribute, but, conservative, he did not reach up to the planets.

The second half of Book I of the *Principia* is an entirely original work, but here NEWTON began to lose hold on his program of deriving everything mathematically from the axioms. I pass over the clever solutions of various problems on the attraction of spheres and spheroids as resting, on the one hand, on statical assumptions not fully stated, while, on the other, making little or no use of the laws of motion.

But surely the masterpiece of Book I is NEWTON's treatment of the problem of three bodies. Here the modern scientist who does not know NEWTON's work at first hand must be initiated into the facts. While *he* may regard NEWTON's laws as equivalent to the differential equations called "NEWTON's equations" in modern textbooks, there is no evidence that NEWTON himself thought of or ever used his principles in any general mathematical form. Earlier in Book I, the problem of two bodies is skilfully reduced to an equivalent problem of one body attracted to a fixed center. Problems of this kind NEWTON could indeed formulate by means of differential equations, expressed in his usual style in terms of components tangent and normal to the path, and then solve. But the three-body problem cannot be so reduced. For it, not only does NEWTON give no solution or approximate solution in the modern sense, but also he shows no sign of any attempt even to set up equations of motion. That he obtained some correct inequalities and groped his way to the major approximate results under this handicap is one more tribute to his peerless grasp of the physical essence of mechanics and to the might of his brain. It does not show, however, that his formulation of the general laws of mechanics was adequate. History proves the contrary. It was not so much additional mathematics that was needed to get further, for NEWTON was a master at approximate solutions, quadratures, and series expansions in definitely set mathematical problems. Rather, the fact that fifty years passed before any improvement over his results on the three-body problem was made shows that he himself had gotten the most out of the subject that his own methods and concepts could produce; to wrest so much from so primitive a formulation of mechanics required the genius of a NEWTON; none of his disciples, who might reasonably have been expected to build on his foundation, could raise the structure an inch higher. Before the next real advance, a half century of abstraction, precision, and generalization of Newtonian concepts was necessary. The first to go substantially beyond NEWTON in the three-body problem was the man who found out how to set up mechanical problems once and for all as definite mathematical problems, and this man was EULER. The year in which the "Newtonian equations" for celestial mechanics were first published is not 1687 but 1749, as we shall see.

While NEWTON's Book I is the achievement of KEPLER's program, NEWTON had a program of his own. Bodies on the earth, as any Aristotelian physicist was ready to say, did not in actual experiments obey GALILEO's orders, and the ideal medium or vacuum in which those proclamations are valid exactly does not, as the Aristotelian was all the readier to say, exist anywhere on this old, practical, physical earth. Seeing clearly that to justify neglect of friction one must first estimate its effects precisely, NEWTON set himself the problem of determining mathematically the nature of motion in resisting media. This, with some excursions, is the subject of Book II.

To make any solid study of fluid resistance, NEWTON had first to learn the laws of fluid motion. Here he had no foregoing pathfinder; here his program of mathematical deduction broke down. Book II is almost entirely original, and much of it is false. New hypotheses start up at every block; concealed assumptions are employed freely, and the stated assumptions sometimes are not used at all. Little from Book II has found its way into either texts or histories; what has, is often misrepresented. To create theory for the spring of air, for the flow of fluid from an orifice, for the progress of waves in water, for the oscillation of water in a tube, for the resistance suffered by bodies in rare or dense fluids, for the propagation of sound in air, for the internal friction of fluids, NEWTON created *new concepts*. That few of his concepts have been retained and few of his solutions are right is less remarkable, considering the entire freshness of the subjects, than that he should have been able, in each case, to grope or beat his way to some definite answer. The brilliantly ingenious but, in the end, largely unsatisfactory Book II laid out the areas and defined the problems for many of the mechanical researches of the next century.

Book III, being astronomical, lies outside the present subject. In it NEWTON showed that the mathematical propositions of Book I, with suitable numerical values, squared with the phenomena of the solar system.

In the three books of the *Principia*, NEWTON displays in the highest every ability of a great theorist: (1) *to organize, derive by mathematics and recast known but separated laws and phenomena;* (2) *to create new concepts;* (3) *to obtain detailed numerical predictions and compare them with measured values.*

But by no means did he give "classical" mechanics its present form, nor were his principles clear and definite enough to do so. If, for example, NEWTON's laws are sufficient, as MACH implies, to determine the motion of fluids, NEWTON himself gave no evidence of knowing it, for in his treatment of fluid flow he made no reference to the principle of momentum, inventing instead the ingenious but false artifice of considering part of the vessel filled with ice around a "cataract" of fluid to which GALILEO's results on falling bodies are presumed to apply. Each other hydrodynamical problem NEWTON attacks and claims to solve by special reasoning, special hypotheses, and special guesses, related vaguely to mechanical ideas and principles, but not in the least derived from a system of mechanics. In fact, neither NEWTON nor any of his disciples or rivals had a grasp of the principle of momentum sufficient to solve a single problem concerning the motion of a deforming body.

The uneasy status of the notion of force in the *Principia* is too well known to need more than a passing remark here. This was the one weakness in NEWTON'S treatment that MACH, following LEIBNIZ and D'ALEMBERT, saw fairly clearly. Is the Second Law a mere definition of force? If so, how does it bring us any nearer to the laws of nature? If not, then what is force, and how do we measure and know it? Is it not but another of the old qualities and tendencies that rational mechanics must reject as meaningless renamings?

While NEWTON went some way toward distinguishing *mass* from *weight*, his concept of *body* was insecure. Sometimes the "bodies" entering his equations are what we now call *mass-points*, but at other times, as in problems of the attraction of a sphere, he considers bodies which fill finite portions of space. *Inertia* or *momentum* he does not define; Laws I and II are usually interpreted as the modern principle of linear momentum, but as a part of the explanation of Law I we read:

"A top, whose parts by their cohesion are continually drawn aside from rectilinear motions, does not stop spinning, except insofar as it is slowed by the air."

Thus NEWTON has rotary as well as linear inertia in mind, but he gives no measure of it, and he leaves the reader without a hint of what principles govern the motion of a wheel on an axle.

This failing is not a matter of mere application or of lack of mathematical process; on the contrary, it is only an inevitable consequence of the block which forced NEWTON already at the middle of Book I, when he came up against the problem of three bodies, to compromise with his program of mathematical proof and to be content with half-guessed and inassessable approximations. Except for certain simple if important special problems, NEWTON *gives no evidence of being able to set up differential equations of motion for mechanical systems*. It is not the function of the historian to guess what NEWTON might have done or could have done, nor is what MACH could do with NEWTON'S principles relevant; the cold fact is, the equations are not in NEWTON'S book. As we shall see, a large part of the literature of mechanics for sixty years following the *Principia* searches various principles with a view to finding the equations of motion for the systems NEWTON had studied and for other systems nowadays thought of as governed by the "Newtonian" equations.

To summarize: In NEWTON's *Principia* occur no equations of motion for systems of more than two free mass-points or more than one constrained mass-point; NEWTON's theories of fluids are largely false; and the spinning top, the bent spring, lie altogether outside NEWTON's range.

So as not to leave too shocked those accustomed to the usual hagiography of science, I close this brief survey of the *Principia* with another, juster epitaph of NEWTON, the one cut upon his tomb:

"Let mortals rejoice that such and so great an ornament to human kind has been."

This estimate is rather strengthened than diminished by knowledge of what NEWTON really did for mechanics: far from *completing* "the formal enunciation of the mechanical principles now generally accepted", he *began* it.

This, then, is the exordium: The achievements, the omissions, and the failures of NEWTON in the *Principia*, 1687. The subject of the following outline is the intercentury to end in 1788 with LAGRANGE's *Méchanique Analitique*, a book hardly less celebrated or, in its effect and acceptance, less definitive—a book as little known and as glibly cited.

2. Scientific Methods and Scenes in the Age of Reason

While by many nowadays mechanics is thought of as a branch of physics, it was not often so regarded in the Age of Reason, and the "classical" mechanics was not discovered by physicists. In the literature of that period, the papers on physics generally concern experiment or speculative rather than mathematical theory. The speculative physicists might almost be regarded as inheritors of the vague, practical, Aristotelian tradition opposed with fury and contempt by GALILEO and DESCARTES; none of them has left a mark on the history of physics. While some fine experimental discoveries were made, they were not of a kind to influence directly the growth of the great mathematical theory now regarded as "classical", for this was created by a handful of "geometers" or "algebrists", as they were called in that day, men who strove to put into mathematical form laws governing the daily physical experience evident to anyone who will open his eyes to it.

These were the cultivators of the new mathematics, the infinitesimal calculus. All had great ability and at least moderate interest in what would now be called pure mathematics. Individual tastes varied; one experimented much; some, little; most, not all; some cultivated pure geometry and the theory of numbers, and some did not. But in mechanics, the aims of all were similar. For none was the mere solving of problems enough; however much a special problem might sometimes excite curiosity and effort, the aim was always a proper statement within the principles of mechanics, a grasp of the case as special at its proper station in the theory as a whole. The clearest expression, of course in part a summary of practice, was given by D'ALEMBERT in 1743:

1. Rational mechanics, like geometry, must be based upon axioms which are obviously true.

2. Further truth in mechanics follows by mathematical proof.

He aimed

"... to envisage in the most abstract and simple manner what can be the particular object of this science; to suppose nothing, to admit nothing in this object, beyond the properties which the science being treated supposes there."

The geometers of the Age of Reason laid as much weight on proof as do the pure mathematicians today. MACH sneered[2] at their "mania for demonstration". The Victorian critics, looking back on the eighteenth century with as much distaste for its standards of reason as for its realistic morals, ridiculed its excesses so as to belittle its triumphs. Indeed, "proofs" of the parallelogram of forces and of the existence of God were published, but these were exuberances in the century that gave us the rational theories of fluids and of rigid bodies, the equations of motion of dynamical systems, the principles of the bending of beams and the general view of elastic vibrations. In those times it was not always clearly realized that *something* must be assumed, or as we say today, postulated—in geometry, too, just as much as in mechanics. But, as D'ALEMBERT wrote, to be content with

---

[2] It has been shown by VAN DER WAERDEN that MACH in criticising ARCHIMEDES' proof ("griechische Beweissucht" and "falsche und verkehrte Strenge") of the law of the lever invented an incorrect substitute of his own rather than taking the pains to follow what ARCHIMEDES really did. See B. L. VAN DER WAERDEN, "La démonstration dans les sciences exactes de l'antiquité," *Bull. Soc. Math. Belg.* 4, 8—20 (1957). There are similar falsifications in MACH's summaries of later mathematical works on mechanics, but his treatment of the whole science in the eighteenth century is in any case so fragmentary and unhistorical as not to deserve detailed study.

merely collecting or organizing knowledge, however true, and solving various problems, however adroitly, was not enough:

"This would ruin the certitude of mechanics and reduce it to being nothing more than an experimental science."

As BERTRAND RUSSELL has said, the method of simply assuming results, once one is persuaded they are true, rather than trying to prove them, has all the advantages of thievery over honest toil. Nowadays geometry and mechanics are split apart; everyone recognizes that geometry without proof is nothing, while mechanics is grouped within physics as a science of empirical origin. In the Age of Reason, not only had this false and noxious cleavage not yet occurred, but the whole program of mechanics was against it: Each generation brought more of the phenomena of nature under the control of exact science. NEWTON's own program was untypical only in the measure of success he obtained by it and the frank ease with which he declared it:

"Therefore *geometry* is founded in mechanical practice and is nothing else than that part of *universal mechanics* that accurately sets forth and demonstrates the art of measuring. Since, however, the manual arts are chiefly employed in moving bodies, it comes about that commonly geometry is referred to magnitude, mechanics to motion. In this sense *rational mechanics* will be the science of the motions that result from arbitrary forces, and of the forces that are required for arbitrary motions, accurately set forth and proved."

Had the geometers of the Age of Reason not somewhat overestimated the power of the mathematical way of thinking, perhaps they would not have achieved with it such brilliant, unparalleled success as in fact they did.

MACH, like his disciples today, confused the drive toward order and precision with metaphysics. He was so violently opposed to metaphysics that he banished from his history of mechanics most of the searching and analysis of the concepts of the subject, leaving the reader with the impression that it grew out of experiments[3].

---

[3] I find that even today there are persons who respect MACH as a historian, although most of these seem to know his book only by hearsay. There are several passages that such persons might find enlightening. For example, though every working mathematician today attributes to GALILEO the proof that the cardinal number of an infinite subsequence of the integers is that of the integers themselves, MACH's remark here is, "GALILEO's atomistic considerations ... lead him to queer discussions about the continuum and to mystical-mathematical speculations about the finite and the infinite, which recall NICHOLAS OF CUSA on the one hand, and on the other, vividly, some modern mathematical investigations, which also are scarcely quite free from mysticism" (§ 22 of Chapter 2 of *Die Mechanik* ..., 9th edition, 1933).

Let it not be thought, however, that the embracing and elegant generality of mechanics came from philosophical speculation. Mechanics is a science of experience; for the theorist, physical experience is balanced against the experience of earlier theories for the phenomena. The history of rational mechanics is neither experimental nor philosophical; it is *mathematical*; it is a history of *special problems*, concrete examples for the solution of which *new principles and methods* had to be created. But the solution of the special problem was never left to stand alone; since there was only one true mechanics, the special case served not as an end in itself but as the *guide to the right conception*. The order and design of the structure were as valuable as the soundness of its members. In 1743 D'ALEMBERT wrote,

"Up to the present, ... more concern has been given to enlarging the building than to illuminating the entrance, to raising it higher than to giving the proper strength to its foundations.

"In this work I have set myself the double object of pushing back the bounds of the science of mechanics and of smoothing the way to it, and my principal aim has been to achieve in some measure the one of these objects by the other; that is, not only to derive the principles of mechanics from the clearest ideas but also to apply them to new uses; to show all at once the uselessness of several principles used heretofore in mechanics and the advantage which may be drawn ... from others; in a word, to extend the principles by reducing them."

With order and precision came elegance and beauty. While these, too, are banished from a positivist philosophy of science, their force is immediately apparent to anyone who knows modern pure mathematicians, so much so, indeed, that according to some current views mathematics is not a science at all, because it is not experimental and because its criteria of excellence are rather aesthetic than utilitarian.

That mathematics and music, pursued in the same spirit, were the special excellences of the Age of Reason, was as plain then as it is now in retrospect. On his third voyage, Gulliver goes to Laputa, an island magnetically suspended in the atmosphere, inhabited by a race given over wholly to mathematics and music,

"... one of their eyes turned inward, and the other directly up to the zenith."

These people are so abstracted in mathematical calculation or musical composition that they have to carry with them everywhere a servant equipped with a rattling bladder full of dried peas or little pebbles, to strike them gently on the mouth or ear when it is time to speak or listen, or to warn them from falling down every precipice or bouncing their heads against every post. At a meal,

"In the first course there was a shoulder of mutton, cut into an equilateral triangle, a piece of beef into a rhomboides, and a pudding into a cycloid. The second course was two ducks, trussed up into the form of fiddles; sausages and puddings resembling flutes and hautboys, and a breast of veal in the shape of a harp. The servants cut our bread into cones, cylinders, parallelograms, and several other mathematical figures ....

"Their ideas are perpetually conversant in lines and figures. If they would, for example, praise the beauty of a woman, or any other animal, they describe it by rhombs, circles, parallelograms, ellipses, and other geometrical terms, or by words of art drawn from music ... ."

Pause to note SWIFT's accuracy. Indeed, he has mastered the spirit of mechanics, not the mechanics of his own day but that of the century previous, of GALILEO. SWIFT's words are a paraphrase of a famous passage in the *Saggiatore:*

"Natural science is written in this great book, which stands ever open before our eyes — it is of the universe I speak — but you cannot understand it unless first you learn to understand its language and to know the characters in which it is written. It is written in mathematical language, and the characters are triangles, circles, and other geometrical figures, without whose means it is impossible for man to learn its content."

SWIFT wrote in 1726, when all this was changing. The new language of mechanics was not geometry but analytical formulae, or, as then it was called, *algebra*; now we call it differential and integral calculus.

Moreover, the association of mathematics and music was not accidental. While SWIFT was attacking the two most prided excellences of his own day, BACH and HÄNDEL were in full creation. Not long afterward an eccentric and semi-learned king of Prussia, himself an expert performer on the traverse flute, was to retain as his private teacher a celebrated virtuoso, as his accompanist a son of BACH, and as workhorse of his academy a disciple of the BERNOULLIS whom he described as a "great algebrist". This latter, LEONARD EULER, had written almost simultaneously treatises on analytical mechanics and on a mathematical theory of music. D'ALEMBERT, the encyclopaedist and litterateur whose judgment FREDERICK II trusted above all others', was a brilliant if not very reliable mathematician; he, too, wrote treatises on dynamics and on the theory of music.

The abstracted inhabitants of SWIFT's isle of Laputa lived poorly:

"Their houses are very ill built, the walls bevil, without one right angle in any apartment, and this defect ariseth from the contempt they bear to practical geometry, which they despise as vulgar and mechanic .... And although they are dexterous enough upon a piece of paper ... yet in the common actions of life I have not seen a more clumsy, awkward, and unhandy people, nor so slow and perplexed in their conceptions upon all other subjects except those of mathematics and music .... Imagination, fancy, and invention they are wholly strangers to, nor have any words in their language by which those ideas can be expressed; the whole compass of their thoughts and minds being shut up within the two forementioned sciences."

That to SWIFT not only mathematics but also music was totally devoid of "imagination, fancy, and invention", is to be remarked.

While hating mathematics and music, SWIFT expressed grudging admiration for the achievements of the inhabitants of Laputa. They are a superior race, the nearly absolute tyrants of a large empire on the earth beneath, and without their skill the island would fall to the ground and break in pieces.

Neither must our colleagues in the natural sciences be allowed to fancy themselves superior to us in SWIFT's esteem. Them he banished to the Grand Academy of Lagado, on the low earth, where in filthy beggary they are engaged in such practical occupations as extracting the sunbeams from out the cucumbers and reconverting humal offal to food—reversing, be it noted, the processes of nature.

Laputa in real life was the mathematical sections of the academies and learned societies of Europe. The establishment of these centers asserts the acceptance of the new, algebraic science. There was no revolution such as that of GALILEO and DESCARTES in the century before. Rather, the course of the eighteenth century researches was perfectly in the spirit of NEWTON's program. In his original preface NEWTON wrote:

"The ancients considered mechanics as twofold: as rational, which proceeds accurately by proofs, and practical. To practical mechanics belong all the manual arts, from which the name mechanics was adopted. But as artificers do not work with perfect accuracy, it comes to pass that mechanics is so distinguished from geometry that what is accurate is called geometrical; what is less accurate, mechanical. However, the errors are not of the art, but of the artificers. He that works with less accuracy is the less perfect mechanic, and if any could work with perfect accuracy, he would be the most perfect mechanic of all."

Far from being the materialist MACH and the physicists would make of him, NEWTON shared fully the "mania for demonstration". That he was not entirely successful grew from the grandeur of his own undertaking. Perhaps mindful of the imperfections in his brilliantly original Book II, he wrote at the same time:

"I wish we could derive the rest of the appearances of nature by the same kind of reasoning from mechanical principles ... ."

As the basis of his "System of the World" in Book III, NEWTON set nine "Hypotheses". The last five, which are assertions about the planets and the moon, he labelled "Phenomena" in the second edition (1713), where he called the first three "Regulae philosophandi", a modern rendering of which would be, "Rules of inference in natural science". In these "Rules", if anywhere, should we seek the experimental basis of his mechanics, but we should seek in vain.

"Hypothesis 1. We are to admit no more causes of natural things than such as may be both true and sufficient to explain their appearances. Indeed, Nature is simple and affects not the pomp of superfluous causes.

"Hypothesis 2. And thus for natural effects of the same kind there are the same causes.

"As breathing in men and beasts; the fall of stones in Europe and in America; the light in cooking fire and in the sun; the reflection of light on the earth and on the planets.

"Hypothesis 3. Every body may be transformed into a body of any other kind, and may take on successively all intermediate degrees of qualities."

This last is hard to understand and is not mentioned by historians with an empirical bent. What it means, roughly, is that physical phenomena occur smoothly, not by jumps; it might be called the "non-quantum hypothesis". Its purpose, we may conjecture, is to assert physical relevance for the differentiations and integrations which occur on almost every page of the *Principia*[4], and in such a context it is familiar to every student of the eighteenth century as "LEIBNIZ's law of continuity", frequently invoked by the continental geometers even after EULER had proved decisively that it is untenable.

In the edition of 1713, NEWTON receded from the boldly speculative position he had taken when younger; he was then sixty-nine, his creative days were past, and he was more prone to tighten the grip of his system than to strengthen its foundation. He replaced Hypothesis 3 by the entirely different

"Rule 3. The qualities of bodies which admit no increase or decrease, and which present themselves in all bodies within reach of experiment, are to be taken as qualities of all bodies whatsoever."

The long, following gloss begins "For since the qualities of bodies lie unknown to us except through experiments...," and here the positivist will erect his ears and listen intently, but with disappointment, for NEWTON's purpose is the opposite of experimental. Far from limiting us to quantities measured experimentally, NEWTON orders us to take the qualities we observe in experimental bodies and *generalize* them to bodies on which we *cannot* experiment[5]. He is an-

---

[4] The historians still repeat old saws from the last century about NEWTON's "concealing his methods" and presenting by "Greek geometry" proofs of propositions he had "discovered by use of calculus". While indeed NEWTON did not use the notation of fluxions in the *Principia,* a modern mathematician, according little respect to those who confuse notations with notions, finds the *Principia* a book dense with the theory and application of the infinitesimal calculus.

[5] In the edition of 1726, following the controversies over gravitation, NEWTON added a fourth rule, to the effect that when we generalize from known phenomena, our assertions must be taken as true until proved false by new phenomena, "so the argument of induction may not

Fig. 35. James Bernoulli (1655—1705), portrait in 1686 by his brother, Nicholas Bernoulli, in the Alte Aula, Basel

swering the skeptics who reproached him for merely mathematical speculation about the forces reigning in places he had never visited, such as the sun and moon.

3. The Discoveries of James Bernoulli

The creation of rational mechanics is due as much to JAMES BERNOULLI as to NEWTON, though JAMES BERNOULLI's work nowadays is known to few. The areas of mechanics not touched by NEWTON were those which JAMES BERNOULLI cultivated, and the approaches and methods of the two great founders of modern mechanics were different in principle as well as detail. While NEWTON emphasized motion resulting from action at a distance, JAMES BERNOULLI developed the idea of contiguous action and the relation of dynamics to statics.

When the problem of the catenary curve was solved by LEIBNIZ, HUYGENS, and JOHN BERNOULLI in 1690, all three used special methods. The solutions of LEIBNIZ and JOHN BERNOULLI rested upon a theorem of PARDIES (1673) concerning the center of gravity of an arc; this theorem, in turn, presumed that the action of any part of a heavy arc on its neighbor is equipollent to a tangential force acting at the junction. JAMES BERNOULLI attacked the general problem of equilibrium of a string subject to arbitrary loading along its length. Retaining PARDIES' concept of contact force, he formalized it as the *tension* in any flexible line, and he succeeded in obtaining the *general differential equations of equilibrium* determining the shape and the tension from the applied loading. Moreover, he found four different methods by which these equations may be derived: balance of normal and tangential forces, balance of forces in two fixed directions, balance of moments, and the principle of virtual work. While this research was done in 1691—1704, it was not published until 1744; meanwhile a part of it had been rediscovered by HERMANN, BERNOULLI's pupil, and published by him in 1716.

In 1684 LEIBNIZ had given the first analysis of the tension in the interior fibres of a loaded beam; on the assumption that this tension varies linearly across the

---

be evaded by hypotheses." This is an illogical defense, because different generalizations can be consistent with the same observations. NEWTON is implying that since universal gravitation explains everything satisfactorily, no other hypothesis is to be admitted until unversal gravitation is proved false. In other words, the *first* adequate theory has the right of predominance over equally adequate aftercomers.

IOANNES BERNOVLLIVS

section he had concluded, in a special case, that the bending moment is proportional to the moment of inertia of the section. JAMES BERNOULLI found by experiment that the linear relation between extension and stretching force, which HOOKE and MARIOTTE had inferred and which LEIBNIZ had adopted, does not hold. He therefore, from that day to the end of his life, rejected it as general principle, though he was content to refer to it as a special case. On the basis of a model of the section of a bent bar as a lever hinged about a point on the concave side and retained by a spring on the upper side, he obtained the first theory of bending. If the spring is linear, the differential equation of the *elastica* results (1694), for the bending moment is then proportional to the curvature of the band.

HUYGENS, like BEECKMAN and HOOKE before him, saw that in bending the inner fibres were contracted and the outer ones extended. When this objection was made to JAMES BERNOULLI, he sought to construct a theory whereby the location of the neutral or unextended fibre would be determined. He attempted to integrate over the cross-section of the beam, as LEIBNIZ had done in the case when bending was negligible, but he never succeeded in doing so correctly, partly because he always refused to adopt any specializing hypothesis regarding the material. Finally he proposed a general principle which was not right; from it he concluded that the position of the neutral line was immaterial, but this is false. The problem he attempted, to derive a one-dimensional theory of beams from a fully general state of interior elastic tension, remains unsolved today, and his best work on it is not yet published.

In the course of these researches JAMES BERNOULLI evolved a general concept of elastic response in one dimension. While BEECKMAN had seen that it was the ratio (change of length)/(length), or *strain*, which should enter an elastic law, HOOKE had been content to measure change of length as a function of stretching force. JAMES BERNOULLI saw that a relation giving the ratio (force)/(area), or *mean stress*, as a function of strain characterizes a material rather than a particular specimen of material (1704). This is the earliest occurrence of a true *stress-strain relation* and a *material property* of a deformable medium. Only BERNOULLI's reluctance to put any faith in HOOKE's law kept him from introducing the so-called

---

Fig. 36. JOHN BERNOULLI (1667—1748) after an engraving done in 1741—1742 by JOHANN JAKOB HAID and AUG. VINDELIC, following a painting by JOHANN RUDOLF HUBER. BERNOULLI points to a figure explaining his solution of the brachistochron.

YOUNG's modulus; his discussion implies the existence of what in recent times has been called the "tangent modulus" of a non-linear stress-strain relation.

The dynamics of rigid bodies, altogether avoided by NEWTON, had been initiated earlier by the famous solution of HUYGENS for the physical pendulum (1673). In effect, HUYGENS assumed that the translational kinetic energy acquired by the parts of the body in falling was sufficient that each of these, if suddenly released, could mount to a height such that the center of gravity of the body would rise to its original level. This ingenious device yields a correct solution to the problem of oscillation of a rigid body about a fixed axis, but, beyond its being only tenuously related to other mechanical ideas, it is not valid or even meaningful in general motion of a rigid body. JAMES BERNOULLI sought a general principle of mechanics whereby HUYGENS' solution would follow. After a faulty attempt in 1686, JAMES BERNOULLI in 1691 created a new approach to mechanics, which he developed in a profound paper of 1703, second only to the *Principia* itself in influence on the later growth of the discipline. In this paper he resolves the acceleration of each mass into two parts; that normal to the path is produced by the constraint of rigidity, while that tangent to the path is again divided into two; one of these is the acceleration of gravity, and the other is the acceleration produced by the mutual interference of the masses because they are joined together. The system of forces of this latter kind, says BERNOULLI in effect, must be equilibrated. Applying the principle of the lever then yields HUYGENS' solution immediately.

Three major ideas of mechanics are foreshadowed in JAMES BERNOULLI's method:

1. To determine the motion of a constrained system, introduce forces which maintain the constraints.
2. The accelerations of bodies if reversed in sign are equivalent to static forces (per unit mass).
3. Not only equilibrium of forces but also equilibrium of moments is necessary in dynamic as well as static problems.

It would be an exaggeration to say that BERNOULLI stated any of these principles or separated them from one another. Rather, they appear mingled and half-implied in his work. The organization of science is as necessary, and as difficult to achieve, as the discovery of root ideas; to locate the vent of the oyster is not to pry it open. Even a modern expert on mechanics, with all the benefit of hindsight, will find the works of NEWTON and JAMES BERNOULLI hard to follow.

To recognize, distinguish, and formulate their concepts and methods in general terms required half a century of trial and search by the finest geometers, as we shall see.

4. Early Efforts to Clarify, Extend, and Organize the Principles of Mechanics

Prior to NEWTON's work, other approaches to mechanics were growing, with various success. The connection of force to motion is the primary problem of Western mechanics, but force is a vague word meaning different things to different persons. We still use it today in many undefined senses: force of habit, force of mind, force of argument, military force, as well as mechanical force. Mechanical force was familiar to everyone in the seventeenth century, too, as muscular sensation and as the cause or accompaniment of motion. But what is the *measure* of force? How is force to be identified with a quantity as well as a quality? NEWTON avoided the question by referring directly to force as if it needed no definiton; in then setting it equal to mass times acceleration in simple dynamical problems, he might as well not have introduced it at all, since in effect he assumed special forms for the acceleration and then solved the resulting kinematical problems. In solving the same class of problems in essentially the same manner today, we acknowledge the efficiency of the means to the end, but we do not face the question which made NEWTON's approach inacceptable to most of his contemporaries, to whom it seemed a brilliant artifice for escaping the real nature of mechanical laws. They sought a simple, *general* statement about the motion of all bodies, from which the motion of a particular body in particular circumstances would follow mathematically.

DESCARTES (1644) had taken mass times velocity as the measure of force and had asserted that there was neither more nor less of it now than when the universe was created. In modern terms, he asserted the conservation of linear momentum. Impatient with special cases, DESCARTES did not apply his principle sufficiently to give a good idea of its meaning. Since he did not claim it applicable to all problems, we may be unfair in remarking that it does not hold for the simple pendulum, but that example suffices to show that the principle is not general. Justly criticizing the views of DESCARTES, LEIBNIZ (1686, 1695), who by no means despised special problems, introduced the concepts of *live force* and *dead force*.

The dead force is only the old force of position, known to the schoolmen and nowadays called potential energy, but the live force is mass times velocity squared, twice what is now called kinetic energy. LEIBNIZ asserted that loss of dead force resulted in corresponding gain of live force. This principle was taken up by JOHN and DANIEL BERNOULLI and made the basis of brilliant successes with special problems. Most if not all of the dynamical problems solved by NEWTON may be solved just as well by LEIBNIZ's principle, combined with use of DESCARTES' principle when it applies. Moreover, the principles of linear momentum and of energy are statements about the *whole* of the system, not requiring it, as do NEWTON's laws, to be cut up into parts. But neither suffices to determine the motion of any but the simplest systems.

While NEWTON's approach led ultimately to mechanics as we know it today, most of the life work of EULER was required in order to clarify and develop the Newtonian concepts, to supplement them by equally important new ideas, and to demonstrate how real problems can be solved. EULER (1707—1783) was the dominating theoretical physicist of the eighteenth century. While his work is undervalued in the usual, vague historical books and attributions, the short factual history in the old *Handbuch der Physik* lists twice as many specific discoveries for EULER as for any other one physicist, earlier or later, and even at that most of his work is omitted. When EULER was nineteen, still a student, he outlined a great treatise on mechanics in six parts, only three of which he lived to finish in the next fifty years, although his program was virtually achieved by the hundreds of papers he published on all branches of mechanics and was supplemented by treatises on naval science, ballistics, and astronomy. The great bulk of EULER's publication is not the only impediment to a just historical estimate of what he did. He put most of mechanics into its modern form; from his books and papers, if indirectly, we have all learned the subject, and his way of doing things is so clear and natural as to seem obvious. In fact, it was he who made mechanics simple and easy, and for the straightforward it is unnecessary to give references. In return, the scientist of today who consults EULER's later writings will find them perfectly modern, while other works of that period require effort and some historical generosity to be appreciated.

Part 1 of EULER's program, his *Mechanics*, appeared in 1736, when he was twenty-nine. It is the first treatise on *analytical* mechanics, in which all problems are set and solved by purely mathematical process. At this time as throughout his life, EULER followed NEWTON in taking force itself, in the sense used in statics, as a primitive idea. The *Mechanics* made three additions to the principles. First, while NEWTON had used the word "body" vaguely and in at least three different meanings, EULER realized that the statements of NEWTON are generally correct only when applied to masses concentrated at isolated points; he introduced the precise concept of *mass-point*, and his is the first treatise devoted expressly and exclusively to it. Second, he is the first to recognize explicitly the *acceleration* and to study it as a kinematical quantity defined in a motion along any curve. Third, he employs the concept of a *vector* or "geometrical quantity", a directed magnitude, as pertaining not only to static force, for which it was familiar, but also to velocity, acceleration, and other quantities. The range of problems considered in this, EULER's first treatise on mechanics, is narrower than in NEWTON's *Principia*. While motion in three dimensions is treated precisely and systematically, no body more complicated than a single mass-point is mentioned. Following NEWTON, EULER here always resolves acceleration along directions defined intrinsically with respect to the curving path of the mass-point, though in some cases he resolves applied forces relative to directions fixed in space. In this first book, EULER consolidates a part of rational mechanics, but only a small part.

5. The Proper Frequencies and Simple Modes of Vibrating Systems before the Discovery of Equations of Motion

The first mathematical analysis of the vibrating string leading to any great result was given by TAYLOR in 1713. He derived the frequency of the fundamental mode, which he in effect assumed to be of sinusoidal form. The corresponding laws of proportion had been obtained earlier by MERSENNE (1625) and GALILEO (1638), but not from theory, and the numerical factor had not been determined. TAYLOR applied NEWTON's second law to a differential element of the string; despite this promising start, he did not recognize the result as a differential equation of motion, and his work rests on confusing and partially erroneous special assumptions.

In 1727 JOHN BERNOULLI calculated the restoring force in a taut massless string loaded by n equal and equidistant masses. This model had been studied earlier by HUYGENS, who had found the frequency when $n=1$. BERNOULLI, using a quasi-static device, obtained the fundamental frequency when $n=2, 3, 4, 5, 6$, or 7.

That the overtones of a string have frequencies $k\nu$, $k = 1, 2, 3, \ldots$, where ν is the fundamental frequency, had been found experimentally by MERSENNE (1633); that the kth overtone corresponds to a form of vibration with $k-1$ nodes, was demonstrated experimentally by NOBLE and PIGOT (c. 1674) and by later writers. While these facts were fairly well known, neither TAYLOR nor JOHN BERNOULLI heeded them. Their theoretical work is confined, tacitly, to the fundamental mode.

DANIEL BERNOULLI, in work beginning about 1733, was the first to understand the nature of small vibrations in general and to obtain a partial theory for them. The crucial problem was that of a massless cord loaded by two weights; this system is often called the compound or bifilar pendulum. DANIEL BERNOULLI, using a quasi-static device applied to a general principle invented by him to determine the effect of constraints, calculated the forms and frequencies of the two modes and indicated that when there are n weights, there are n different modes, each with its own frequency; he showed that there were infinitely many modes for the continuous heavy cord, and he determined the forms and frequencies of the first two. While giving evidence of knowing the corresponding theory for the taut string, of course much simpler to treat, he did not trouble to publish anything about it until many years later. EULER at once generalized these results in various ways; while his dynamical principle is different from BERNOULLI's, it is equally special.

In 1734 DANIEL BERNOULLI and EULER simultaneously found the differential equation for the simple modes of transverse vibration of elastic bars but could integrate it only in series. The resulting scaling law makes definite the old suggestion of LEIBNIZ (1684) that the elastic and acoustic properties of bodies are connected. In 1739 EULER gave the first analysis of a simple harmonic oscillator driven by a harmonic generator; he emphasized the phenomenon of resonance, which had never appeared in a mathematical theory before, suggesting practical application for driving a naturally oscillatory machine by an oscillatory engine. Then he learned how to integrate explicitly linear ordinary differential equations with constant coefficients. This made the problem of transverse vibrations of bars much easier. By 1743 DANIEL BERNOULLI and EULER had given the conditions to be satisfied at a clamped, pinned, or free end; had recognized the six possible

---

Fig. 37. DANIEL BERNOULLI (1700—1782), portrait by JOHANN NIKLAUS GROOTH in 1760, in the Alte Aula, Basel

kinds of vibration; and for four of the kinds had calculated, to more than sufficient accuracy, the entire sequences of proper frequencies and the nodal ratios for the first few modes. EULER's work was more complete and elegant; DANIEL BERNOULLI's was closer to the physical phenomena. In particular, BERNOULLI verified many of the results by simple experiments, and in 1742 he began to see some theoretical reason that two or more simple modes could be excited simultaneously in the same body.

The experience gained with special theories for vibratory phenomena was of great help toward discovery of the general principles of mechanics, as we shall see.

6. Fluid Mechanics before the Discovery of Differential Equations of Equilibrium and Motion

While remarkable theorems of hydrostatics had been proved by ARCHIMEDES and STEVIN, and certain isolated results concerning fluid motion were known, it was two special problems treated for the first time in NEWTON's *Principia* that gave rise to the great theories of the next century. First was the problem of the figure of the earth, which NEWTON identified with the figure of equilibrium of a rotating mass of fluid subject to its own mutual gravitation; he concluded that the earth was flatter at the poles than at the equator. The Cartesians somehow inferred from their systems of vortices that it should be flatter at the equator. In each treatment, both the application of physical principle and the mathematical detail were dubious, but the general opposition of the two views of mechanics was made clear, and a simple issue of this kind could be discussed with happy partisanship by numerous well informed persons who understood nothing about either theory. An experimental test on a large scale was proposed: Expeditions were sent to the corners of the earth to measure its curvature. The expedition to Lapland returned sooner, with its determination that the earth was indeed flatter at the poles, and MAUPERTUIS, its leader, was called in triumph "the great flattener". While in the popular view the Newtonian system was by this one blow proved correct, the measured eccentricity did not agree with NEWTON's value, and the geometers were moved to read more critically the passage in which NEWTON derived his result. They found his argument insecure. NEWTON had not used either the momentum principle or any principle of hydrostatics, for these were not yet known in sufficient generality to be applicable. As an independent hypothesis, he had asserted that two perpendicular columns of fluid, one axial and the other equatorial, meeting at the center of the spinning figure must have equal weight. Other principles were proposed by HUYGENS and MACLAURIN; the latter succeeded in proving in 1741 that any oblate spheroid was a figure of equilibrium and in relating the angular speed to the eccentricity. Thus began a search for the true and general laws of hydrostatics.

In 1743 appeared CLAIRAUT's book, *The Figure of the Earth*, where CLAIRAUT set himself the problem of determining the possible figures of equilibrium of fluid masses spinning subject to an arbitrary law of force. In this theory is the first occurrence of a *general vector field*. While CLAIRAUT could not solve the problem, he concluded that in order for a system of forces to be compatible with equilibrium, it was necessary that the work done in driving those forces through any closed circuit be zero; that is, the differential element of work must be perfect. CLAIRAUT thus obtained a condition to be satisfied by the forces, but he did not reach the general equations of hydrostatics.

The second main problem of fluid mechanics to grow out of NEWTON's unsuccessful attempts was that of efflux from a vessel. The first real progress on this particular problem was made by BORDA at the end of the century, but long before this, other trials, while failing to gain their immediate object, had opened the field of hydrodynamics. DANIEL BERNOULLI (1727) was the first to apply the principle of momentum correctly to a continuous medium; he found the force exerted by a jet impinging upon an inclined plate; the result has the same form as NEWTON's formula for the resistance of a body moving in a "rare fluid" or very rarefied gas, but the physical circumstances are different.

DANIEL BERNOULLI's *Hydrodynamica* (1733—1738) contains the first successful attempt to analyse the pressure and velocity of a fluid in motion. By *pressure*, DANIEL BERNOULLI means the force exerted by the fluid upon the walls of the tube or vessel containing it, and as a measure of this pressure he takes the height of a column of stagnant fluid, which had long been used to measure the pressure of a fluid in equilibrium. His result has the form of the now well known "BERNOULLI equation" for steady flow, but it would be misleading to attribute to DANIEL BERNOULLI the ideas with which that equation is now associated. In his book is no attempt to get equations of motion or to describe the deformation of a continuous medium. Treating the fluid as a block, BERNOULLI imagines that the wall of the vessel downstream from the point where the pressure is sought suddenly dissolves into thin air; the block of fluid formerly occupying it then experiences an impulse, which BERNOULLI by use of a principle of energy relates to the pressure that the wall must in fact exert in order to hold the fluid in.

7. The Earliest Differential Equations of Motion for Systems

The first differential equations of motion for systems having more than two mass-bearing points were published in 1743 by JOHN BERNOULLI and by D'ALEMBERT. The systems considered are more complex than one might expect; as often happens, the simplest cases were not the first to be treated successfully.

By 1739 JOHN BERNOULLI had found the *general equations of hydraulics* for incompressible inviscid fluids. While the main innovation in this work will be described in § 11, here we remark that BERNOULLI uses a Newtonian principle: The accelerating force on an infinitesimal slice of fluid equals the mass of the slice times its acceleration. After the work of JAMES BERNOULLI on the center of oscillation (§ 3), and apart from the incomplete work of TAYLOR on the vibrating string, here is the earliest occurrence of this now familiar argument. However, while BERNOULLI derives the quantities that would suffice to write down the partial differential equation of motion, he does not quite do so, but rather integrates along the tube. The result he obtains is a generalization of the hydraulic equation of DANIEL BERNOULLI.

This work is published in a volume dated 1742 but issued in 1743. In the same volume are JOHN BERNOULLI's treatments of the bifilar pendulum and the more general weighted cord. Here he introduces explicitly the *tensions* in the parts of the cord. For the case of two weights, he uses the balance of forces against accelerations to derive the *differential equations of finite motion*. In the case of n weights, JOHN BERNOULLI comes close to the general equations but misses them by use of a special device. His treatment is the first to refer all positions to a *single, rectangular Cartesian system*.

In a work of D'ALEMBERT printed in this same year, 1743, the differential equations of small motion for the weighted cord are derived by a general method. At the age of twenty-four, D'ALEMBERT published a book with a title indicative of what mechanics was in the Age of Reason: "Treatise on Dynamics, in which the laws of equilibrium and motion of bodies are reduced to the smallest possible number and are proved in a new way, and where is given a general principle for finding the motion of several bodies which react mutually in any way." This is the book to which we have several times referred in connection with the program of the age. In regard to practice, it is less successful. Contrary to the usual claims, neither did D'ALEMBERT reduce dynamics to statics, nor did he, here or anywhere, propose either of the two forms of the laws of dynamics now usually called "D'ALEMBERT's principle", these being due to EULER and to LAGRANGE, respectively, at a later period. D'ALEMBERT was the first to give a general rule for obtaining equations of motion of constrained systems. After decomposing the motion into two parts, one being "natural" and the other due to the presence of the constraints, he asserts that the forces corresponding to the accelerations due to the constraints form a system in static equilibrium. Thus his principle is a development of one of the ideas in JAMES BERNOULLI's great paper of 1703; it is still closer to the principle stated even more obscurely by DANIEL BERNOULLI in his treatment of the hanging cord in 1732—1733 (published 1740). Like the older assertions of DESCARTES and LEIBNIZ, it is a statement about the system as a whole, not about its parts, and it is insufficient to solve the general problems of dynamics; D'ALEMBERT tacitly invoked other principles as well, but he got results; moreover, he was the first to derive a *partial differential equation* as the statement of a law of motion, the particular case being that of a heavy hanging cord.

While EULER was soon to become the champion of NEWTON'S approach to mechanics, D'ALEMBERT started a new and opposed way of thinking. If the motion is known, he observed, then what we call forces are merely manifestations which may be calculated from it.

"Why should we appeal to that principle used by everybody nowadays, that the accelerating or retarding force is proportional to the element [*i.e.* differential] of velocity, a principle resting only on that vague and obscure axiom that the effect is proportional to the cause? We shall not investigate at all whether this principle be a necessary truth; we shall assert only that the proofs of it given up to now do not seem very convincing to us; neither shall we adopt it, like some other geometers, as a purely contingent truth, for that would ruin the certainty of mechanics and reduce it to being nothing but an experimental science. Rather, we shall be content to remark that the principle [of accelerating force], be it true or be it dubious, be it clear or be it obscure, is useless to mechanics and ought therefore to be banished from it."

While NEWTON's name is not mentioned, it is NEWTON's entire view of mechanics that D'ALEMBERT is banishing. If, for POPE and MACH, NEWTON had made all light, D'ALEMBERT regarded "forces inherent to bodies in motion", the central concept of the Newtonian view, as "obscure and metaphysical beings, capable of nothing but spreading darkness over a science clear by itself".

Immediately upon seeing JOHN BERNOULLI's work on hydraulics, EULER acclaimed it for its use of the "true and genuine method", namely, the Newtonian balance of force against acceleration. EULER's new treatments of constrained systems are close to JOHN BERNOULLI's cases in spirit but far clearer and more general. In 1744 EULER set up the differential equations of finite motion for the taut loaded string and for n rigid bars linked together, described in terms of a single rectangular Cartesian co-ordinate system. The only mechanical principle used is the "Newtonian" one, and the integrals of linear momentum and kinetic energy are derived by integration. This is *the earliest occurrence of exact differential equations of motion for a system of n bodies*, and also the earliest example of what we are now taught by textbooks of mechanics as the "Newtonian" method. As a corollary, the first recognition of the status of the integrals of linear momentum and kinetic energy for the system is to be found here also. By a brilliant passage to the limit in certain corollaries of the fundamental equations, EULER derived a complicated set of partial differential equations governing *finite motion of the continuous string*, but he was unable to draw any conclusion from them. This work appeared in print in 1751.

In 1746 D'ALEMBERT derived the *linear wave equation* as governing the small vibration of a string. He did not use his own principle of mechanics but instead, in this case, adopted the Newtonian approach; in effect, he took up and linearized an equation TAYLOR had derived but failed to exploit. D'ALEMBERT's wave equation is not the first partial differential equation of motion to be obtained, but the first to attract any attention.

8. Definitive Formulation of the Principle of Linear Momentum: Euler's "First Principles of Mechanics" ("Newton's equations") (1750)

The first formulation of the general problem of celestial mechanics, and in particular of the problem of three bodies, by means of a system of differential equations occurs in a paper of EULER, written in 1747 and published in 1749. With this paper begins the systematic development of Newtonian principles applied to the solar system. The method used is the same as in EULER's papers on con-

Fig. 38. LEONHARD EULER (1707—1783), portrait in 1753 by EMANUEL HANDMANN, in the Kunstmuseum Basel

strained systems completed just before, but of course the mechanical problem is simpler and more easily surveyed. As often happens, correct results were obtained before the general method was fully comprehended, and the generality was seen by induction from the successes with concrete special cases. That, sixty years after the *Principia*, a general approach to dynamical systems had finally been found, came as an anti-climax; once it was seen, no one could be surprised; indeed, it was "obvious". As EULER himself wrote,

"The foundation ... is nothing else than the known principle of mechanics, $du = p\,dt$, where p is the accelerating power and u is the velocity .... But some reflection is necessary before we can see that this principle holds equally for each partial motion into which the true motion is thought of as reduced. Moreover, this lemma includes all the principles ordinarily used in the determination of curvilinear motion."

The modern student may find it hard to understand how sixty years of experience with special cases had to follow before this simple conclusion, which he is taught to accept unquestioningly in a first course in physics, was seen. It is unlikely, however, that he has a better grasp of mechanical principles than did NEWTON or EULER. No-one doubted the correctness of "NEWTON's second law", at least as a rule for problem-solving, but what no-one saw, until it was shown, was that among all the various mechanical principles then used it was this one which was *general: It applies to every part of every system*, and more than this, *it suffices to get all the equations* determining the motion of many systems. As LAGRANGE explained when writing a history of mechanics in 1788,

"But if one looks for the motion of several bodies which react upon each other ..., then the question is of a higher order, and the preceding principles [*i.e.* those known before 1742] are insufficient to solve it."

EULER himself was not yet so far along in 1747. He had reached clarity on discrete systems, but for solid or fluid media he still failed of a general and unifying method. After experience using a similar approach to the vibrating string and to hydraulics, in 1750 he finally saw that the principle of linear momentum applied to *mechanical systems of all kinds*, whether discrete or continuous. His paper called "Discovery of a new principle of mechanics", published in 1752, presents the equations

$$F_x = M a_x, \quad F_y = M a_y, \quad F_z = M a_z,$$

where the mass M may be either finite or infinitesimal, as the axioms which "include all the laws of mechanics." Later he called them "the first principles of mechanics". These are the famous "Newtonian equations", here proposed for the first time as general, explicit equations for mechanical problems of all kinds.

The discovery of this principle seems so easy, from the Newtonian ideas, that it has never been attributed to anyone but NEWTON; such is the universal ignorance of the true history of mechanics. It is an incontestable fact that more than sixty years of research using more complicated methods even for rather simple problems took place before this "new principle" was seen.

The fruit of the "new principle" was immediate. In the very paper in which it was announced, EULER succeeded in deriving the equations of motion of a rigid body. In his notebooks, which even today remain unpublished, he derived at once and by inspection from the long known formulae for the restoring force, the partial differential equations for small transverse vibrations of a rod and for various flexible systems.

But in asserting that this principle "contains by itself all the principles that can lead to the knowledge of the motion of all bodies, of what nature soever they be," EULER was wrong. It required the accumulated mechanical experience of much of the rest of his life before he saw that a second principle was needed, and before, as we shall indicate below, he was able to find it.

9. Rigid Bodies, and the Distinction of Mass from Inertia

While the problem of oscillation of a heavy rigid body about a fixed axis had been solved correctly by HUYGENS, and while a more satisfactory method containing the germ of several later principles had been created by JAMES BERNOULLI in 1703, in 1750 it could not be said that the general motion of a rigid body was understood at all. Even for motion about a fixed axis, the reaction of the body upon its support could not be calculated, and no method for determining the behavior of a spinning top was known.

EULER's "first principles" changed the scene overnight. As just mentioned, in the paper where these principles are published EULER obtained the general *equations of motion of a rigid mody* about its center of gravity. He applied the "first principles" to the elements of mass making up the body, at the same time replacing the acceleration of the element by its expression in terms of the *angular velocity vector*, which makes its first appearance here. Taking moments about the center of gravity then yields, after some reduction, the differential equations of motion known as "EULER's equations" for a rigid body, subject to assigned torque about its center of mass. In the process arise naturally the six components of what is now called the "tensor of inertia". The "moment of inertia" had occurred, of course, in the old researches and had been named by EULER much earlier. The paper begins with the proof that in order for a body to spin freely about an axis, not only must that axis pass through the center of gravity, but also the two appropriate products of inertia must vanish, generalizing the corresponding results for laminae which EULER had discovered five years before. Thus NEWTON's example of the spinning cylinder is put at its proper and special place: While inertia as an expression for the property of continuing motion is indeed appropriate for translational motion, it applies only to restricted kinds of rotation.

In 1755 SEGNER proved that every rigid body has at least three axes of permanent rotation, and that these are mutually perpendicular; for special bodies, there may be infinitely many axes. In work of 1758—1760 EULER revised the whole theory of rigid bodies. Beginning with the most general deformable body, he defined and named its "center of mass" or "center of inertia", possibly distinguished from the center of gravity when this latter exists, and asserted that in any body the center of mass moved in the same way as would a mass-point located there, having the same mass as the body, and subject to the resultant force acting on the body. For a rigid body, EULER showed more clearly how the inertia, or resistance to change of motion, was determined not by the mass but by the *tensor of inertia*, the properties of which he analysed briefly but exhaustively. This material he developed still more clearly in his great treatise on rigid bodies, published in 1765.

Thus the distinction of mass from weight, begun by NEWTON, was completed, and at the same time inertia and mass, confused in the older work, were separated. The position of "NEWTON's second law" was fixed as appropriate only to infinitely small bodies or to the centers of mass of finite bodies.

10. Problems of Vibration and Wave Motion seen from the Equations of Motion

As soon as differential equations of motion were discovered, the whole theory of small oscillations could be approached in a unified way. The first modern work on vibrations is EULER's analysis (1748) of the motion of n equal masses coupled

Fig. 39. J. LE ROND D'ALEMBERT (1717—1783), after an engraving published as the frontispiece to Volume 2 of the third edition of DIDEROT's Encyclopédie, 1778

J. D'ALEMBERT.

by like springs; all results may be interpreted also in the context of the taut loaded string. EULER found the general solution as a sum of n particular solutions, each of which corresponds to a simple mode with a definite frequency of oscillation. EULER calculated these frequencies explicitly and found the values of the arbitrary constants in the case when only the end mass is displaced initially.

While D'ALEMBERT (1746) obtained the solution of the partial differential equation for the vibrating string in terms of arbitrary functions, he hedged it about with so many unnecessary conditions as to make it useless. These were denied by EULER (1748), who maintained that the partial differential equation itself sufficed as a full statement of the mechanical problem. In 1734—1739 DANIEL BERNOULLI in his studies of various oscillating systems had concluded that small vibrations which a body might execute separately it might also execute simultaneously, and in 1753 he wrote a paper asserting that the most general vibratory motion of any body could be composed by superposition of such simple modes. Since BERNOULLI never consented to begin from differential equations of motion, mathematical proof of such a proposition was out of the question; rather, he regarded it as a new law of physics. He was never able to substantiate his contention by solving important special initial-value problems, mainly because the expansion theorems necessary to carry out the details were not discovered until the next century. LAGRANGE entered the controversy in 1759 by calculating the explicit solution of the initial-value problem for the loaded string, generalizing slightly EULER's results of 1748, and attempted to obtain the solution in arbitrary functions for the continuous string by a passage to the limit. As D'ALEMBERT pointed out, the work is faulty. LAGRANGE claimed to disprove BERNOULLI's contentions, but here, too, LAGRANGE was in error.

The verbose subsequent polemic contributed little to mechanics, but in the course of it EULER was impelled to study the wave equation more closely. Virtually everything known today about the uniform vibrating string was found by him, directly from the functional solution. In particular, he showed that disturbances were propagated to the right and left at the speed corresponding to the fundamental period, and he discovered the laws of reflection from the end points. This work proved LEIBNIZ's law of continuity untenable; by destroying the old prejudice in favor of "equations" (*i.e.*, roughly, analytic functions), it opened the way to the general concept of wave motion.

Euler recognized the simple modes of bars and other vibrating systems as being special solutions, corresponding to special initial conditions, of the appropriate partial differential equations, but he was unable to get the general solutions. By considering a membrane as a network of fibres, he obtained the correct partial differential equation for its small motions (1759); he calculated the simple modes and some of the proper frequencies for rectangular and for circular drums. His theories of curved rods are faulty from lack of a developed geometry of the strain of a curve; similar lack of differential geometry, combined with D'Alembert's false idea that a plate may be regarded as a bundle of rods, invalidates James II Bernoulli's theory of plates (1787), which was motivated by Chladni's experimental determination of nodal patterns and proper frequencies (1787).

11. The Concept of Normal Stress, and the Principles of Hydrodynamics

In John Bernoulli's *Hydraulics* (1739), as we have seen, is the first successful use of the balance of forces to determine the motion of a deformable body. It would be wrong to consider this a mere application of Newtonian principles. In order to render those principles definite, Bernoulli saw that it was necessary to make some hypothesis about the action of fluid upon fluid. Guided, perhaps, by the old researches on generalized catenaries, where he had been the first to isolate correctly the forces acting upon an infinitesimal element of a cord (1691 to 1692, published 1743), he assumed that the fluid on each side of an infinitesimal normal slice pressed normally upon that slice, with a varying force which was itself a major unknown. He introduced the *internal force* of hydraulics. This is neither the wall pressure of Daniel Bernoulli nor the weight of a column of stagnant water as in the older researches; including both these ideas as special cases, it is, like the tension in a flexible line, a major independent concept of mechanics, a great advance in principle, for it achieves the separation of pressure from weight and replaces vague notions of action at a distance by a concrete view of contiguous action in a certain kind of material. The *resultant accelerating force* on the slice is the difference of the internal forces acting upon its two sides.

John Bernoulli's explanation of these ideas was poor, and his use of them limited to immediate extensions of the older results of his son. Euler took up the new hydraulics, made it clear in principle, and showed its application to many important cases. In a sequence of researches done in 1749—1752, he determined the flow of fluid in pumps and turbines, calculated the reactions in the machines, and found conditions sufficient to prevent cavitation within them.

D'ALEMBERT had entered the field of fluid mechanics with a treatise, published in 1744, concerning the problems which had been solved before by DANIEL BERNOULLI. This work, like the hydraulic researches of JOHN BERNOULLI and EULER, is limited to motions in one dimension where the velocity is supposed uniform over the cross-section of the tube. Theories of this kind cannot predict the forces exerted by fluids on submerged obstacles, and a prize was announced by the Academy of Berlin on the subject of the resistance of fluids. D'ALEMBERT's essay, completed in 1749, contains the first analysis of the *velocity field*, allowing the velocity of the fluid particles to vary arbitrarily from one place to another. The Academy, refusing to award the prize, returned the manuscripts and suggested that the contestants compare their results with measured data. D'ALEMBERT angrily withdrew, publishing his essay in 1752. It contains in major special cases the partial differential equation expressing the conservation of mass in the flowing fluid, the condition of potential flow, and what may now be recognized as the condition of conservation of vorticity in more general flows. As in his studies of punctual systems, D'ALEMBERT avoided the use of force whenever possible; in hydrodynamics, this is difficult to do correctly, and D'ALEMBERT's pioneering effort is incomplete as well as remarkably obscure in method and principle. Comparison of the results with experimental data was impossible and would have been premature in any case, for we know now that the data then available were taken in what are today called turbulent flows, not to be understood even roughly for another century; according to D'ALEMBERT's theory, an obstacle in a steady flow suffers no resistance at all, as had been shown by EULER in 1745, giving rise to misunderstandings over the next 150 years. More important than would have been the explanation of any particular experiment is the concept of a *field theory* of material behavior, which is initiated with this work of D'ALEMBERT.

The theory was taken up at once and rendered simple, clear, and general by EULER, who laid out the foundations of hydrodynamics in 1752—1759. The separation of the *kinematical* from the *dynamical* aspects of fluid motion, begun in JOHN BERNOULLI's *Hydraulics* and extended in EULER's hydraulic papers, is more important in the theory of the three-dimensional field, and EULER succeeded in constructing a definite and clear mathematical description of some aspects of the change of shape of a continuous body. From this point begins the full understanding of the *conservation of mass*, independently of changes of shape or volume.

EULER's greatest triumph in the theory of deformable bodies is his progress toward the theory of stress. In his earliest hydrostatic researches (1738, published 1749), he had seen that for incompressible fluids at rest on the earth's surface, the single axiom that the pressure of the fluid is normal to the surface on which it acts, and the same for all forms and orientations of that surface, sufficed to prove everything in the subject. His hydraulic researches had rested upon JOHN BERNOULLI's idea that fluid, even in motion, pressed normally upon itself, but only for particular fluid elements and particular kinds of flows. As if by a sudden intuition seeing that all these ideas could be combined, EULER at once creates the fully general concept of *internal pressure*, according to which the force exerted by the surrounding fluid upon any imagined interior boundary surface within the fluid, whatever its place and form, is equipollent to the action of a *field of normal pressures upon the surface*.

This idea is simpler than the predecessors it replaced, but, like many simple ideas, at first it seemed abstract, and it was slow to be understood. Nowadays it is often presented to beginners as if it were obvious; it cannot be obvious, because, as shown by later theories of fluid behavior, it is not always true. EULER saw at once that his statement of the theory included everything done before in fluid mechanics and rendered all former approaches obsolete:

"The generality which I conceive, far from shutting off our light, will reveal to us rather the laws of nature in all their brilliance, giving still more reason to wonder at their beauty and their simplicity."

Nothing of this kind had ever been achieved in mechanics before. Even LAGRANGE, late in his life, grew to see the greatness of what EULER had done:

"By this discovery, the entire mechanics of fluids was reduced to a single point of analysis, and if the equations which include it were integrable, one could determine completely the circumstances of motion and of action of a fluid moved by any forces. Unfortunately, they are so rebellious that up to the present time only a few very limited cases have been worked out."

EULER, however, like D'ALEMBERT before him, despite numerous brilliant studies was unable to conclude from his equations a single new result fit for comparison with experiment. Judged from a positivist philosophy, EULER's hydrodynamic researches are misconceived and unsuccessful: Their basic assumptions cannot be established experimentally, nor did EULER obtain from them numbers which can be read on a dial. Yet, after EULER's death, special solutions of his equations have given us the theories of the tides, the winds, the ship, and the airplane, and every year new practical as well as physical discoveries are found by their aid.

EULER's success in this most difficult matter lay in his *analysis of concepts*. After years of trial, sometimes adopting some semi-empirical compromise with experimental data, EULER saw that the experiments had to be set aside for a time. They concerned phenomena too complicated for treatment then; some remain not fully understood today. By creating a *simple* field model for fluids, defined by a set of partial differential equations, EULER opened to us a new range of vision in physical science. It is the range we all work in today. In this great insight, looking within the interior moving fluid, where neither eye nor experiment may reach, he called upon the "imagination, fancy, and invention" which SWIFT could find neither in music nor in mathematics.

12. The Laws of Elasticity, and the Concept of Shear Stress

Apart from the work on small vibrations, studies of elasticity were restricted to static deformation and thus pursued their course, unaffected by the major revision of dynamical principles at the middle of the century. Four major problems of principle originated from the work of JAMES BERNOULLI: (1) to derive the law of bending in an elastic band from the law of extension of the longitudinal fibres, (2) to determine the position of the neutral fibre in a bent beam and the nature of the forces acting upon a cross-section, (3) to unify the theory of flexible lines, resting upon the equilibrium of forces, with the theory of elastic bands, resting upon the equilibrium of moments, (4) to specify the elastic response of materials.

The first and fourth of these were solved by EULER in the course of a remarkable note on bells written while he was a student of JOHN BERNOULLI in Basel (1727) but, since its main subject was an incorrect theory of the oscillation of curved rods, left unpublished for nearly 150 years. To begin with, for a beam of any form, EULER assumed a linear variation of the tensions over the cross-section and integrated to obtain the total bending moment, thus uniting the procedure of LEIBNIZ, who had neglected bending, with that of JAMES BERNOULLI, who had not taken proper account of the variation of the tension. The result is the now familiar BERNOULLI-EULER law giving the bending moment as the product of the so-called YOUNG's modulus by the moment of inertia and the increase of curvature. As is clear from this statement, to this end EULER has to introduce the *modulus of extension*, which characterizes an elastic material rather than a particular specimen and is nowadays named after YOUNG[6].

In papers published in 1760 and 1780, EULER presented these ideas more compactly and explained them better. In particular, he defined explicitly the modulus of extension, which had been used with only passing remark in his earlier work. Correct scaling laws for the dependence of frequencies of vibration upon the shape of the cross-section, under debate for some time, followed immediately from the introduction of this modulus.

While it had been remarked that a column loaded in compression usually bends before it breaks, no tests of this phenomenon were made until the great program of experiment on the strength of materials carried out by MUSSCHENBROEK (1729). MUSSCHENBROEK found that the breaking strength is inversely as the square of the length of the column; he reported also a dependence upon breadth and depth which may be correct for breaking but cannot be so for buckling, the elastic bending which precedes it. EULER's rigorous determination of all forms that may be assumed in finite deflection of a straight or circular band subject to terminal load is the most brilliant achievement of the theory of elasticity in the century, but it cast light upon the principles only through his incidental discovery of the formula for buckling (1743). His result confirms MUSSCHENBROEK's experimental rule in its dependence upon the length of the column; when the modulus of bending is properly related to the modulus of elasticity, EULER's result gives also the correct dependence upon the shape of the cross-section. LAGRANGE (1770) observed that there was an infinite sequence of buckling loads, at each of which an additional bent form becomes possible, but he did not conjecture which form was actually assumed; he tried to find the shape of column strongest for given bulk, but his analysis is wrong (1773). EULER's determination of the height at which a uniformly heavy column buckles (1776, 1778) is a *tour-de-force* of analysis and mechanics but does not advance the basic concepts. EULER found that a stress equal to the buckling stress in compression of a wooden bar would produce, if reversed, a strain of only 0,07%. From this point begins the understanding that enormous loads may produce only minute elastic strains;

---

[6] As is well known, the modulus introduced by YOUNG (1807) does not enjoy this independence and is not the modern "YOUNG's modulus", being the constant of proportionality between force and strain, which was in standard use throughout the century preceding YOUNG's work.

all older work of any importance concerned finite deformation, and the distinction between large strain and large deflection, essential for the present century, was not seen.

PARENT (1713) extended and corrected JAMES BERNOULLI's analysis of the neutral fibre. He was the first to see how to apply the condition of equilibrium of forces as well as equilibrium of moments to the actions at the cross-section of a beam. Allowing for different moduli of compression and extension, he showed that the areas under the curves representing the tensions and compressions had to be equal. This condition restricts the position of the neutral fibre; for a continuous linear relation, it shows that the neutral fibre must be the central one. PARENT saw also that forces tangent to the cross-section, corresponding to what are now called *shear stresses*, were possible. EULER throughout his life spoke of the neutral fibre as being on the concave side; this persistent error had no effect on his results concerning elastic curves and buckling, because he never specialized them sufficiently to have any need of using it. JOHN III BERNOULLI showed some understanding of the problem of the neutral line (1776), but the first advance beyond PARENT's work was made by COULOMB (1773), who wrote down all conditions of equilibrium for the forces acting upon the cross-section of a loaded beam. COULOMB proved that shear stress was not only possible but *necessary*; in fact, almost everything he did rested upon hypotheses about it. While GALILEO had associated rupture with a given tensile force, and MARIOTTE with given elongation, COULOMB took a shear stress of given magnitude as his criterion for failure. He showed that the greatest shear stress in compression occurred along the planes inclined at 45° to the line of action of the load.

The great work of PARENT and COULOMB did not go so far as a theory of elasticity; like the studies of GALILEO and LEIBNIZ, and unlike the more developed analyses of JAMES BERNOULLI and EULER, it was purely statical, neglecting deformation, and too primitive in regard to mathematics to yield a sound foundation upon which theories could be built.

Before COULOMB's paper was published, EULER had finished a complementary research. One of his earliest essays (1728—1730) had succeeded in using the method of moments to derive a single equation including both the perfectly flexible line and the flexurally elastic band. The work of JAMES BERNOULLI, published later, made it clear that in the case of a perfectly flexible line, any of the general approaches could be applied, and each led to the same result. For bands, this did not seem to be true. In various trials, all his life long EULER sought to unify the flexible line with the elastic band in every respect, and particularly in regard to the equilibrium of forces. In 1771 he saw, at last, that what was lacking was a proper idea of the force exerted by one part of a general line upon its neighbor. In addition to the tension which alone is present in a perfectly flexible line, there is a normal force, or *shear resultant*. When this is taken into account, the condition of equilibrium of forces may be applied. The result has a threefold importance beyond EULER's immediate aim: (1) Extending the ideas of JAMES BERNOULLI, EULER separated the general principles of mechanics from the constitutive equations for particular materials. His *general equations of equilibrium* for a plane deformable line of any kind are the first example since his "first principles" of a general law of mechanics, independent of ideas defining fluids, flexible bodies, elastic bodies, *etc.*, and alike valid for all. (2) Distinguished from the purely tangential action in the flexible line and from the purely normal pressure of EULER's hydrodynamics, and analogous to the shear stress introduced just afterward by COULOMB, the *general line-stress vector* which appears here, unrestricted in direction or magnitude except by the conditions of equilibrium, is the precursor of the stress tensor. (3) It is clearly understood that the condition of balance of moments is *independent* of that of balance of forces. This independence, obscured in much of analytical dynamics but essential for continuum mechanics, is the subject of the next section.

By use of a framework model (1764), EULER had inferred that in torsion the torque was proportional to the sine of the twist, but, intent upon a theory of plates, he had not made any use of this result. COULOMB (1777—1784) carried out extensive experiments on the twisting of silk threads and metal wires, whence he found that the torque was proportional to the twist for small twists. Among his results is the remark that large tension changes the torsional modulus before destroying the linearity of the response. COULOMB made no attempt at a theory of torsion.

While all the ideas necessary for the general theory of stress had been proposed separately by 1773, there was no sign of that other necessary ingredient of a full theory of elasticity, namely, the theory of strain. There was no sign of it, that is, in works on elasticity, where it is needed. In a treatise on the mechanics of perfect fluids, where it was of little importance, EULER (1766) had introduced and interpreted the tensor of rate of deformation or stretching. Only a change of wording was needed to yield the description of infinitesimal strain later constructed by CAUCHY, but EULER's work on this subject was never noticed.

13. The Principle of Moment of Momentum

It is clear enough that in statics the equilibrium of moments is not assured by the equilibrium of forces, nor *vice versa*. In dynamics, the principle of moment of momentum developed late, and much of the earlier work concerning it gives the impression that the two principles were somehow hoped to be equivalent, so that there would be but a single law of motion. This illusion is fostered in the teaching of mechanics by physicists today.

It will be recalled that HUYGENS' solution for the physical pendulum does not use the principle of moment of momentum at all. Developing an idea latent in JAMES BERNOULLI's great paper of 1703, EULER and DANIEL BERNOULLI in their early work on rigid, linked, or flexible systems frequently invoked the principle of moment of momentum, often disguised by an additional quasi-static assumption. Its importance began to be seen gradually, so that only with difficulty can a particular date be fixed as that in which it rose to the level of a principle of mechanics. I believe the correct date is 1751, the year of publication of EULER's paper of 1744 which has been mentioned above as the first in which the "Newtonian equations" are recognized as sufficient to give all the mechanical principles determining the motion of a complex system, the case in point being the taut loaded string. The latter part of this same paper obtains the equations of motion for a system of *n* rigid rods linked together and subject to arbitrary forces at the junctions. Here the "Newtonian equations" do not suffice. In addition, EULER sets up as an independent principle the balance of moment of momentum about the center of mass of each rod.

EULER's development of the concept of rotary inertia has been mentioned in § 9, where his first method of deriving the equations of motion of a rigid body was described. The kinematical and dynamical aspects of the problem were not clearly separated there. EULER sought later an approach to mechanics as a whole that would yield directly and easily the equations of motion of a rigid body as a special case. This he achieved in a memoir published in 1776, where he laid down the following laws as applicable to *every part of every body*, whether punctual or space-filling, whether rigid or deformable:

Law 1. The total force acting upon the body equals the rate of change of the total momentum.

Law 2. The total torque acting upon the body equals the rate of change of the total moment of momentum, where both the torque and the moment are taken with respect to the same fixed point.

These laws, which may be called EULER's *laws of mechanics*, imply not only "NEWTON's laws" for mass-points but also all the other principles of classical mechanics and are just as convenient for continuous bodies as for discrete systems. The first law is equivalent to EULER's "first principles" of 1750 and yields the general theorem on the center of mass. The second law follows from it in some cases; for example, if all forces are absolutely continuous functions of mass and if all torques are the moments of forces; but in more general systems, such as those in which shear stress is present, the second law is independent of the first. For a rigid body, EULER was able to derive his former equations of motion directly and easily from the second law.

The law of moment of momentum is subtle, often misunderstood even today[7].

---

[7] In presentations of mechanics for physicists it is usually derived as a consequence of "NEWTON's laws" for elements of mass which are supposed to attract each other with mutual forces which are central and pairwise equilibrated. Although the formal steps of this derivation are correct, the result is too special for continuum mechanics as well as methodologically wrong for rigid mechanics:

1. Any forces between the particles of a rigid body never manifest themselves, by definition, in any motion. The condition of rigidity applied to EULER's second law suffices to determine the motion. To hypothecate mutual forces is to luxuriate in superfluous causes, which are to be excised by OCKHAM's razor.

2. To introduce mutual forces in a rigid body drags in action at a distance in a case when it is unnecessary to do so.

3. In continuum mechanics the total force acting upon a finite body arises principally from the stress tensor, which represents the contiguous action of material on material. There is no physical reason to assume that forces arise only from action at a distance, and no purpose served by doing so. If rigid-body mechanics is viewed as a special case of continuum mechanics, the stress within the rigid body is indeterminate and need never be mentioned, but a uniform process based upon EULER's second law remains possible.

No work of EULER, or of any other geometer of the eighteenth century, approaches rigid bodies by hypothecating forces acting at a distance as in modern books. To discover his equations, EULER assumed that since a body does not take on motion spontaneously, the forces among its particles "mutually destroy each other" and hence may be neglected in calculating the torque; the right answer comes out, but we are justified in doubting the argument. His final treatment does not make any presumption or restriction regarding the presence or nature of internal forces.

14. The Invariance of the Laws of Mechanics, and Lagrange's Méchanique Analitique (1788)

While the principle of "Galilean invariance" is now well known, during the eighteenth century it was felt rather than stated, and discovery of the properties of invariance of the principles of mechanics grew rather from concrete cases than from general studies. In the 1740's DANIEL BERNOULLI, EULER, and CLAIRAUT set up some tricky special problems, such as to determine the motion of a mass-point going down a wedge which is free to slide upon a table, or of a mass-point moving within a curved rigid tube cast in arbitrary initial motion. The principle to be used was stated most clearly by CLAIRAUT in a paper published in 1745; in effect, it is the modern principle of relative motion, according to which a body seen from a non-inertial frame experiences an "apparent force" per unit mass equal to the negative of the acceleration of that frame relative to an inertial frame. In special cases involving motion relative to the non-inertial frame, however, CLAIRAUT did not calculate the relative acceleration correctly. EULER adopted the principle, stated it more clearly, and calculated correctly the solutions of a number of extremely difficult problems. In particular, in his researches on hydraulic machines he obtained what is now called the "CORIOLIS acceleration", expressed in terms of angular variables appropriate to the problem. In an attempt to calculate that acceleration in rectangular Cartesian co-ordinates, for general motion, he fell into error and got but one half of the right amount (1755).

A special potential function was introduced by DANIEL BERNOULLI in connection with the elastic band (1739). JAMES RICCATI (c. 1754) suggested that the work done in deforming an elastic body was available, at least in part, for the recovery of the deformation when the forces are removed, but he gave nothing mathematical based upon this important idea. EULER (1743) showed that DANIEL BERNOULLI's principle of minimum potential energy led to the equation of the elastic curve, but this is not the same as the principle of conservation of energy.

Two principles proposed by MAUPERTUIS as general statements of the laws of mechanics attracted much notice because they were single and invariant statements about the whole of a system, not breaking it up into parts. The static *law of rest* (1740) is a generalization of the old principle that the center of gravity falls as

---

Fig. 40. J.-L. LAGRANGE (1736—1813)

low as it can; this principle, known to the ancients, had been shown by JAMES BERNOULLI to yield the solution to the catenary problem. MAUPERTUIS introduced the potentials of central forces varying as the nth power of the distance. His dynamic *principle of least action* (1744) takes the "action" of a moving body as the integral of the momentum with respect to distance along the path and asserts that nature wishes to economize as much as possible in dispensing this quantity[8]. He desired to imitate and subsume FERMAT's principle in optics, and he succeeded in deriving a number of simple mechanical results from his own principle, but he was not a sufficiently able mathematician to put it into general form or to get anything new from it; his formulation is neither clear nor correct.

A special case of the principle of least action was discovered independently by EULER at this time (1743); EULER's statement is clear, and he correctly limits the comparison paths to those consistent with the same total energy. In researches done in 1748 EULER succeeded in putting the law of rest (known nowadays as "DIRICHLET's principle") into fairly general form. In particular, he showed how to calculate the work done by a couple, thus making it possible to apply the principle to an elastic band subject to distributed load.

A general formal statement of the principle of least action for systems of mass-points was first obtained by LAGRANGE (1761), though without the proper restrictions. This is the first case in which an adequate statement of the laws of a fairly extensive branch of mechanics was gotten without the use of an *a priori* concept of force. In his work with continuous systems at this time, LAGRANGE by formal manipulations obtained other variational principles, various in scope. Potential functions occur in this work, but LAGRANGE never explains any concept, being content to verify the formalism. [The velocity-potential of hydrodynamics and the partial differential equation it satisfies ("LAPLACE's equation") had been derived in EULER's work on hydrodynamics (1752), and the "Newtonian" gravitational potential was to appear first in work by LAGRANGE in 1779.]

In the *Méchanique Analitique* (1788) LAGRANGE succeeded, at last, in formulating a variational principle uniformly valid for most kinds of mechanical systems then known. His formula, called "D'ALEMBERT's principle" in most books today, expresses the *principle of virtual work* with the assigned forces supplemented by

---

[8] A closely related idea had been put forward by LEIBNIZ. For a review of the controversy over the principle of least action, see J. O. FLECKENSTEIN's Introduction to L. Euleri *Opera Omnia* (2) **5**.

"inertial forces" arising from the accelerations, and with the assumption that the constraints do no work. For systems of mass-points, in 1782 LAGRANGE had introduced "generalized co-ordinates" and had achieved the invariant formulation of analytical dynamics now called "LAGRANGE's equations", from which his treatise has gained its main fame.

Much has been said in praise of these equations, which belong to the only part of the mechanics of the eighteenth century that is reasonably well known today. Their importance for later work in analytical mechanics and in modern physics is clear. Less clear, perhaps, is that abstractness of formulation conceals the main conceptual problems of mechanics: The role of inertial frames and the concept of rigidity, essential to the classical idea of "observer", are hidden in the invariance of the algebra.

It is true that LAGRANGE's equations have the same form in all descriptions, but it is not true that a system of differential equations in LAGRANGE's form necessarily belongs to a dynamical system satisfying EULER's laws in some frame, let alone an inertial frame. By obscuring the forces, LAGRANGE's equations conceal the invariance group of classical mechanics, which is immediately plain from EULER's equations[9]. Moreover, LAGRANGE's equations do not reflect the space-time geometry of classical mechanics, the main property of which is the possibility of adding together vectors localized at different points. Without this distant parallelism, we may speak of energy, but we cannot even form those other basic quantities of classical mechanics, momentum, center of mass, and moment of momentum. From LAGRANGE's equations, we cannot tell whether a system has momentum or not; EULER's equations show at once that it does, and this is borne out by the fact that the general integrals of linear momentum first appear in treatments based upon EULER's formulation. In any case, LAGRANGE's equations are relevant only to special kinds of mechanical systems and are far less general than EULER's laws or than LAGRANGE's principle of virtual work.

Something regarding the *Méchanique Analitique* itself must also be said. Its object was expressed by LAGRANGE as follows:

"There are already several treatises on mechanics, but the plan of this one is entirely new. I have set myself the task of reducing the theory of this science, and the art of solving the problems

[9] *Note added for the reprinting.* Indeed, that there are *two* basic laws is a reflection of the theorem that every automorphism of Euclidean space may be decomposed into a translation and a rotation. *Cf.* the basic paper by W. NOLL, "La mécanique classique, basée sur un axiome d'objectivité," *La Méthode Axiomatique* ... (1959), Paris, Gauthier-Villars, 1963.

concerned with it, to general formulae, the simple development of which gives all the equations necessary to the solution of each problem."

Its method is purely algebraic:

"No drawings are to be found in this work. The methods which I present require neither constructions nor geometrical or mechanical arguments, but only algebraic operations, subject to a regular and uniform progress."

Like the first half of Book I of NEWTON's *Principia*, it is largely a retrospective work, ordering within a single formal scheme a selection from the discoveries of the immediately preceding generations. In the main, the results presented are slight generalizations of some of EULER's, set within LAGRANGE's severely formal style. Unlike Book I of the *Principia*, it contains no major discovery, although some of its contents are taken from earlier original work by LAGRANGE himself. Neither is there anything corresponding to Books II and III of the *Principia*, nothing looking forward rather than backward, and no comparison of theoretical and observed values. Mechanics appears as a physically closed subject, a branch of the theory of differential equations.

At the end of the century there was a dismaying tendency to turn away from the basic problems, both in mechanics and in pure analysis. Directly contrary to the great tradition set by JAMES BERNOULLI and EULER, this formalism grew rapidly in the French school and is reflected in the *Méchanique Analitique*. Much of the misjudgment that historians and physicists have passed upon the work of the eighteenth century comes from unwillingness to look behind and around the *Méchanique Analitique* to the great work of EULER and the BERNOULLIS which it left unmentioned. As its title implies, the *Méchanique Analitique* is not a treatise on rational mechanics, but rather a monograph on one method of deriving differential equations of motion, mainly in the special branch now called, after it, analytical mechanics. This may account for its failure to include any concrete solutions for the vibrating string[10], the tones and nodes of bars and membranes, the shapes of elastic curves, the flow of fluids in tubes, *etc.*, or to mention such fundamental concepts as relative motion, stress, strain, elastic modulus, neutral line, instability of equilibrium, wave propagation, *etc.*, although all these solutions and concepts may be found in the literature of the generation preceding LAGRANGE's.

Granted its more modest scope, estimates of LAGRANGE's performance must remain a matter of taste. In music, in painting, in literature, tastes have changed

[10] Something about it was added in the second edition.

in the past century. Why should they not also change in mechanics? The historians delight in repeating HAMILTON's praise of the *Méchanique Analitique* as "a kind of scientific poem", but it is unlikely that many persons today would find HAMILTON's recommendations in non-scientific poetry congenial.

15. Retrospect: Experience, Theory and Experiment in the Age of Reason

The true nature of mechanics in the Age of Reason I have attempted to display in outline, to the best of my ability to read and survey the hundreds of mathematical papers and books in which this luxuriant growth of the aims, methods and results of mathematical physics, never before or since equalled, was triumphantly recorded. He who is trained in physics today will ask, what were the fundamental experiments upon which "classical" mechanics was founded?

I have never been able to find any. In the books physicists write on the history of physical experiment, little mention is made of the period between NEWTON and LAGRANGE, and much of the little that is said turns out to be a description, after all, of theory or of *a posteriori* verification at best. Two examples may be noted here. First, while there had been experiment enough in regard to the oscillation of pendulums in HUYGENS' day, when it came to the general motion of a rigid body there is no evidence that any experiment was even suggested; once an exact solution was obtained, the only function of measurement could be to estimate the magnitude of the frictional forces or the defect from perfect rigidity; and in the 200 years since EULER derived his equations, no direct experimental test of them has ever been proposed, so far as I know, but they are applied and verified indirectly in hundreds of devices later designed by use of them. For a second example, consider MUSSCHENBROEK's discovery of the law of buckling; there is no evidence that EULER knew of it when he derived his buckling formula, and comparisons were not made for another twenty years; moreover, the structural engineers for more than a century following seemed not to know of MUSSCHENBROEK's tests, for they sneered at EULER's result as a piece of useless pure mathematics. It is unlikely that today there is a bridge designed anywhere without some use of the MUSSCHENBROEK-EULER formula.

What was, then, the method? Rational mechanics was a science of *experience*, but no more than geometry was it *experimental*. While some great mechanical experiments were done in the Age of Reason, they had only occasional bearing on the growth of the theories we now regard as classical. Experiment and theory result from different kinds of reaction to experience. If, ideally, they should complement and check one another, yet even today, with all our superior knowledge not only of facts but also of scientific method, it is difficult enough to relate them, why should it have been easier 300 years ago? It was not. A factual view of the history of mechanics must concede that rational mechanics and experimental mechanics, both arising from human beings' intelligent reaction to mechanical experience, grew up separately.

Not only private, individual experimental researches were performed in the eighteenth century; there were also large, cooperative projects. As today, they cost more than real science, and they attracted administrators. But the effect of all this expense on what we now consider the achievement of the period was nil. The method used in the great researches was entirely mathematical, but the result was not what would now be called pure mathematics. *Experience* was the guide; *experience*, physical experience and the experience of accumulated previous theory. If we were to seek a word for what was done, it would not be physics and it would not be pure mathematics; least of all would it be applied mathematics: It would be *rational mechanics*.

Acknowledgment

This paper was written during my tenure of a Fellowship of the U.S. National Science Foundation. It includes material given in a SigmaXi lecture at Indiana University (1951), public lectures at Brown University (1955) and at the State University of Iowa (1956), and a lecture to the Research Society of the Socony-Mobil Laboratory at Dallas (1960). It was first published in the *Archive for History of Exact Sciences*, Volume 1, pp. 3—36 (1960). A partial translation into Italian was printed as "La scienza del moto dai 'Principia Mathematica Naturalis Philosophiae' di Newton alla 'Méchanique Analitique' di Lagrange," *Accad. Naz. Sci. Lett. Arti di Modena, Atti e Memorie* (VI) **2**, 3—32 (1960).

In the reprinting I have corrected some errors and added a few phrases and footnotes.

References

The following historical works, easily accessible to the general reader, give quotations, details, and citation of the original works used for some parts of the foregoing article. For other parts, as explained in the Introduction, no adequate detailed history is presently available. Some of the points made in this essay are presented in greater detail, with references to chapter and verse, in the fourth and fifth essays in this volume.

JOUGUET, E., *Lectures de mécanique,* 2 vols. Paris: Gauthier-Villars 1908, 1909, $x+210+284$ pp.
DUGAS, R., *La mécanique au XVIIe siècle.* Neuchatel: Ed. du Griffon 1954. 620 pp.
TRUESDELL, C., "Rational fluid mechanics, 1687—1765." L. Euleri *Opera Omnia* (2) **12**, IX—CXXV (1954).
TRUESDELL, C., "I. The first three sections of EULER's treatise on fluid mechanics (1766); II. The theory of aerial sound (1687—1788); III. Rational fluid mechanics (1765—1788)." L. Euleri *Opera Omnia* (2) **13**, VII—CXVIII (1956).
TRUESDELL, C., "The rational mechanics of flexible or elastic bodies, 1638—1788." L. Euleri *Opera Omnia* (2) $\mathbf{11}_2$, 428 pp. (1960).

III. Reactions of Late Baroque Mechanics to Success, Conjecture, Error, and Failure in Newton's Principia

1. Newton's Principia, Book I

NEWTON'S *Principia* is cited with awe by historians of science and even writers of physics texts; its worth is attested by the price a copy of the first edition commands from bibliophiles; but all these mostly join in expressing their awe and their valuation by venerating the book from without, not venturing to profane the shrine by opening its covers to peer inside. It is a very hard book to read. If a page from the middle of it were copied on a typewriter and handed to a Professor of the History of Science who had just detailed NEWTON'S magnificent discoveries to his class, he might well fail to recognize it as coming from the pen of his hero. It is densely mathematical. On many pages there is not a single word concerning what historians and physicists now call physics (Fig. 41). The new professionalized History of Science joins Social Science in excluding, if politely and often tacitly, mathematics from the sciences. Unlike the historians of today,

Fig. 41. A typical purely mathematical passage in the *Principia*, reproduced from a copy, annotated in HALLEY'S hand, now in the library of the University of Texas

tia GJ minor esse potest quam assignata quævis. Est autem (ex natura circulorum per puncta ABG, Abg transeuntium) AB *quad.* æquale $AG \times BD$ & Ab *quad.* æquale $Ag \times bd$, adeoq; ratio AB *quad.* ad Ab *quad.* componitur ex rationibus AG ad Ag & BD ad bd. Sed quoniam JG assumi potest minor longitudine quavis assignata, fieri potest ut ratio AG ad Ag minus differat a ratione æqualitatis quam pro differentia quavis assignata, adeoq; ut ratio AB *quad.* ad Ab *quad.* minus differat a ratione BD ad bd quam pro differentia quavis assignata. Est ergo, per Lemma I, ratio ultima AB *quad.* ad Ab *quad.* æqualis rationi ultimæ BD ad bd. *Q. E. D.*

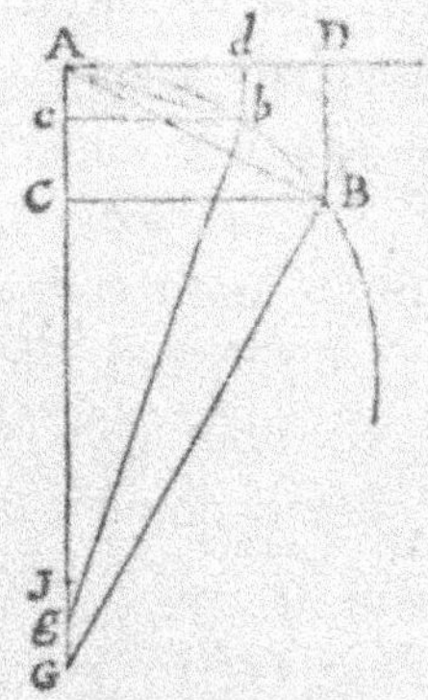

Cas. 2. Inclinetur jam BD ad AD in angulo quovis dato, & eadem semper erit ratio ultima BD ad bd quæ prius, adeoq; eadem ac AB *quad.* ad Ab *quad.* *Q. E. D.*

Cas. 3. Et quamvis angulus D non detur, ~~tamen anguli D, d~~ ad æqualitatem semper vergent & propius accedent ad invicem quam pro differentia quavis assignata, adeoq; ultimo æquales erunt, per Lem. I. & propterea lineæ BD, bd in eadem ratione ad invicem ac prius. *Q. E. D.*

Corol. 1. Unde cum tangentes AD, Ad, arcus AB, Ab & eorum sinus BC, bc fiant ultimo chordis AB, Ab æquales; erunt etiam illorum quadrata ultimo ut subtensæ BD, bd.

Corol. 4. Triangula rectilinea ADB, Adb sunt ultimo in triplicata ratione laterum AD, Ad, inq; sesquiplicata laterum DB, db: Utpote in composita ratione laterum AD & DB, Ad & db existentia. Sic & triangula ABC, Abc sunt ultimo in triplicata ratione laterum BC, bc.

Corol. 5. Et quoniam DB, db sunt ultimo parallelæ & in duplicata ratione ipsarum AD, Ad; erunt areæ ultimæ curvilineæ

F A

Newton's contemporaries and successors did not revere him as a seer and a prophet. They respected him as a mathematician of gigantic power.

Now a mathematician has a matchless advantage over general scientists, historians, politicians, and exponents of other professions: He can be wrong. *A fortiori*, he can also be right. There are errors in Euclid, and, to within a set certainly of measure zero on the ordinary human scale, what Euclid proved to be true in ancient Greece is true even in the colossal, unprecedented, nucleospatial, totally welfared today. In the advance through the physical, social, historical, and other sciences, the demarcation between truth and falsehood grows vaguer, until in some areas truth can be rezoned as falsehood and falsehood enshrined into truth by consensus of "acknowledged experts and authorities" or even popular vote. One professor discussing the doctrines of Karl Marx may label them as grave errors; a second, equally qualified, may present them as problematic, partly true and partly not so; while a third, living in a different part of the world, may proclaim them as the quintessence of human knowledge. In the mathematical sciences as taught by the colleagues of these same three social scientists, there is no disagreement as to what is true and what is not. A mistake made by a mathematician, even a great one, is not a "difference of point of view" or "another interpretation of the data" or a "dictate of a conflicting ideology", it is a mistake. The greatest of all mathematicians, those who have discovered the greatest quantities of mathematical truths, are also those who have published the greatest numbers of lacunary proofs, insufficiently qualified assertions, and flat mistakes. By attempting to make natural philosophy into a part of mathematics, Newton relinquished the diplomatic immunity granted to non-mathematical philosophers, chemists, psychologists, *etc.*, and entered the area where an error is an error even if it is Newton's error; in fact, all the more so because it is Newton's error.

The mistakes made by a great mathematician are of two kinds: first, trivial slips that anyone can correct, and, second, titanic failures reflecting the scale of the struggle which the great mathematician waged. Failures of this latter kind are often as important as successes, for they give rise to major discoveries by other mathematicians. One error of a great mathematician has often done more for science than a hundred impeccable little theorems proved by lesser men. Since Newton was as great a mathematician as ever lived, but still a mathematician, we may approach his work with the level, tactless criticism which mathematics demands.

The *Principia* is a book on the mechanics of the solar system seen within a half formulated, half sensed rational mechanics as a whole. In the century following its publication, down even until the work of LAPLACE, the *Principia* influenced nearly all mechanical researches. Two great challenges offered by the *Principia* may be seen at once and distinguished, for the answers to them are almost disjoint:

1. To determine in greater precision the motion of a system of mass-points subject to universal gravitation.
2. To discover and set forth the general rational mechanics which NEWTON sought and to apply it to the motion of ordinary bodies on the earth: deformable masses of finite extent.

Most eighteenth-century researches on theoretical mechanics respond to one or the other of these challenges. Historians of physics, from MACH on into our own day, have studied the history of the first so exclusively as to fall into the delusion that NEWTON himself had solved the second, and that the three laws set at the head of his book really suffice as a basis for mechanics as a whole. No modern specialist in mechanics shares this opinion; even if it were true today, to project it backward to the century from 1687 to 1787 would be most unhistorical, for NEWTON himself made no such claim, either outright or by implication or tacitly in his own work in the second area, and the great savants of the eighteenth century who struggled their lives long to master the mechanics of rigid and flexible and fluid and elastic bodies would consider it poor pay indeed for their labors to be told by some paedagogue that they were merely "applying NEWTON's laws".

A total estimate of the reactions to the *Principia* would have to include a summary of eighteenth century work on the problem of three bodies and on perturbations in the solar system. I am not competent in celestial mechanics and must therefore keep silent about it, even about its history. While today an astronomer does not seem to need to know much about mechanics, it was a different matter in NEWTON's day, since physics had not yet been cut up into compartments, and students of natural science had not yet reached certainty as to what was unimportant to them. The specialist did not yet exist.

At the time he wrote the *Principia*, NEWTON showed little interest in any part of mechanics save the motion of heavenly bodies. He treats these first as points moving subject to two forces alone: universal gravitation and their own inertia. So as to justify a model in which such immense bodies as the sun, earth, and moon are represented as mathematical points, he then proves that the force of attraction between two homogeneously layered spheres is directed along the line of centers and is independent of the diameters. Here, as in several earlier places in the book, his program of deduction from the laws breaks down (Fig. 42). While he once refers to the third law, which in fact he does not need, the arguments rest on a rule for adding infinitely many forces. No such rule was given explicitly by him, but apparently he thought the vector law for pairs of forces sufficed. This law he had stated as Corollary II to the Laws of Motion, but the proof, resting on Corollary I, is false[1]. If we overlook this use of unstated axioms, common and excusable in works of great originality, we can regard NEWTON's program as brilliantly successful.

NEWTON himself was not satisfied. In earthly experience, bodies move in media, and these media resist their motion. The heavenly bodies move in something, too. If we are to complete a unified mechanics based on laws common to the earth and the heavens, we must take account somehow of such resistance as heavenly bodies may encounter. Moreover, the proof that big spheres are dynamically equivalent to concentrated masses then fails. If the resistance on a tennis ball in air is small, it need not be so on a great planet moving at enormous speed in the faraway heavens.

---

[1] In modern terms, with granules of impulse replaced by forces, the proof of Corollary I is roughly as follows. Given two forces $\boldsymbol{f}_1$ and $\boldsymbol{f}_2$, let them produce motions $\boldsymbol{x}_1(t)$ and $\boldsymbol{x}_2(t)$ in a body of mass m. Then by Law II,

$$\boldsymbol{f}_1 = m\ddot{\boldsymbol{x}}_1, \quad \boldsymbol{f}_2 = m\ddot{\boldsymbol{x}}_2.$$

Addition yields

$$\boldsymbol{f}_1 + \boldsymbol{f}_2 = m(\ddot{\boldsymbol{x}}_1 + \ddot{\boldsymbol{x}}_2).$$

From kinematics, accelerations add vectorially, say $\ddot{\boldsymbol{x}}_1 + \ddot{\boldsymbol{x}}_2 = \ddot{\boldsymbol{x}}_3$, but by Law II

$$\boldsymbol{f}_3 = m\ddot{\boldsymbol{x}}_3.$$

Hence $\boldsymbol{f}_1 + \boldsymbol{f}_2 = \boldsymbol{f}_3$. That is, forces add vectorially. This argument applies only to *resultant* forces on bodies. In statics, these resultant forces are all $\mathbf{0}$, and every step in the proof reduces to $\mathbf{0} = \mathbf{0}$. The vectorial composition of the parts of a force with resultant $\mathbf{0}$ is essential in statics, and in particular in the theory of attraction. It cannot be derived from any laws which relate forces only to the motions they produce.

Fig. 42. NEWTON's proof of the proposition that the resultant attraction of two homogeneously layered spheres is as the inverse square of the distance between their centers, from the *Principia*

Sunto Sphæræ quotcunq; concentricæ similares *AB*, *CD*, *EF* &c. quarum interiores additæ exterioribus componant materiam densiorem versus centrum, vel subductæ relinquant tenuiorem; & hæ, per Theor. XXXV, trahent Sphæras alias quotcunq; concentricas similares *GH*, *IK*, *LM*, &c. singulæ singulas, viribus reciproce proportionalibus quadrato distantiæ *SP*. Et componendo vel dividendo, summa virium illarum omnium, vel excessus aliquarum supra alias, hoc est, vis qua Sphæra tota ex concentricis quibuscunq; vel concentricarum differentiis composita *AB*, trahit totam ex concentricis quibuscunq; vel concentricarum differentiis compositam *GH*, erit in eadem ratione. Augeatur numerus Sphærarum concentricarum in infinitum sic, ut materiæ densitas una cum vi attractiva, in progressu a circumferentia ad centrum, secundum Legem quamcunq; crescat vel decrescat: & addita materia non attractiva compleatur ubivis densitas deficiens, eo ut Sphæræ acquirant formam quamvis optatam; & vis qua harum una attrahet alteram erit etiamnum (per argumentum superius) in eadem illa distantiæ quadratæ ratione inversa. Q. E. D.

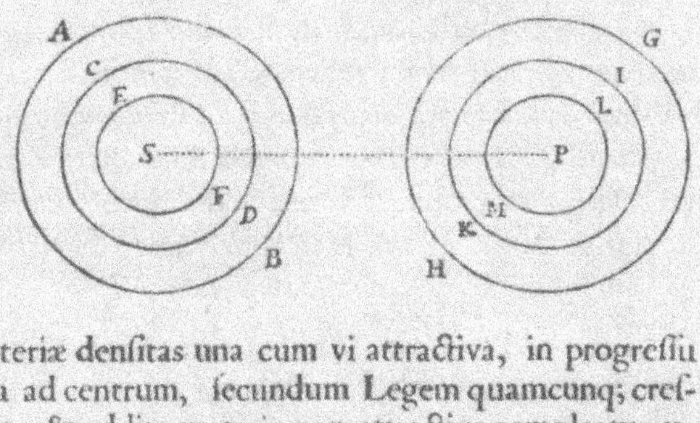

Corol. 1. Hinc si ejusmodi Sphæræ complures sibi invicem per omnia similes se mutuo trahant; attractiones acceleratrices singularum in singulas erunt in æqualibus quibusvis centrorum distantiis ut Sphæræ attrahentes.

Corol. 2. Inq; distantiis quibusvis inæqualibus, ut Sphæræ attrahentes applicatæ ad quadrata distantiarum inter centra.

Corol.

2. Newton's Principia, Book II

NEWTON faced this problem in Book II of the *Principia.* This is the part of the *Principia* that historians and philosophers, apparently, tear out of their personal copies. NEWTON may have hoped to show that resistances according to all simple laws influenced the motion of a "body", ultimately, in much the same way. After finding, on the contrary, that the effects of friction according to various hypothetical laws were different, he was impelled to seek the true nature of the resistance offered to a finite body by a fluid medium. According to his metaphysics, explanation of celestial mechanics was to be sought in earthly mechanics. Obviously, such a metaphysics is not confined, like the mind of a physicist today, to regard inertia and gravitation as the beginning and end of "classical" mechanics. Not having been taught in a physics course to shut out most ordinary facts as being only "unphysical" or "approximate" or "macroscopic", NEWTON sought the laws of the heavenly medium just where he had sought the laws of inertia, namely, in earthly experience. Since the laws of fluid motion on the earth were unknown, NEWTON had to find them. This is not so easy. While there is only one kind of inertia, there are many kinds of fluids, and the variety of their behaviors is bewildering. NEWTON envisaged a fluid alternately as composed of discrete particles "fleeing each other" or as a "continued" medium. His program of rational deduction from the axioms then broke down completely, and in many places he took refuge in merely qualitative remarks, sometimes labelled as "corollaries" although they are no more proved from the propositions than the propositions themselves are proved from the axioms. The arguments are made still more difficult by being not entirely conjectural: Here and there a few lines or even pages of regular mathematical proof of the easier details rise, as they do throughout the writings of GALILEO, to lend the appearance of logic to what seems in fact to be guesswork.

The areas treated are as follows:

1. General and obscure assertions about resistances and motions of parts of a fluid.
2. Resistances of spheres and cylinders in a rare static medium.
3. The motion of a continued fluid falling through a hole in a vessel, based on statics.
4. Resistances of cylinders, spheres, and spheroids in a continued fluid, either of infinite extent or confined by a canal.

5. General remarks on the propagation of pressures through a heap of particles.
6. Oscillation of an incompressible fluid in a *U*-tube.
7. The speed of surface waves and the nature of the fluid motion giving rise to them.
8. The speed of sound in a gas.
9. The resistances of cylinders and spheres spinning in a fluid with internal friction.

In his study of these miscellaneous problems of fluid mechanics, NEWTON seems to have forgotten what his original purpose was. He nowhere applies to the solar system his conclusions about resistance, perhaps because they are so varied and so hypothetical as to defy any such application. He concludes Book II with a proposition intended to destroy DESCARTES' vortex theory of the planets, but the proposition itself is unconnected to what went before and is meaningless without definition of the terms used, and the "proof", like that of the earlier and false proposition on the "cataract", rests upon an unsound principle of solidification.

While in celestial mechanics and the simpler motions of "bodies" NEWTON had centuries and even millenia of sound observation and sound special theory on which to base his "intuition", as the physicists today like to call the application of common sense to previously organized experience, in fluid mechanics he had few guides. Book II is almost entirely original.

In writings on the history of science today, as in all aspects of social intercourse, it becomes increasingly bad taste to call a spade a spade. In the particular application to the history of science, the compulsion to euphemy assumes the form of a solemn refusal to admit that there is such a thing as wrong in science. All that matters is how the scientists of one epoch thought and felt about nature, their own work, and the work of others. In particular, the beginner is enjoined above all not to take sides, like the positivist historians of the last century, and set past science into categories of true and false according as it does or does not agree with what is taught in school science of our own day[2]. However admirable this

[2] Lest this remark seem unfounded, I append a quotation from Mr. COHEN's eulogy of the late A. KOYRÉ, published in *Isis*, Volume 57, p. 161 (1966). Asserting that KOYRÉ was "more than any other responsible for the *new* history of science," Mr. COHEN explains, "He rejected the kind of history in which a modern scholar remarks that this was correct and that was wrong judging each statement, not in the context of the time in which it was composed, but

philosophy may be in promoting peace and mutual love among historians of science, it disregards one aspect of science that is not altogether negligible, namely, that scientists seek *the* truth, not *a* truth. He who refuses to "take sides" in science in effect negates science itself by denying its one and common purpose. He reduces science to just one more social manifestation. In so doing, not only does he by implication belittle the great scientists of the past, but also he sins against history, for in his attempt at historical impartiality he destroys the object, namely science, the history of which he claims to write.

NEWTON's treatment of the efflux of water from a vessel is a case to the point here. In the first edition, he gave an essentially static argument which in fact sets aside his own laws of motion (Fig. 43); after learning that his end result contradicted experiment, he replaced the whole passage in the second edition by an argument far less plausible, based on an *ad hoc* fiction of a cataract of melting ice (Fig. 44). Thus NEWTON himself clearly believed then that his first treatment had been at fault, but he did not say how. Instead, he manufactured a maze to lead to a different answer he at that time considered right. In a language where a spade is called a spade, the first treatment of the efflux problem is wrong, while the second is a bluff. Much the same may be said of the calculation of the velocity of sound, for which NEWTON in the second edition introduced the fiction of the "crassitude of the solid particles of the air" so as to insert what would nowadays

rather grading it from the viewpoint of our own presently accepted theories and collections of facts. Such historians, according to KOYRÉ, were not engaged in conceptual analysis; they were not attempting to understand what the men of science of the past were thinking and even accomplishing with the mathematical, theoretical, and experimental tools then available. Professor KOYRÉ was interested — first and foremost — to find out the degree to which men of science in the past may have been 'right' in the terms of their own times, not ours."

The logical completion of this approach — so far as I know, not yet advocated openly though perhaps already practised — is to ban from the education and the library of the Newtonian scholar all writings on mathematics and physics since NEWTON's death, for fear they might suggest applications and extensions of NEWTON's ideas which cannot have influenced "men of science in the past ..., 'right' in the terms of their own times, not ours."

It is amusing to carry the new historians' impartiality in science a little further. If science, which presumably remains even for them the theory and demonstration of the phenomena of nature, is time-conditioned, social, and institutional, like politics and moral philosophy, are such also the phenomena themselves? If purely mechanical perpetual motion and purely chemical transmutation of elements are impossible today, as we are now taught in school, were they perhaps possible in the times of LEONARDO and NEWTON?

Fig. 43. NEWTON's argument that the reaction of a jet issuing from a vessel is the same as that which would be required to keep the hole stopped up, in the first edition of the *Principia*, 1687

fit $\frac{2d}{e}S$ & latitudo eadem quæ foraminis, posset eo tempore defluendo egredi de vase, hoc est columna $\frac{2d}{e}SF$. Quare motus $\frac{2dd}{ee}SFV$, qui fiet ducendo quantitatem aquæ effluentis in velocitatem suam, hoc est motus omnis tempore effluxus illius genitus, æquabitur motui $AF \times V$. Et si æquales illi motus applicentur ad FV; fiet $\frac{2dd}{ee}S$ æqualis A. Unde est dd ad ee ut A ad $2S$, & d ad e in dimidiata ratione $\frac{1}{2}A$ ad S. Est igitur velocitas, quacum aqua exit e foramine, ad velocitatem quam aqua cadens, & tempore T cadendo describens spatium S acquireret, ut altitudo aquæ foramini perpendiculariter incumbentis, ad medium proportionale inter altitudinem illam duplicatam & spatium illud S, quod corpus tempore T cadendo describeret.

Igitur si motus illi sursum vertantur; quoniam aqua velocitate V ascenderet ad altitudinem illam S de qua deciderat; & altitudines (uti notum est) sint in duplicata ratione velocitatum: aqua effluens ascenderet ad altitudinem $\frac{1}{2}A$. Et propterea quantitas aquæ effluentis, quo tempore corpus cadendo describere posset altitudinem $\frac{1}{2}A$, æqualis erit columnæ aquæ totius AF foramini perpendiculariter imminentis.

Cum autem aqua effluens, motu suo sursum verso, perpendiculariter surgeret ad dimidiam altitudinem aquæ foramini incumbentis; consequens est quod si egrediatur oblique per canalem in latus vasis, describet in spatiis non resistentibus Parabolam cujus latus rectum est altitudo aquæ in vase supra canalis orificium, & cujus diameter horizonti perpendicularis ab orificio illo ducitur, atque ordinatim applicatæ parallelæ sunt axi canalis.

Hæc omnia de Fluido subtilissimo intelligenda sunt. Nam si aqua ex partibus crassioribus constet, hæc tardius effluet quam pro ratione superius assignata, præsertim si foramen angustum sit per quod effluit.

generari quo tempore Globus duas tertias diametri suæ partes, velocitate uniformiter continuata describat, ut densitas Medii ad densitatem Globi, si modo Globus & particulæ Medii sint summe elastica & vi maxima reflectendi polleant: quodque hæc vis sit duplo minor ubi Globus & particulæ Medii sunt infinite dura & vi reflectendi prorsus destituta. In Mediis autem continuis qualia sunt Aqua, Oleum calidum, & Argentum vivum, in quibus Globus non incidit immediate in omnes fluidi particulas resistentiam generantes, sed premit tantum proximas particulas & hæ premunt alias & hæ alias, resistentia est adhuc duplo minor. Globus utique in hujusmodi Mediis fluidissimis resistentiam patitur quæ est ad vim qua totus ejus motus vel tolli possit vel generari quo tempore, motu illo uniformiter continuato, partes octo tertias diametri suæ describat, ut densitas Medii ad densitatem Globi. Id quod in sequentibus conabimur ostendere.

PROPOSITIO XXXVI. PROBLEMA VIII.

Aquæ de vase Cylindrico per foramen in fundo factum effluentis definire motum.

Sit *ACDB* vas cylindricum, *AB* ejus orificium superius, *CD* fundum horizonti parallelum, *EF* foramen circulare in medio fundi, *G* centrum foraminis, & *GH* axis cylindri horizonti perpendicularis. Et concipe cylindrum glaciei *APQB* ejusdem esse latitudinis cum cavitate vasis, & axem eundem habere, & uniformi cum motu perpetuo descendere, & partes ejus quam primum attingunt superficiem *AB* liquescere, & in aquam conversas gravitate sua defluere in vas, & cataractam vel columnam aquæ *ABNFEM* cadendo formare, & per foramen *EF* transire, idemque adæquate implere. Ea vero sit uniformis velocitas glaciei descendentis ut & aquæ contiguæ in circulo *AB*, quam aqua cadendo & casu suo describendo altitudinem

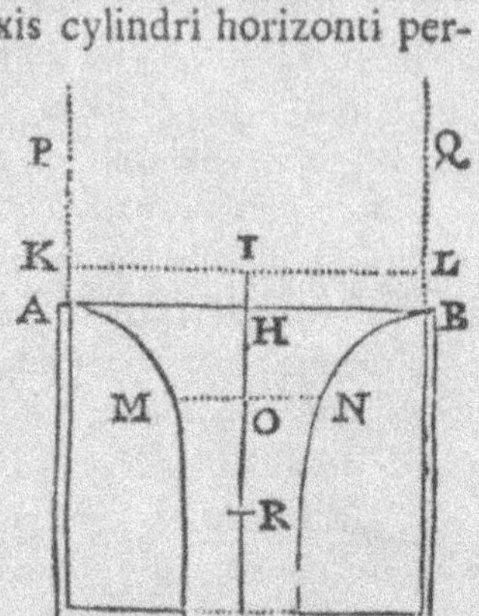

IH acquirere potest; & jaceant *IH* & *HG* in directum, & per punctum *I* ducatur recta *KL* horizonti parallela lateribus glaciei.

be called a "fudge factor" yielding the desired numerical value from a recalcitrant theory.

It was these two flimsy and unmathematical props, the "cataract" and the "solid particles of the air", no better than the vortices of the Cartesians, that drew the strongest criticism upon the *Principia*. However successful was Book I, Book II was a failure as an essay toward a unified, mathematical mechanics.

With its bewildering alternation of mathematical proof, brilliant hypothesis, pure guessing, bluff, and plain error, Book II has long been praised, in whole or in part, and praised justly, as affording the greatest signs of NEWTON's genius. To the geometers of the day, it offered an immediate challenge: to correct the errors, to replace the guesswork by clear hypotheses, to embed the hypotheses at their just stations in a rational mechanics, to brush away the bluff by mathematical proof, to create new concepts so as to succeed where NEWTON had failed. It is not an exaggeration to say that rational mechanics, and hence mathematical physics as a whole and the general picture of nature accepted today, grew from this challenge as it was accepted by the Basel school of mathematicians: the three great BERNOULLIS and EULER, on the basis of whose work LAGRANGE, FOURIER, POISSON, NAVIER, CAUCHY, GREEN, STOKES, KELVIN, HELMHOLTZ, KIRCHHOFF, MAXWELL, and GIBBS constructed what is now called classical physics. Between 1700 and 1750 the Basel school met occasional competition from TAYLOR, MACLAURIN, CLAIRAUT, and D'ALEMBERT, though of these only the last did work of any breadth.

3. Success and Tradition in Mechanics before Newton

It is wrong to picture the Basel geometers as disciples of NEWTON. Rather, they began as doubting opponents, who accepted, grudgingly, only some isolated results. NEWTON's achievement, great though it was, was far from enough to cause the revolution in scientific thought historians of the last century and ours have been pleased to imagine. In the first place, three major mechanical systems which had been treated with greater or lesser success by others were left out by NEWTON. These are:

1. *Rigid bodies*. NEWTON's famous remark about the inertia of a spinning top did

---

Fig. 44. Page 303 of the second edition of the *Principia* (1713), in which NEWTON introduces the "cataract" of ice so as to get a value of the speed $\sqrt{2}$ times as great as in the edition of 1687

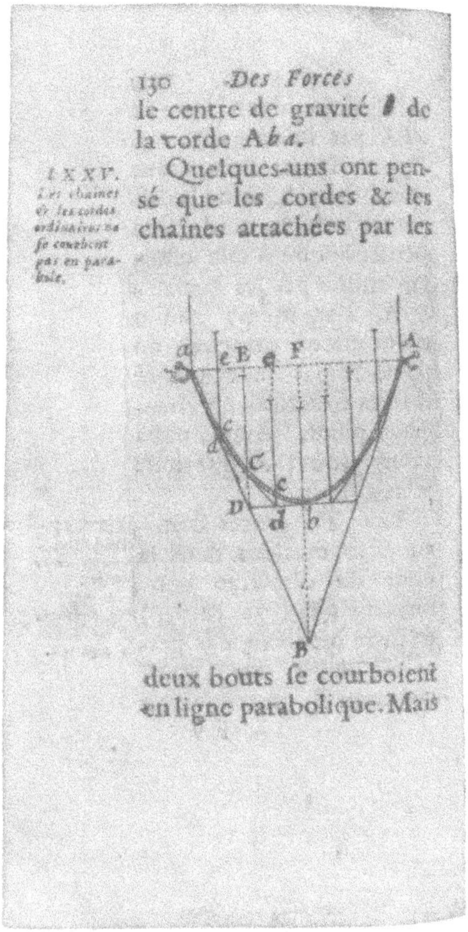

130 *Des Forces*

le centre de gravité *I* de la corde *Aba*.

LXXV. *Les chaines & les cordes ordinaires ne se courbent pas en parabole.*

Quelques-uns ont pensé que les cordes & les chaînes attachées par les deux bouts se courboient en ligne parabolique. Mais

Fig. 46. PARDIES' solution of the problem of the suspension bridge (1673)

Fig. 45. CHRISTIAAN HUYGENS (1629—1695), after a painting by KASPAR NETSCHER in 1671, Collection Haags Gemeentemuseum — The Hague

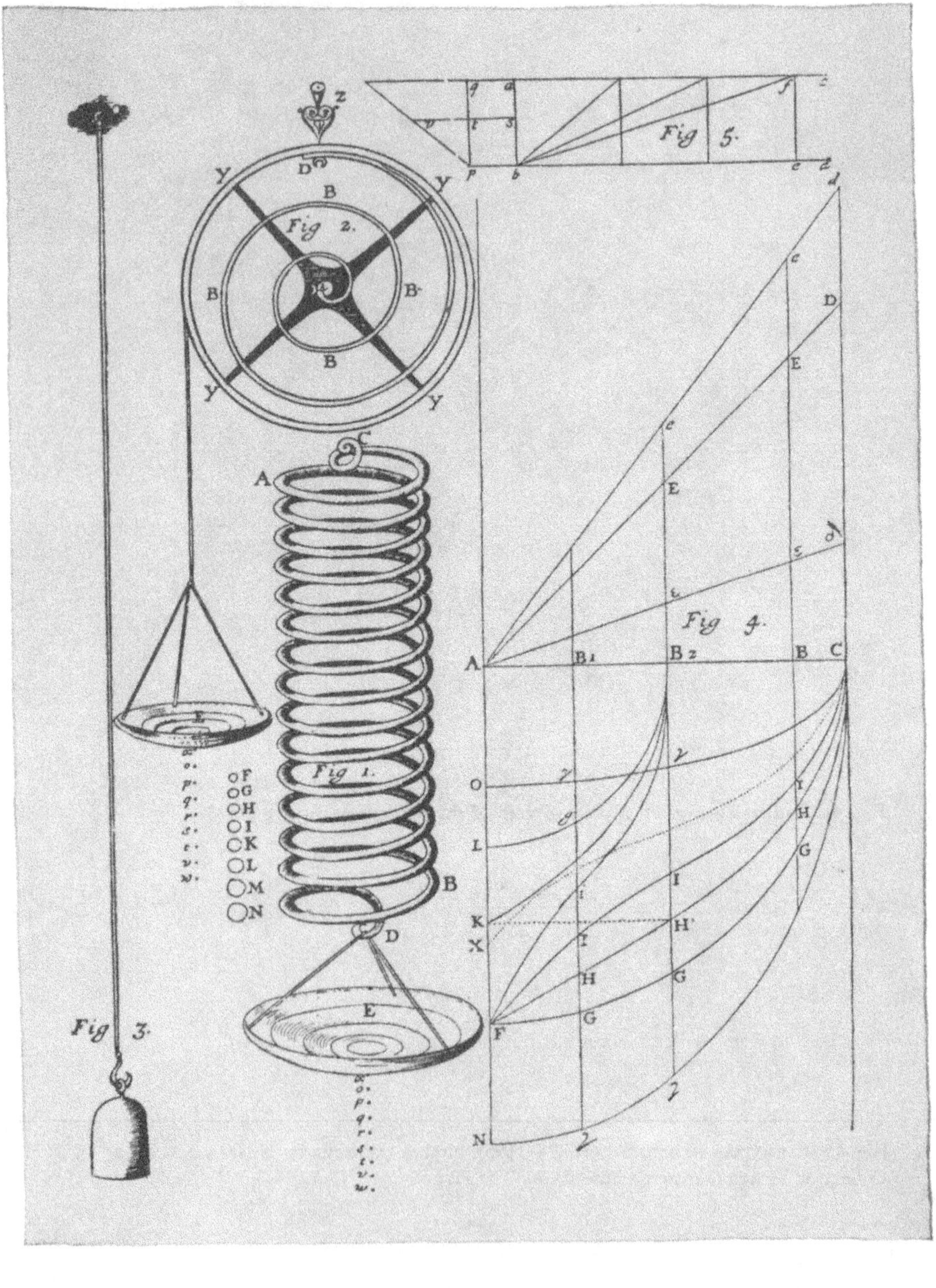
Fig 1.
Fig 2.
Fig 3.
Fig 4.
Fig 5.

not lead him to attempt any specific theory of rotation, nor did he mention the problem of the pendulous oscillation of a finite body, a correct solution to which, based on special assumptions, had been published by HUYGENS in 1673.
2. *Flexible bodies.* The problems of the vibrating string and the catenary curve had been widely discussed even before NEWTON's birth. If most of the work of BEECKMAN and HUYGENS remained unpublished, this fact did not keep it from being noised abroad. MERSENNE had published his empirical laws of vibration in 1625; PARDIES, his beautiful solution of the suspension bridge problem in 1673 (Fig. 46). NEWTON must have known much if not all of this, but his writings do not mention anything in this area of study.
3. *Elastic bodies.* While GALILEO had discussed the rupture of solid bodies, and while various authors had published attempts on the theory of elasticity, NEWTON's nearest approach comes in his description of the pressure of a gas composed of particles. NEWTON must have known not only HOOKE's vociferations on elasticity (Fig. 47), made excessively public in 1678, but also the magnificent note of 1684 in which LEIBNIZ initiated the mathematical theory of stress and gave the first application of integral calculus to problems concerning contact forces.

NEWTON's work in mechanics opened a new stream of thought, which immediately became the most powerful of all. However, it by no means stopped or absorbed the others at once. These others were:
1. *Statics.*
a) *The Parallelogram of Forces and the Law of the Lever.* While by example NEWTON taught us to use statically conceived forces in kinetic problems, his explanation of forces suggests that they can be detected only when they give rise to motions. If this were true, there would be no science of statics. It may seem a paradox that although it was NEWTON above all who showed us how to turn knowledge gained from equilibria so as to solve problems of motion, his handling of mechanics was weakest where statics was involved. In 1687, the year of the *Principia*, VARIGNON published his *Project of a New Mechanics* (Fig. 48), in which the principle

---

Fig. 47 (p. 152). The plate from HOOKE's pamphlet, *Lectures De Potentia Restitutiva, or of Spring, Explaining the Power of Springing Bodies*, 1678

Fig. 48 (p. 154). Plate 1 from VARIGNON's *Projet d'une Nouvelle Méchanique*, 1687, one of many illustrating the parallelogram of forces

Fig. 49 (p. 155). The argument for the parallelogram of forces in STEVIN's *De Beghinselen der Weeghconst*, 1586, as republished in French in 1634

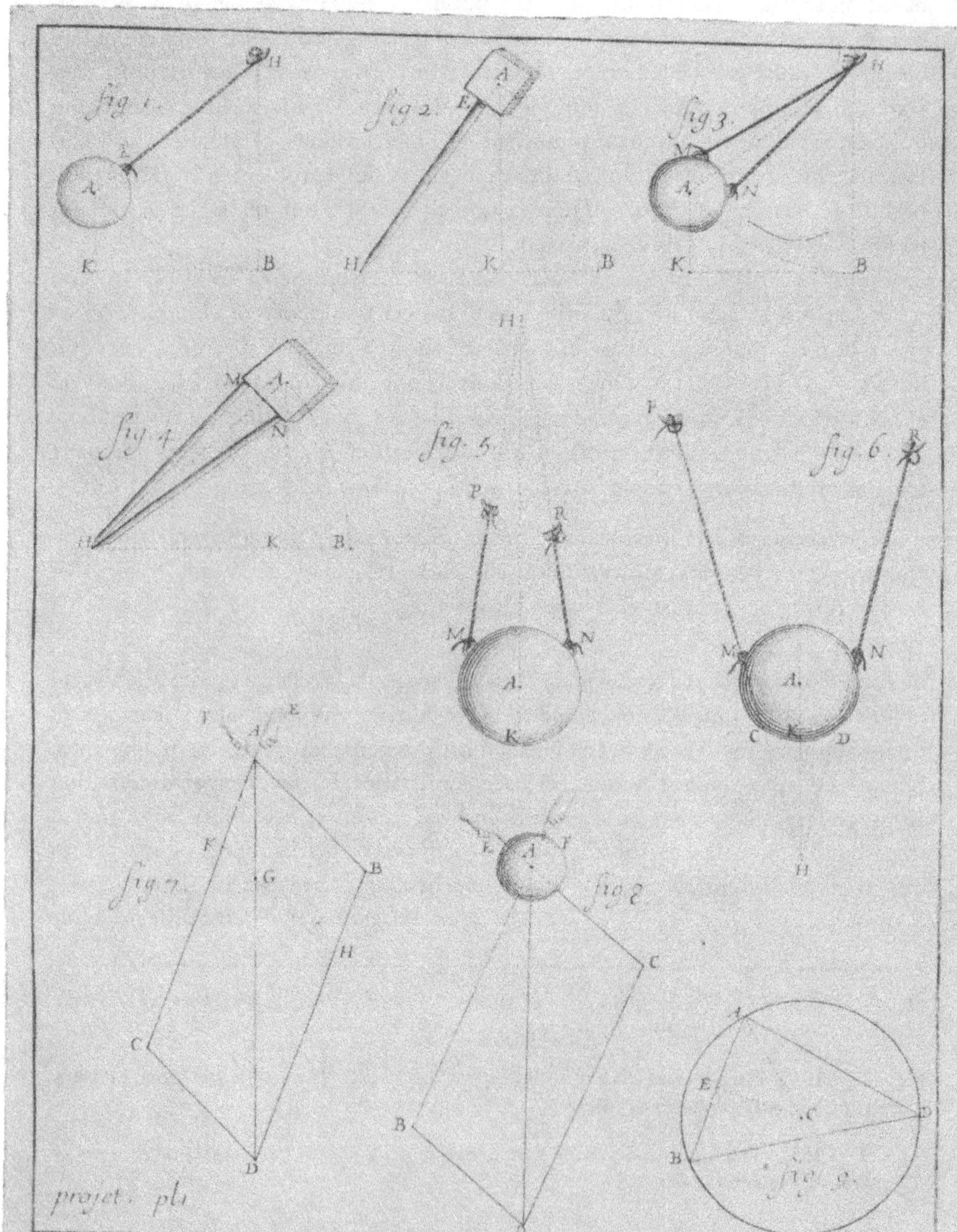
fig 1.
fig 2.
fig 3.
fig 4.
fig. 5.
fig. 6.
fig 7.
fig 8.
fig. 9.
projet. pl1

sur P, repose aussi sur I, & partant comme G O à G P, ainsi seront les pesanteurs qui reposent sur I & H.

Conclusion. Une colomne donc reposant, &c.

Corollaire.

Il est manifeste par ce que devant, que voulant recognoistre la raison de la pesanteur reposant sur I, à celle qui repose sur H, qu'on doit à ceste fin mener les perpendiculaires K L, M N, coupant l'axe és poincts O, P; & que la raison de G O à G P seroit la requise; ainsi donc, lors que la pesanteur de la colomne est notoire, qu'aussi seront les pesanteurs de celles qui reposent sur chacun poinct, tel que H, I.

JUSQUES ICY ONT ESTE declarées les proprietez des pesanteurs directes; suivent les proprietez & qualitez des obliques, desquelles le fondement general est compris au Theoreme suivant.

Theoreme XI. Proposition XIX.

Si un triangle, a son plan perpendiculaire à l'horizon, & sa base parallele à iceluy; & sur un chacun des deux autres costez un poids spherique, de pesanteur egale; comme le costé dextre du triangle, au senestre; ainsi la puissance du poids senestre, à celle du poids dextre.

Le donné. Soit A B C un triangle ayant son plan perpendiculaire à l'horizon, & la base AC parallele à iceluy horizon: & soit sur le costé A B (qui est double à B C) un poids en globe D, & sur B C un autre E, egaux en pesanteur & en grandeur.

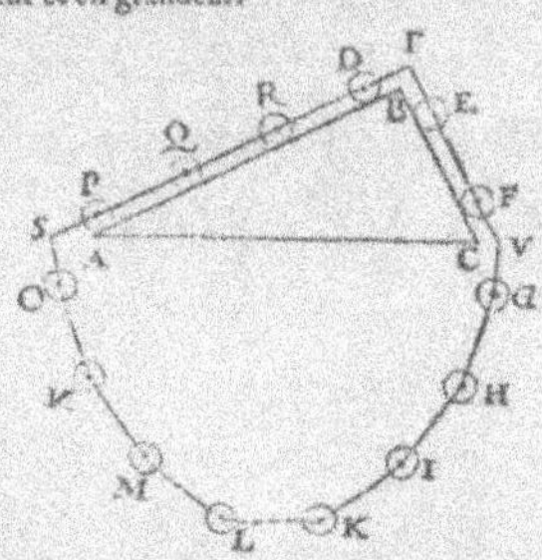

Le requis. Il faut demonstrer que comme le costé A B 2 au costé B C 1, ainsi la puissance ou pouvoir du poids E à celle de D.

Preparation. Soit accommodé à l'entour du triangle un entour de 14 globes, egaux en pesanteur, en grandeur, & equidistans, comme D, E, F, G, H, I, K, L, M, N, O, P, Q, R, enfilez d'une ligne passant par leurs centres, ainsi qu'ils puissent tourner sur leurs susdits centres, & qu'il y puisse avoir 2 globes sur le costé B C, & 4 sur B A, alors comme ligne à ligne, ainsi le nombre des globes au nombre des globes: qu'aussi en S, T, V, soyent trois poincts fermes, dessus lesquels la ligne, ou le filet puisse couler, & que les deux parties au dessus du triangle soyent paralleles aux costez d'iceluy A B, B C; tellement que le tout puisse tourner librement & sans accrochement, sur lesdits costez A B, B C.

Demonstration.

Si le pouvoir des poids D, R, Q, P, n'estoit egal au pouvoir des deux globes E, F, l'un costé sera plus pesant que l'autre, donc (s'il est possible) que les 4 D, R, Q, P, soyent plus pesans que les deux E, F; mais les 4 O, N, M, L, sont egaux aux 4 G, H, I, K; parquoy le costé des 8 globes D, R, Q, P, O, N, M, L, sera plus pesant selon leur disposition, que non pas les 6, E, F, G, H, I, K, & puis que la partie plus pesante emporte la plus legere, les 8 globes descendront, & les autres 6 monteront: Qu'il soit ainsi donc, & que D vienne, où O est presentement, & ainsi des autres; voire que E, F, G, H, viennent, où sont maintenant P, Q, R, D, aussi I, K, où sont maintenant E, F: Ce neantmoins l'entour des globes aura la mesme disposition qu'auparavant, & par mesme raison les 8 globes auront le dessus en pesanteur, & en tombant feront revenir 8 autres en leurs places, & ainsi ce mouvement n'auroit aucune fin, ce qui est absurde. Et de mesme sera la demonstration de l'autre costé: La partie donc de l'entour D, R, Q, P, O, N, M, L, sera en equilibre avec la partie E, F, G, H, I, K; que si on oste des deux costez, les pesanteurs egales, & qui ont mesme disposition, comme sont les 4 globes O, N, M, L, d'une part, & les 4, G, H, I, K, d'autre part; les 4 restans D, R, Q, P, seront & demeureront en equilibre avec les 2 E, F; parquoy E aura un pouvoir double au pouvoir de D; comme donc le costé B A 2, au costé B C 1, ainsi le pouvoir de E, au pouvoir de D.

Conclusion. Si un triangle donc a son plan, &c.

Corollaire I.

Soit A B C un triangle comme devant, & A B double à B C, & soit D un globe sur A B, double à E, qui est sur B C; en F soit un poinct ferme, par dessus lequel la ligne DFE puisse couler sans empeschement, ainsi que D F, FE soyent paralleles aux costez du triangle A B C, procedantes des centres des globes, il appert que D, E seront encor en equilibre, puis que cy-dessus P, Q, R, D, l'estoyent à E, F; parquoy comme A B à B C, ainsi le globe D au globe E.

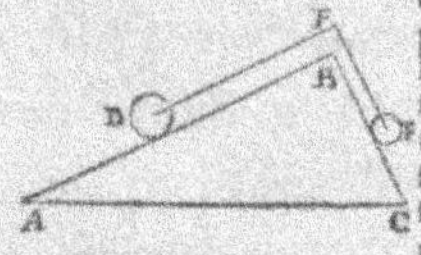

Corollaire II.

Soit maintenant l'un des costez du triangle, comme BC (qui est moitié de l'autre A B) perpendiculaire à AC, comme cy joignant, le globe D, qui est double à E, sera encor en equilibre avec E, car comme le costé AB à BC, ainsi le globe D au globe E.

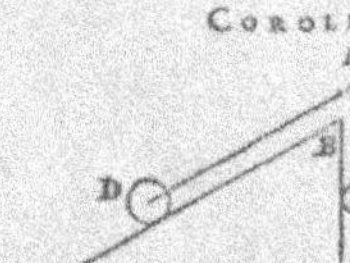

Corollaire III.

Soient derechef les mesmes posées, mais au lieu du poinct ferme F, soit adaptée une poulie comme icy, ainsi que D F demeure parallele à A B; & que E soit un

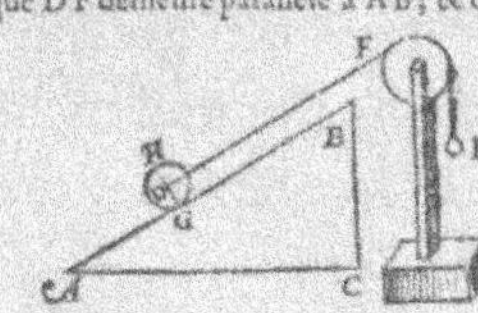

poids de quelle figure que ce puisse estre, egal en pesanteur à celuy de devant; iceluy avec D, seront encor en equilibre, parquoy comme A B à B C, ainsi D à E.

Corol-

of the parallelogram of forces, inferred from a rather Aristotelian viewpoint, is applied to the solution of a class of problems untouched by NEWTON. While the book is pedestrian, it bears witness that the didactic tradition of STEVIN (Fig. 49), long prior and hence independent, was still very much alive, as in fact it is today. The law of the lever had been known since antiquity. The numerous unsuccessful attempts, including VARIGNON's, to derive it from the equilibrium of forces are not germane to my subject.

b) *Virtual Work.* While the principle of virtual work may be of Peripatetic origin, a correct formulation of it had been sketched by DESCARTES, and it was to remain on the periphery of mechanics until LAGRANGE made it his principal axiom in 1788.

c) *The Action of Contact Forces.* The concept of interior forces exerted by the parts of a body on each other is difficult to trace, but it was skirted by GALILEO and used explicitly in the treatise of PARDIES.

2. Dynamics

a) *The Principle of Energy.* A particular principle of energy had been laid down as an axiom and used with brilliant success by HUYGENS. Another was published by LEIBNIZ in 1686, the year the *Principia* was licensed.

b) *The Principle of Inertial Force.* Although it is customary to attribute this principle to NEWTON or D'ALEMBERT, there is no evidence that NEWTON had a grasp of it sufficient for application to finite bodies, and certainly it was known and used widely before D'ALEMBERT's treatise of 1743, where, in fact, it can be located only with difficulty.

4. Some of the Bernoullis' Work

While the parallelogram of forces, the law of the lever, the principle of virtual work, the action of contact forces, and the principle of energy had extensive earlier histories and were to be absorbed in or united to the Newtonian stream, the principle of inertial force grew independently from the same antecedents as did NEWTON's mechanics, found its first expression in 1686, and finally incorporated NEWTONS' dynamics and every other. It was JAMES BERNOULLI's idea to set aside HUYGENS' special principle of energy and determine the oscillation of a pendulous body by an appeal to statics. The law of the lever is taken as valid in motion as well as at rest, provided the motion itself be regarded as giving rise to forces per unit mass equal to the accelerations reversed. In his note of 1686 JAMES BER-

NOULLI does not get this principle quite right, but he is the first to come near it. He proposed and applied it correctly, though only in the context of the pendulum, in his great paper of 1703 (see Figs. 87 and 88, pp. 250, 251, and 253).

Though it was many years before this great unifying principle was grasped in general form, we see today that to make it useful, a far more developed, explicit statics was necessary. This, too, we owe in the main to JAMES BERNOULLI. It is well known that the catenary curve was determined in 1690 by JOHN BERNOULLI, HUYGENS, and LEIBNIZ in a public contest. It is less well known that all three used special devices for arriving at the equations to be solved. NEWTON, and with him the British school, took no part in this work, nor is it even in the slightest detail influenced by NEWTON's book, published only three years before. Immediately thereafter, in 1691—1704, JAMES BERNOULLI, who had been unable to match his younger brother's brilliance and who had failed to appear among the solvers of the great problem at the appointed time, pondered the general nature of perfectly flexible lines in equilibrium, subject to arbitrary loading. Here the devices of LEIBNIZ and JOHN BERNOULLI would not work, and of course the old approach of HUYGENS was hopelessly elaborate. In the end, JAMES BERNOULLI found four independent proofs of the general differential equations:

a) By the balance of forces acting on an infinitesimal element.
b) By the balance of moments acting on an infinitesimal element.
c) By the principle of virtual work.
d) By the principle of minimum potential energy.

The first two are the most important, for they point toward the general theory of stress, and they show that when applied to a particular and sufficiently special system, the two independent basic principles of mechanics may fall into equivalence.

To get the equation of motion of the vibrating string, anyone today knows that he need only apply "NEWTON's second law" to an infinitesimal element, or, more elegantly, apply to the static equation of the catenary JAMES BERNOULLI's principle of reversed accelerations. In his famous paper of 1713, TAYLOR followed the

---

Fig. 50 (p. 158). BROOKE TAYLOR's treatment of the vibrating string in the *Philosophical Transactions* No. 337, 1714

Fig. 51 (p. 159). D'ALEMBERT's paper on the vibrating string, showing the first appearance of the partial differential equation, in the *Mémoires de l'Académie des Sciences*, Berlin, for 1747

Lemma 2.

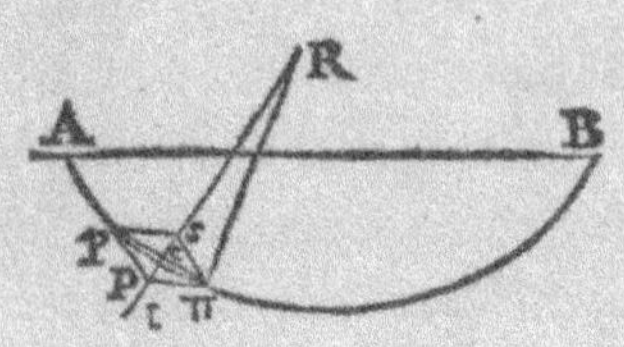

In aliquo articulo vibrationis suæ induat Nervus tensus, inter puncta A & B, formam curvæ cujusvis A p π B. Tum dico quod sit incrementum velocitatis puncti alicujus P, seu acceleratio oriunda a vi tensionis Nervi, ut curvatura Nervi in eodem puncto.

Demonstr. Finge Nervum constare ex particulis rigidis æqualibus infinite parvis p P & P π, &c. & ad punctum P erige perpendicularem P R = radio curvaturæ in P, cui occurrant tangentes p t & π t in t, iis parallelæ π s & p s in s, & chorda p π in c. Tum, per Principia Mechanicæ, vis absoluta, quâ urgentur particulæ ambæ p P & P π versûs R, erit ad vim tensionis fili, ut s t ad p t; & hujus vis dimidium, quo urgetur particula una p P, erit ad Nervi tensionem, ut c t ad t p, hoc est, (ob triangula similia c t p, t p R) ut t p vel P p ad R t vel P R. Quare, ob tensionis vim datam, erit vis acceleratrix absoluta ut $\frac{P\,p}{P\,R}$. Sed est acceleratio genita in ratione compositâ ex rationibus vis absolutæ directè & materiæ movendæ inversè; atq; est materia movenda ipsa particula P p. Quare est acceleratio ut $\frac{1}{P\,R}$, hoc est ut Curvatura in P. Est enim Curvatura reciprocè ut radius circuli osculatorii. Q. E. D.

$\alpha\, dt^2 : 2a = F\beta\, dt^2 : p\theta^2$, donc $\alpha = \frac{2aF\beta}{p\vartheta^2} = \frac{2apml\beta}{p\vartheta^2} =$

$\beta \quad \frac{2aml}{\vartheta^2}$.

VI. Nous remarquerons d'abord, que l'on peut repreſenter le tems donné θ par une ligne conſtante de telle grandeur que l'on voudra : il faudra ſeulement avoir ſoin de prendre, pour exprimer les parties variables & indeterminées du tems, des lignes t qui ſoyent à la ligne qu'on aura priſe pour marquer θ, dans le rapport de ces parties variables du tems au tems conſtant & donné, pendant lequel un corps peſant parcourt l'eſpace a. On pourra donc ſuppoſer θ telle, que $\vartheta^2 = 2aml$: & en ce cas on aura $\alpha = \beta$. Donc puisque $dp = \alpha\, dt + v\, ds$, il faut que dq ou $v\, dt + \beta\, ds$ ſoit $= v\, dt + \alpha\, ds$.

VII. Pour déterminer par ces conditions les quantités α & v, on remarquera, que comme $dp = \alpha\, dt + v\, ds$, & $dq = v\, dt + \alpha\, ds$, on aura $dp + dq = (\alpha + v) \cdot (dt + ds)$; & $dp - dq = (\alpha - v) \cdot (dt - ds)$. d'où il s'enſuit

1°, que $\alpha + v$ eſt egale à une fonction de $t + s$, & que $\alpha - v$ eſt egal a une fonction de $t - s$.

2°. Que par conſequent on aura $p = \frac{\Phi(t+s) + \Delta(t-s)}{2}$ ou ſimplement $= \Phi(t+s) + \Delta(t-s)$; & $q = \Phi(t+s) - \Delta(t-s)$, d'où l'on tire P M ou $s\,(p\, dt + q\, ds) = \psi(t+s) + \Gamma(t-s)$, $\psi(t+s)$ & $\Gamma(t-s)$ exprimant des fonctions encore inconnuës de $t + s$ & de $t - s$.

l'equation generale de la courbe eſt donc

$$y = \psi(t+s) + \Gamma(t-s).$$

VIII.

former path and obtained, if in obscure form, one of the two partial differential equations of finite motion (Fig. 50), but he had no idea what to do with it, and he fell into a morass of guesswork to obtain only the fundamental mode. As late as 1727 JOHN BERNOULLI, following HUYGENS in using the discretely loaded string as a model, could do little better. The differential equation was first recognized as the basic and sufficient statement of the mechanical principles for this particular system by D'ALEMBERT in 1746 (Fig. 51). In the sixty years since the *Principia* enough mathematics had been created that a fully mathematical statement of the problem could be recognized as such and put to use.

After JAMES BERNOULLI's exhaustive analysis of the general catenary, he was in a position to attack the far more difficult problem of the elastic beam. Here the equilibrium of forces and the equilibrium of moments are not at all the same thing, and the action of a part of the beam on its neighbor must be represented not only by a contact force but also by a contact couple. In a series of deep and difficult researches in 1691—1694, JAMES BERNOULLI went far toward understanding these facts, and he obtained the correct differential equations of the elastica. His basic principle was the equilibrium of moments.

While the work of JAMES BERNOULLI shows not the slightest influence from NEWTON's, the English approach found a great champion in DANIEL BERNOULLI, son of JOHN and nephew of JAMES. It was DANIEL BERNOULLI who first applied to a deforming mass NEWTON's idea that the resultant force on a body, whatever its form and nature and internal motion, is the rate of change of its total momentum. In 1727 he calculated in this way the force exerted on a wall by a jet of water playing steadily upon it. We could describe this work as an alteration of one of NEWTON's problems in such a way as to render just NEWTON's own approach.

However, when DANIEL BERNOULLI came to the motion of water in tubes, he could find no key in his uncle's work or in NEWTON's, and he turned instead to a primitive and unconvincing principle of impulse and energy, reminiscent of NEWTON's groping toward the law of linear momentum.

While DANIEL BERNOULLI discovered his law of hydraulics in 1730, it was not published until 1738. His old father, JOHN BERNOULLI, thereupon attacked the method as being indirect and replaced it by another, "resting solely on dynamical

---

Fig. 52. JAMES BERNOULLI's figures for explaining his treatment of the elastica, from the *Acta Eruditorum* for June, 1694

TAB. VI. ad A. 1694. pag. 264. M. Jun.

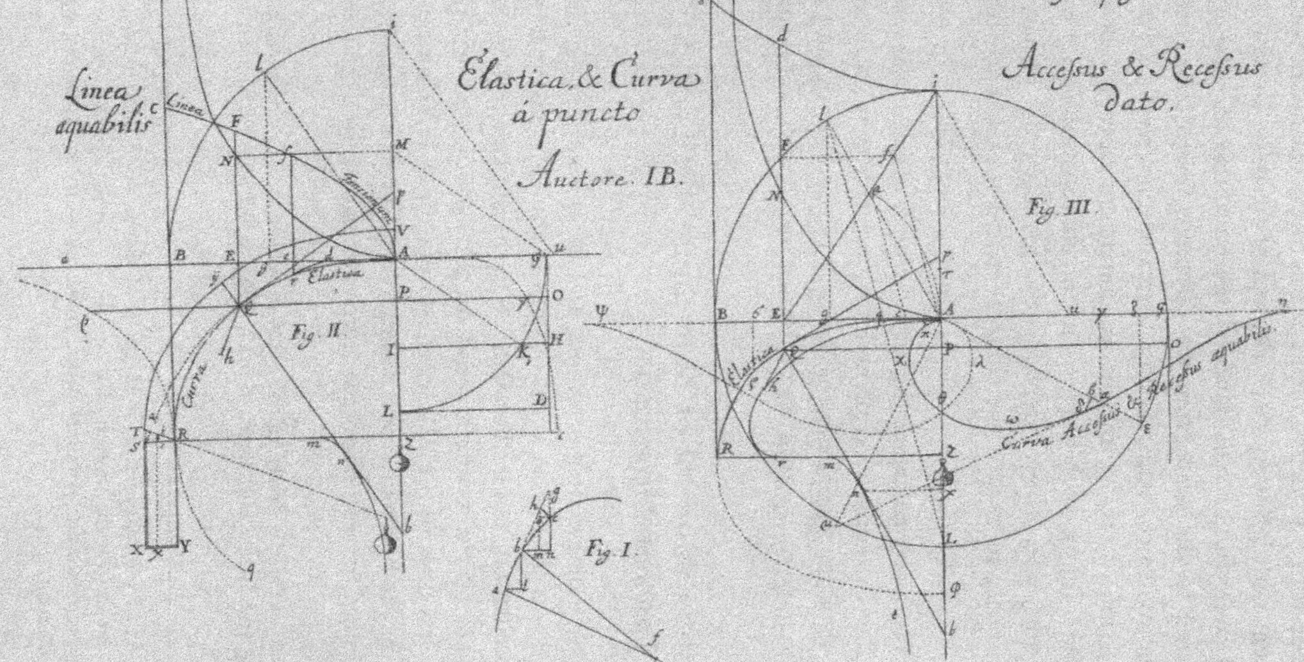

principles, denied by no-one." These principles are, 1. calculation of the difference of internal pressures on opposite sides of a slice of fluid, and 2. the linear momentum principle. This work is simultaneous with D'ALEMBERT's on the vibrating string and shares with it priority in application of the balance of forces to all the parts of a system with infinitely many degrees of freedom. In work done earlier but published later, EULER had obtained the equations of finite motion for a string by passing to the limit from the discretely loaded one. Priority is not important. The evidence, while failing to reveal NEWTON and his disciples as being in possession of tools sufficient for competent treatment of hydraulic or flexible motion, confirms such progress in the concepts of statics by the 1740's that three savants were able, independently, to apply the principle of linear momentum correctly to two of the simplest models of extended bodies on the earth.

Advance so far had been gained through the study of particular cases and particular systems like the compound pendulum; once this point had been reached, generalization to more spatial dimensions was relatively easy, and of course it was EULER who gave us the theories of flexible membranes and of perfect fluids.

5. Rigid Motion, Moment of Momentum, the Elastica, and Euler's Laws of Mechanics

The principle of moment of momentum and its applications lie deeper. Here, I remind you, there is not one word of help in the *Principia* or in the work of HUYGENS on oscillation, and in the early 1700's the difficult and cryptic papers of JAMES BERNOULLI on the pendulum and on the elastica were the only reliable guides. In the first draft of the *Mechanica*, written about 1730 but not published[3] until last year, EULER broke off just after beginning to try to determine the motion of a body first forced to rotate about two perpendicular axes and then freed (Fig. 53).

[3] EULER, *Mechanica seu Scientia Motus*, pp. 93—224 in *Manuscripta Euleriana II*, ed. G. K. MIKHAILOV, Moscow and Leningrad, Izdat. Nauka, 1965. See Sectio III and especially the last paragraph, § 648.

Fig. 53. Conclusion of the manuscript of EULER's *Mechanica seu Scientia Motus*, showing his unsuccessful attempt to treat the motion of a rigid body, written about 1730, first published in 1965 (reproduced here from the Archives of the Academy of Sciences, Leningrad, by courtesy of Dr. G. K. MIKHAILOV).

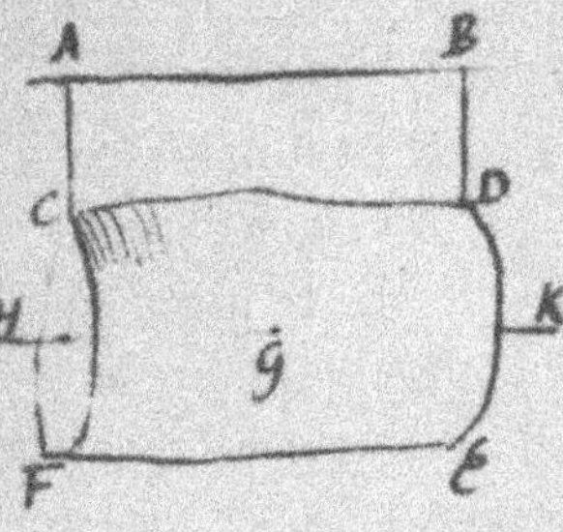

§. 647. Atq etiam ex hoc colligitur, si
corpus CDEF, revolvatur circa axem fixum
AB, idq subito liberum efficiatur, qualem
tum corpus motum sit habiturum. Cen-
trum centrum gravitatis G uniformi-
ter movebitur in directum celeritate, qua
momento, quo ab axe AB rescindebatur ha-
buit, ita autem movebitur ut etiam
recta HK, quae per G transit et paralle-
la est axi AB etiam motu sibi paralle-
lo in directum progrediatur. Interea vero
totum corpus circa HK revolvetur aequali-
liter, motu rotatorio aequali ei, quem
antea circa AB habuerat. Scilicet te-
pus revolutionis circa HK aequale erit tem-
pori revolutionis, quod ante circa AB habuit.

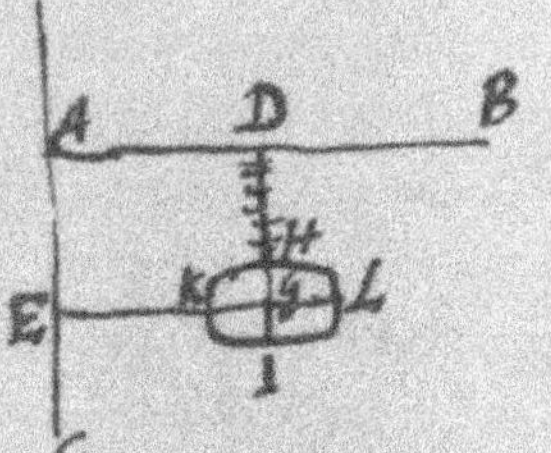

§. 648. Praeterea intelligitur si corpus HK
LI circa duos axes AB et AC fixos revol-
vatur subitoq sibi ipsum relinquatur, centrum
gravitatis G ea celeritate, quam eo ipso
momento habuit, uniformiter in directum
esse progressurum, atq interea duplicem
etiam motum rotatorium habiturum, alte-
rum circa axem HI per G transeuntem sem-
per axi AC parallelum, cujus tempus periodi-
cum aequale erit tempori periodico circa AC.
alterum circa axem KL etiam per G transe-
untem semperq parallelum axi AB, circa
quem revolutiones eodem quoq tempore
absolvet, quo ante circa AB.

In the book[4] as finished in 1734, everything concerning rigid bodies has been suppressed. In their correspondence in the 1730's EULER, JOHN BERNOULLI, and DANIEL BERNOULLI all complained that the "ordinary principles" of mechanics were insufficient to determine the motion of a rigid body, and none of them was able to find the new general principle needed, although all three did succeed, by special devices, in obtaining differential equations for physical systems somewhat but not much more general than HUYGENS' pendulous body. In particular, neither they nor anyone else knew how to describe, let alone determine, a rigid motion whose axis is not restricted to lie in one plane. The first non-planar case of rigid motion that had to be faced squarely came up in EULER's researches on the stability of ships. So as to discuss the problem of small vibrations of a floating body, he had no choice but, like NEWTON before him in the problem of efflux, to abandon the known principles of mechanics and resort to conjecture. His "hypothesis"[5] was a happy one: that every rigid body has three orthogonal axes about any of which it may oscillate freely in infinitesimal motion (Figs. 54a, b). We shall come back to this hypothesis in a moment.

The development of the general theory of rigid motion was slower than the brilliant successes of HUYGENS and EULER in these key examples might lead us to expect. The treatment of various special cases by EULER and DANIEL BERNOULLI gradually built up enough experience with different mechanical ideas to indicate which one was the easiest to apply. No-one today will be surprised to learn that the quickest and surest was JAMES BERNOULLI's principle of regarding reversed accelerations as forces per unit mass. This principle was adopted also by D'ALEMBERT in 1743, if in a rather obscure form. As always, EULER's treatment was clearest and most direct, and it was he who first made it plain that in certain special systems, namely, those consisting of rigid bars linked together, the principle of moment of momentum must be laid down as an independent basic law.

[4] EULER, *Mechanica sive motus scientia analytice exposita*, Petropoli, 1736, 2 vols. = *Opera omnia* II, **1** and **2**.

[5] EULER, *Scientia Navalis* ... I (completed 1738), Petropoli, 1749 = *Opera omnia* II, **18** (1968). See § 184.

Fig. 54a. The first conjecture of the principal axes of inertia, in EULER's *Scientia Navalis*, completed about 1738, published in 1749

Fig. 54b (p. 166). The same, continued

ita vt OC ſit diſtantia centri oſcillationis ab axe. Conſideremus autem corpus extra ſitum verticalem detruſum, eiusque centrum grauitatis in G, ita vt ipſi angulus GOC ſit deſcribendus, donec in ſitum aequilibrii pertingat. Vis autem, quae corpus ad hunc motum angularem ſollicitat, eſt pondus corporis quo in directione GH deorſum vrgetur. Sit nunc maſſa ſeu pondus corporis $= M$ eiusque momentum inertiae reſpectu axis per centrum grauitatis G tranſeuntis et axi oſcillationis paralleli $= S$ erit huius corporis momentum reſpectu axis oſcillationis $= S + M.OC^2$. Momentum autem grauitatis ad motum angularem circa O generandum eſt $M.GO.\text{ſin}.O$; ideoque vis gyratoria erit $= \frac{M.GO.\text{ſin}\,O}{S+M.OC^2}$. Contemplemur nunc pendulum ſimplex og aequali angulo goc a ſitu verticali oc diſtans cui in g ponduſculum infinite paruum p ſit alligatum circa o oſcillans, erit vis gyratoria, qua ponduſculum p ad angulum goc abſoluendum animatur $= \frac{\text{ſin}.O}{OC}$. Si ergo haec vis gyratoria aequalis fuerit priori, pendulum ſimplex og et compoſitum OG ſimul in ſitum verticalem peruenient, quia vtrique aequalis angulus eſt percurrendus. Faciamus ergo $\frac{M.CO.\text{ſin}\,O}{S+M.OC^2} = \frac{\text{ſin}.o}{oc}$ prodibit $oc = \frac{S+M.CO^2}{M.CO}$ quae eſt longitudo penduli ſimplicis iſochroni, ſeu diſtantia centri oſcillationis in pendulo compoſito OC ab axe oſcillationis, erit ergo centrum oſcillationis in Z, vt ſit $OZ = CO + \frac{S}{M.CO}$; Vnde apparet centrum oſcillationis perpetuo infra centrum grauitatis G cadere, eſſeque interuallum $CZ = \frac{S}{M.CO}$.

Hypotheſis.

Tab. XI. fig. 2. 184. In omnibus corporibus aquae innatantibus praecipue vero in nauibus concipere licet tres axes per centrum

trum grauitatis G transeuntes inter se normales, primum verticalem scilicet CGP, secundum horizontalem AGB spinae RS parallelum, in plano diametrali situm ARSB et tertium EGF pariter horizontalem, si quidem nauis fuerit in statu aequilibrii, et ad priorem AGB normalem. Deinde ponere licet corpus huiusmodi a viribus sollicitantibus circa vnumquemque horum axium ita conuerti posse, vt motus gyratorius circa vnum horum axium non turbetur a motibus gyratoriis circa reliquos.

Scholion 1.

185. Ex superioribus satis intelligitur corpus circa alium axem liberum et immotum gyrari non posse, nisi circa quem omnes vires centrifugae se destruant. Quamobrem si vires sollicitantes corpus circa alium axem rotare conentur, motus orietur maxime irregularis, cum etiam axis inclinetur; quem motum definire difficillimum etiamnum est. Huic igitur incommodo medela afferetur, si talis motus irregularis resolui posset in duos vel tres motus rotatorios circa axes fixos simul factos; tum enim cognito motu circa quemque axem seorsim, motus totus inde facile colligeretur. Quamuis autem haec resolutio accurate non succedat, tamen si ad praxin respiciamus, tuto satis adhiberi poterit, si tres illi axes inter se fuerint normales; tum enim motus circa vnum minime a motibus circa reliquos turbabitur. Praeterea vero si hi axes ita sint comparati vt corpus circa quemli' et seorsim immotum gyrari queat, resolutio ista eo magis veritati erit consentanea. In nauibus autem, ad quas hanc tractationem praecipue accomodare institui, huiusmodi tres axes

 vel

The statements I have just made are misleading in that they are phrased by hindsight. To us, today, it is clear that the principle of moment of momentum, as a consequence of JAMES BERNOULLI's law of reversed accelerations applied to the ancient law of the lever, was used again and again in the mechanical researches of the early eighteenth century, especially those of EULER. Since, however, the principle of moment of momentum had not yet been stated, the geometers of that day did not know they were applying it. Indeed, they were discovering it, discovering it by the commonest method for discovering general laws of theoretical physics, namely, generalization from special cases by discarding the differences and retaining the common features.

Three problems were critical and central for the development of a general mechanics. All three consisted in discovery of differential equations of motion for particular kinds of space-filling bodies:

1. A rigid body.
2. A perfect fluid.
3. An elastic bar.

All three of these problems were solved by EULER. The first stage of his thought culminated in the idea that the principle of linear momentum could be expressed, once and for all, as a set of differential equations. Not until 1747 did EULER see, and he was the first to see, that for *all* discrete systems the equations of motion are of the form (Fig. 55)

$$m_k \ddot{\boldsymbol{x}}_k = \boldsymbol{F}_k,$$

where the force $\boldsymbol{F}_k$ acting on the kth body is prescribed. While physicists call these "NEWTON's equations", they occur nowhere in the work of NEWTON or of anyone else prior to 1747. It is true that we, today, can easily read them into NEWTON's words, but we do so by hindsight.

Once he had this idea, it did not take EULER long to propose as the one and general law of mechanics (Fig. 56)

$$\ddot{\boldsymbol{x}}\, dm = d\boldsymbol{F},$$

Fig. 55 (p. 168). The first statement of the "Newtonian equations" for discrete systems, in EULER's memoir on celestial mechanics in the *Mémoires de l'Académie des Sciences*, Berlin, for 1747

Fig. 56 (p. 169). EULER's "New Principle of Mechanics", *i.e.* the general principle of linear momentum, in the *Mémoires de l'Académie des Sciences*, Berlin, for 1750

paralleles refpectivement aux directions des trois coordonnées CP, PQ & QM. Que ces forces abfolues ou motrices foient

$$Mx = X;\ My = Y\ \&\ Mz = Z.$$

& pofant la maffe du Corps $= M$, ces mêmes forces acceleratrices feront:

$$Mx = \frac{X}{M};\ My = \frac{Y}{M}\ \&\ Mz = \frac{Z}{M}.$$

Cela posé, prenant l'element du tems dt pour conftant, le changement inftantané du mouvement du Corps fera exprimé par ces trois équations :

$$\text{I. } \frac{2ddx}{dt^2} = \frac{X}{M};\ \text{II. } \frac{2ddy}{dt^2} = \frac{Y}{M};\ \text{III. } \frac{2ddz}{dt^2} = \frac{Z}{M}$$

d'où l'on pourra tirer pour chaque tems ecoulé t les valeurs x, y, z, & par conféquent l'endroit où le Corps fe trouvera. C. Q. F. T.

Coroll. 1.

XIX. La viteffe du Corps fuivant la direction Mx fera $= \frac{dx}{dt}$, fuivant $My = \frac{dy}{dt}$ & celle fuivant $Mz = \frac{dz}{dt}$, & cela enforte que les quarrés de ces formules favoir $\frac{dx^2}{dt^2}$; $\frac{dy^2}{dt^2}$; $\frac{dz^2}{dt^2}$ expriment les hauteurs mêmes, qui conviennent à ces viteffes. Et c'eft à caufe de ce rapport, que les formules differentio-differentielles font multipliées par le nombre 2.

Coroll. 2.

XX. Si les forces X, Y, Z, qui agiffent fur le Corps M, evanouïffent, on aura ces équations :

$$\text{I. } ddx = 0;\ \text{II. } ddy = 0;\ \&\ \text{III. } ddz = 0.$$

qui étant integrées à caufe de dt conftant, donnent

I.

cette viteſſe ſoit egale à celle qu'un corps grave acquiert en tombant de la hauteur v, il faut prendre $\frac{dx^2}{dt^2} = v$, ou l'elément du tems ſera $dt = \frac{dx}{\sqrt{v}}$; d'où l'on connoit le rapport entre le tems t & l'eſpace x.

XXII. Comme cette formule ne détermine que l'eloignement ou l'approchement du corps par rapport à un plan fixe quelconque, pour trouver le vrai lieu du corps à chaque inſtant, on n'aura qu'à le rapporter en même tems à trois plans fixes, qui ſoient perpendiculaires entr'eux. Donc, comme x marque la diſtance du corps à un de ces plans, ſoient y & z ſes diſtances aux deux autres plans : & aprés avoir décompoſé toutes les forces qui agiſſent ſur le corps, ſuivant des directions perpendiculaires à ces trois plans, ſoit P la force perpendiculaire qui en réſulte ſur le premier, Q ſur le ſecond, & R ſur le troiſième. Suppoſons que toutes ces forces tendent à eloigner le corps de ces trois plans; car en cas qu'elles tendent à le rapprocher, on n'auroit qu'à faire les forces negatives. Cela poſé, le mouvement du corps ſera contenu dans les trois formules ſuivantes:

I. $2 M ddx = P dt^2$; II. $2 M ddy = Q dt^2$; III. $2 M ddz = R dt^2$.

XXIII. Si le corps n'eſt ſollicité par aucune force, de ſorte que $P = o$, $Q = o$, $R = o$, les trois formules trouvées, à cauſe de dt conſtant, ſe réduiront par l'intégration à celles-cy:

$$M\,dx = A\,dt;\quad M\,dy = B\,dt;\quad \&\ M\,dz = C\,dt.$$

d'où l'on voit d'abord, que dans ce cas le corps ſe mouvra dans une ligne droite, avec un mouvement uniforme; & partant ces formules renferment en ſoi la premiere loi du mouvement, en vertu de laquelle tout corps étant en repos y demeure; or étant en mouvement le corps continuë uniformément ſelon la même direction, à moins qu'il ne ſoit ſollicité par quelque force de dehors. Mais il eſt clair que nos

or, in the modern notation of Stieltjes integrals,

$$\int_{\mathscr{P}} \ddot{\boldsymbol{x}}\, dm = \boldsymbol{F}(\mathscr{P}),$$

where $\boldsymbol{F}(\mathscr{P})$ is the resultant force acting upon the arbitrary part $\mathscr{P}$ of the body.

If we ask, how can this idea have been missed, we can answer only by profitless and unconfirmable guessing about how people thought and felt in days long past. However, there can be no doubt of the fact, and we ought not to be so childish as to follow MACH and the standard historical writers on physics in their implication that everyone knew the Newtonian equations, whether or not these had yet been written down[6]. The key example is the problem of general motion of a rigid body, which, as I mentioned, EULER in the 1730's was the first

[6] It is enlightening to read what EULER himself wrote in 1747 when he took up, by no means for the first time, the general problem of the planetary system. "To solve these problems, I shall use a method a little different from that which others who have written on this subject have used. First, they tried to determine the true velocity of the body, from which they sought the motion at each instant; and from this velocity compared with the space traversed they determined the place where the body should appear at each instant. So as to avoid this rather tedious operation, and since in astronomy the true velocity of the heavenly bodies is never wished, I have found means of reaching straightaway an equation connecting the time elapsed with the apparent place of the planet .... Finally, according to my method, I am not obliged to take account of the curvature of the line that the body describes, and by this means I avoid much laborious research, above all when the motion of the body does not take place even approximately in the same plane. This great advantage is included in the following *Lemma* ....

Lemma

"XVIII. If a body M is acted upon by arbitrary forces, to determine the instantaneous change that these forces produce in the motion of the body."

The outcome, as the reader may see for himself in Fig. 55, is the "Newtonian equations", there numbered I, II, III. EULER apparently considers the *principle* these equations express to be so well known that it requires no reference and no proof. He merely affirms the equations.

What then are the "method a little different" and the "great advantage"? As EULER writes, these lie in the *general conception*, which is new in that it requires no geometric restrictions or devices. To apply in three dimensions the intrinsic conception of NEWTON, EULER had had to construct a differential geometry of skew curves, and a large part of his *Mechanica* is taken up by cumbrous manipulations concerning what we should now call curvature and torsion. The "method a little different" sweeps all this away.

Clearly, EULER did not regard his own innovation here a great one; he attributes to it only mathematical convenience. Just as clearly, it was this small innovation that showed him within three years the way to what he unhesitatingly called a *new principle* of mechanics: The linear momentum principle for each element of volume in an arbitrary space-filling body.

person to face squarely, and which he could at first solve only by resort to hypothesis, a hypothesis he cautiously limited to small motion. However, the moment he had his "New Principle of Mechanics", as he called it in 1750, he confirmed its usefulness by applying it to the motion of a rigid body so as to obtain the general differential equations which, unlike most of his discoveries, are named after him today. Shortly thereafter he was able to prove as a theorem his hypothesis of fifteen years earlier that in any body there is an orthogonal triad of axes of free rotation and to liberate it from any restriction to small motion.

We all know that the equations of motion of a rigid body express the balance of moment of momentum, not of linear momentum, so we might conjecture that EULER, like a modern physicist, must have made some hypothesis about mutual forces such as to maintain rigidity. Here hindsight misleads, for in the Age of Reason no real, producing geometer (and here I wittingly exclude BOSCOVICH) ever introduced hypothetical forces beyond gravity in the interior of a body unless compelled to. EULER proposed instead an exclusion principle: Since a body does not spontaneously assume any motion in virtue of whatever internal forces there may be within it, these do not contribute to any of its motions as a whole. We know today that this principle is false in general, but for rigid bodies, it leads to the correct equations. We may say with hindsight that to discover the general equations of rigid motion EULER used a special device to *avoid* the principle of moment of momentum, just as HUYGENS had used a different special device to avoid it when he determined the center of oscillation. It should be unnecessary to remark, though unfortunately it is not, that the general motion of a rigid body affords a mechanical problem far too subtle for any savant of the seventeenth century, no matter how great.

The problem of the elastica is of a still higher order of conceptual difficulty. In one of his earliest researches, EULER in 1728 had succeeded in unifying the catenary and the elastica in a single theory as far as equilibrium was concerned. Obviously, the basic principle had to be the balance of moments. These two systems are of such a kind that balance of forces, while of course it holds, need not be considered. To get equations of motion is a different matter. While the nature of the contact force could be left aside in the problem of equilibrium, for the problem of motion it could not.

Here EULER had a second earlier success to guide or misguide him, namely, his hydrodynamics. After having simplified and generalized the hydraulic theories of DANIEL and JOHN BERNOULLI in 1750—1751, EULER was in a position to extend

the ideas to three-dimensional fluid bodies. D'ALEMBERT in a devious and obscure essay submitted for a prize in 1749 had woven a number of correct equations for rotationally symmetric or plane flow into a tissue of philosophy, conjecture, doubt, sophomorisms, and error. Just at this time EULER was attacking the same and more general problems. As always, his approach was brilliantly simple and easy. He introduced the concept of fluid pressure, the action of the fluid on one side of an imagined boundary on that on the other. By assuming this contact force to be perpendicular to the surface, no matter what the state of motion of the fluid, and by applying his differential form of the principle of linear momentum, EULER derived the basic equations of hydrodynamics in a few pages. The Bernoullian tradition had taught him how to calculate the resultant force on a one-dimensional element, and his own calculus of partial derivatives enabled him to extend the idea to three dimensions.

The simple modes of infinitesimal vibration for straight elastic bars had been found by DANIEL BERNOULLI and EULER in their youthful days, even before there were equations of motion. The partial differential equation of small motion was obtained straightaway by EULER the moment he had the general principle of linear momentum. Finite motion offers a problem of a different order of difficulty.

After having studied this problem on and off for fifty years, EULER found the key to it, as simple as it is powerful. This key is to regard the moment of momentum principle as *an independent basic law of mechanics*, and to decline to specify the contact force and contact couple, either in magnitude or in direction. In this way EULER in 1771 was able to write down the general equations of mechanics for a plane deformable line. After doing this, he found it easy to state the general principle of moment of momentum:

$$\int_{\mathscr{P}} \boldsymbol{p} \times \ddot{\boldsymbol{x}}\, dm = \boldsymbol{L}(\mathscr{P}),$$

where $\boldsymbol{p}$ is the position vector and $\boldsymbol{L}$ is the total torque, including couples if need be, applied to the part $\mathscr{P}$ of the body. In his final memoir of 1775, he laid down the two principles as fundamental and independent: On every part of every body, the total force equals the rate of change of total momentum, and the total torque equals the rate of change of total moment of momentum, where those torques that are moments of forces are taken with respect to the origin of $\boldsymbol{p}$, that origin being a fixed point. These are EULER's *laws of motion*. The nature of the forces acting, and in particular the nature of contact forces, varies from one kind of body to another. This same distinction had been made by JAMES BERNOULLI in 1705,

but he had not attained correct general principles. Again with hindsight, we may say that at the beginning of the eighteenth century the mathematics of three-dimensional space had not been developed sufficiently to make possible clear and precise physical thought about space-filling bodies in general. With EULER's memoir of 1775, the whole program of rational mechanics becomes clear. There are the two general laws, *common to all bodies* but insufficient to specify their motions. The *differences* between bodies are represented by *constitutive equations*, which specify the nature of their response to their surroundings. The constitutive equations studied in the eighteenth century are those defining the discrete system, the rigid body, the perfectly flexible line, the perfectly flexible sheet, the elastica, and the perfect fluid, with a few others which are less important. All these fit easily into the general scheme laid down by EULER in 1775. In fact, this scheme remained general enough for all of mechanics for at least 100 years.

6. The Méchanique Analitique of Lagrange, and Later Reactions

Something must be said about LAGRANGE's *Méchanique Analitique*, published in 1788. In it, the basic axiom of statics is the principle of virtual work, stated explicitly and in some generality (Fig. 57). Dynamics follows by the BERNOULLI-EULER principle of reversed accelerations. While today this generalized axiom is often attributed to D'ALEMBERT, LAGRANGE himself claims, and justly so, that it replaces the much more complicated principle D'ALEMBERT had proposed as a general basis for mechanics.

Almost as much nonsense has been written about the *Méchanique Analitique* as about the *Principia* and the *Two New Sciences*. Although LAGRANGE's book is far easier to read with understanding than is NEWTON's or GALILEO's, still it does not seem to be easy enough for most historians of science to penetrate the contents. There are few errors, few novelties, and many routine manipulations in it. While it contains interesting historical parentheses, the presentation of mechanics is strictly algebraic, with no explanation of concepts, no illustrations either by diagrams or by developed examples, and no attempt to justify any limit process by rigorous mathematics. It does not enter at all a number of the fields opened by NEWTON, and it leaves unmentioned most of the deeper and harder problems of mechanics solved by the Basel geometers in the century preceding it. In particular, it does not include the general principle of moment of momentum. It could not do so, because the principle of virtual work does not yield the principle of

tems les eſpaces parcourus par le corps m ſuivant les lignes p, q, r, &c. Donc $mP \times \delta p$, $mQ \times \delta q$, $mR \times \delta r$, &c, ſeront les *momens* des forces mP, mQ, mR, &c, agiſſantes ſuivant ces mêmes lignes, p, q, r, &c.

Or la formule générale de l'équilibre conſiſte en ce que la ſomme des *momens* de toutes les forces du ſyſtême doit être nulle (Part. I, Sect. 2, art. 2); donc on aura la formule cherchée en égalant à zéro la ſomme de toutes les quantités

$$m\left(\frac{d^2x}{dt^2}\delta x + \frac{d^2y}{dt^2}\delta y + \frac{d^2z}{dt^2}\delta z\right)$$
$$+ m(P\delta p + Q\delta q + R\delta r + \&c),$$

relatives à chacun des corps du ſyſtême propoſé.

7. Donc ſi on dénote cette ſomme par le ſigne intégral S, qui doit embraſſer tous les corps du ſyſtême, on aura

$$S\left(\frac{d^2x}{dt^2}\delta x + \frac{d^2y}{dt^2}\delta y + \frac{d^2z}{dt^2}\delta z + P\delta p + Q\delta q + R\delta r + \&c.\right)m$$

pour la formule générale du mouvement d'un ſyſtême quelconque de corps, regardés comme des points, & animés par des forces accélératrices quelconques P, Q, R, &c.

Pour faire uſage de cette formule, on ſuivra les mêmes regles que pour la formule de l'équilibre; ainſi il faudra appliquer ici tout ce qui a été dit dans la ſeconde Section de la premiere Partie, depuis l'article 3 juſqu'à la fin, en obſervant que les différentielles marquées par la note ou caractériſtique δ dans la formule précédente répondent aux différentielles marquées par la caractériſtique ordinaire d dans la formule de l'équilibre, & ſe déterminent par les mêmes regles & les mêmes opérations.

moment of momentum until the nature of the contact forces is made somewhat explicit. CAUCHY had still to be born and to create the general concept of stress, by which all theories of space-filling bodies are united.

Other great problems set before the world as challenges in the *Principia* remained unsolved, some even untouched, a century after it was published. The theory of infinitesimal surface waves waited for LAPLACE, KELLAND, and AIRY, who derived it from EULER's equations of fluid dynamics. NEWTON's ideas on fluid friction were to be taken up by NAVIER and STOKES; while the latter was to see the error of NEWTON's solution for the spinning cylinder and correct it, the problem of the spinning sphere remains unsolved today. The flow of fluid from an orifice, after a special result found ingeniously by BORDA in 1766, was to be studied successfully on the basis of EULER's hydrodynamics by KIRCHHOFF and HELMHOLTZ, but only in the plane, and the three-dimensional theory is scarcely touched today. While D'ALEMBERT and EULER had derived wave equations for flexible solids and gases, the first satisfactory explanation of the observed velocity of sound was found in the early years of the nineteenth century, and a theory inclusive enough to allow loud sounds to reinforce while weak sounds die out was first discovered only last year by COLEMAN & GURTIN.

7. Nineteenth-Century Misconceptions of Newton's Work and Program. The Position of Faraday

In being asked to speak about reactions to the *Principia*, I am grateful for an assignment licensing an approach simpler than that most of the other speakers have chosen. Here, unpublished documents are almost irrelevant, since only historians of today, not scientists of NEWTON's time or any other, have had a chance to react to them. So as to survey the reactions to NEWTON's published work, I must first correct common misconceptions as to what NEWTON's achievement and program in physics really were. One such misconception, put forward in the eighteenth century by a few interpreters not to be numbered among creative scientists and then adopted by almost all physicists and historians from the nineteenth century on, is that NEWTON insisted all of natural philosophy should be founded on calculations of central forces.

Fig. 57. The general principle of virtual work as stated by LAGRANGE in the *Méchanique Analitique*, 1788

Whatever may have been the speculations NEWTON chose not to publish, the works he did release do not justify imputation to him of so narrow a view of nature. In the first place, it would have to rest largely on one sentence in the preface to the *Principia* and on the comments at the end of the Queries in the *Opticks*. The exaggerated weight often laid upon these few pages may reflect the fact that while a mathematician may need a day to get through a page of the *Principia*, then he has it beyond quibble, but anybody can read the Queries in ten minutes and go on arguing about them forever. The Queries are not science; they show NEWTON from a side he withheld usually and would have been better advised to have withheld without exception; at best, they are prophecies of science. Their interest lies not in themselves but in the mere fact that it was NEWTON who wrote them. Second, even if we must tarry over these fleshless speculations, we cannot justly read into them atomism in the narrow sense later given to the word. NEWTON's fancied atoms were "solid, massy, hard, impenetrable, movable". Of these five adjectives, three are not punctual. The repulsion experienced by two solid, hard, impenetrable grains, however small, when they strike each other, is not an action at a distance but a contact force. Contact forces, if crudely handled by NEWTON, are nevertheless nowhere excluded by him. His treatments of water waves, efflux, and resistance of a continued fluid rest on concepts, vague though they be, of contact forces, not between corpuscles but in an infinitely divisible medium. NEWTON's particles are "even so very hard, as never to wear or break in pieces." Breaking cannot be denied unless there are pieces into which to break. The "particles" are "particulae", little parts, not mass-points; they fill a small region, which contains infinitely many points. I do not deny that NEWTON here and there proclaimed an atomistic view of nature, but I see equally that in his own *practice* he was far from adhering without exception to the 39 Articles the nineteenth-century headline writers and the text compilers of today strap him to with the hagiography of a Parson WEEMS. Even in the explicit statement of the third law of motion in the *Principia*, where only forces between pairs of "bodies" are mentioned, NEWTON does not demand that they be central. In chanting that electromagnetism destroyed NEWTON's views, the physicists force upon him a program that may have been once theirs but was not his.

Writings on the history of science all too often proclaim and develop an antithesis between "Newtonian" and "anti-Newtonian" science in the nineteenth century, using the supremacy of central interatomic forces as the shibboleth. I do not deny that central forces had their passionate partisans and opponents, but among these enthusiasts I do not find any mathematical theorists but those of the second rate, or, oftener, mere journalists of science. The great mathematicians who gave us "classical" physics do not fall into either category. CAUCHY and POISSON created theories of solids alternatively as assemblies of point centers and as plena, and they proved that the two models were equivalent, more or less. MAXWELL created both the kinetic theory of gases and the electromagnetic theory of the aether; he showed, incidentally, that in his kinetic theory no result inconsistent with the conception of a gas as a continuum could ever be derived by correct mathematics. These men tower above the idle speculators who claimed all and proved nothing for their atoms or subtle fluids. CAUCHY, POISSON, and MAXWELL were neither "Newtonian" nor "anti-Newtonian" in the historians' sense. Like EULER before them, they bore NEWTON's mantle. While EULER is regarded often, and with some justice, as an upholder of Cartesian metaphysics, in his *practice* of mechanics he was the truest of all NEWTON's ultimate disciples, for he took over, ennobled by abstraction, and fixed by explicit and exact application all of NEWTON's fruitful ideas and merged them with JAMES BERNOULLI's to form what we know now as "classical mechanics". Indeed, what physicists today call "Newtonian mechanics" has little direct relation to NEWTON's own work but is rather a combination of EULER's mechanics with LAGRANGE's.

The heritage of science cannot be equated to its history. We do injustice to a great scientist if we merely chronicle his views and actions, or even if we cast the arithmetic sum of all his scientific work. If we excise judgment, we insult the goddess we claim to worship. While he who writes the history of CHARLES II need not be an expert tennis player and keep a costly international stall of mistresses, he who approaches NEWTON without having himself experienced the neurosis of scientific thought, the perverted solitary ecstasy of creation, and the aristocratic duelli of polemics at the summit[7], is not likely to get nearer than did poor puerile BREWSTER and RIGAUD to the real science of NEWTON. Even if the difference is merely that a historian can less painfully imagine himself in the slippers of a CHARLES crowned by a garland of doxies than in the "absolute space,... always like and immobile" of a NEWTON with only fluxions of that equable fluent, "absolute, true, and mathematical time" as his companions, the difference is essential. CHARLES is dead beyond recall. A scientist does not so die. NEWTON is alive today, just as are the HADAMARD and the BIRKHOFF some of us

[7] *Cf.* the remarks of Mr. MANUEL, which, set against Mr. HILL's, may serve to remind us that in the age of CHARLES II and NEWTON, blessed by Apollo and Venus, the "normal" of iatropsychological philosophy did not exist, either in word or in concept.

met in the flesh, the HILBERT all of us know through his disciples. NEWTON is alive, not in the barren speculations of the Queries but in the solid, permanent, mathematical pages of the *Principia*, in the program he there laid down for us and illustrated by brilliant examples, and in the great new concept he there taught us to use.

First, the program. Often quoted but invariably misread is NEWTON's own statement, "Utinam caetera Naturae phaenomena ex principiis Mechanicis eodem argumentandi genere derivare liceret." It is right to translate "ex principiis Mechanicis" as "from mechanical principles", but only in the vague, inclusive sense the word "mechanics" had in the seventeenth century, already explained by Mr. HILL. Many, on the contrary, read this passage as if NEWTON had meant "from mechanical principles" in the sense of the nineteenth-century physicists' mechanics, namely, "from my own three laws of motion". As one of its definitions of "mechanic", with a quotation from CONGREVE in 1697, the *Oxford Dictionary* gives "working like a machine, having a machine-like action"; another, supported by a quotation from 1663, "acting or performed without exercise of thought or volition"; and "practical as opposed to speculative", with the following quotation from a mathematical book of 1570: "A speculative Mechanicien... differreth nothyng from a Mechanicall Mathematicien." The physicists' and historians' use of "mechanical" as distinguished from "optical", "electromagnetic", *etc.*, is too recent even to be mentioned in the *Dictionary*, which comes near it only in the antithesis to "chemical", quoting an example a full century after NEWTON's time. NEWTON himself, while he regarded force as a "mechanical" idea, mentioned "electric" and "magnetic" forces. The electromagnetic "vacuum" of MAXWELL, HERTZ, and LORENTZ is just as consonant with NEWTON's "ex principiis Mechanicis eodem argumentandi genere" as were the ponderable aethers of NEWTON's own early and late speculations[8].

[8] NEWTON continues in this passage, "For many things move me to suspect somewhat that they all may depend upon certain forces by which the particles of bodies through causes not yet known are driven toward one another and cohere according to regular figures, or are driven away and withdraw from one another ..." The phrase "secundum figuras regulares" is usually translated "*in* regular figures", as if NEWTON had suggested geometric arrays of atoms, while in fact it is not so specific and could refer, as the phrase "geometrical figures" often did at that time, merely to use of specific mathematical functions. The classical "figura" means not only "form, shape, figure" but also "quality, kind, species, nature, manner", and in the writing of GALILEO and others "geometrical figures" are said to describe or specify a situation where a modern author would use the term "function". Thus, I think, NEWTON's own attempts to describe flows of a "continued" fluid in Book II involve coherence of "the particles of bodies...

The aspect of NEWTON's program least regarded in historical studies is that on which he himself laid greatest weight, namely, that it is *mathematical*, not only mathematical but *rigorously* mathematical. The word "mathematical" appears in the title and frequently in the text, always with a capital letter. In the preface NEWTON, after reminding us of the old distinction between rational and practical mechanics, continues: "Since, however, artificers are wont to operate with small accuracy, *mechanics* has come to be distinguished thus from *geometry*, that whatever is accurate is classed as *geometry*, whatever is less accurate, as *mechanics*. Notwithstanding, the errors are of the artificers, not of the art. He who operates less accurately is the less perfect mechanic, and if any one could operate with greatest accuracy, he would be the most perfect mechanic of all." Going on to explain the nature of geometry, NEWTON concludes, "*Geometry*, therefore, is founded in mechanical practice and is nothing else than that part of *universal mechanics* which sets forth and proves the art of measuring accurately... *Rational mechanics* will be the science of the motions that result from any forces whatever, and of the forces that are required for any motions whatever, accurately set forth and proved." Those who read NEWTON's "ex principiis mechanicis eodem argumentandi genere" as referring only to the one following sentence regarding forces among the particles disregard not only the page before it, from which I have just given extracts, but also the ten lines above it, where in describing his book NEWTON speaks of "our mathematical principles of philosophy", "propositions mathematically proved", and "mathematical propositions" — three references to mathematical proof. "The same kind of reasoning" means, in plain English, *proved by rigorous mathematics from mathematical axioms*. NEWTON's preface, if read straight through, makes his program plain: to construct a *mathematical* natural philosophy, in which geometry is only the simplest part but a perfect specimen of method.

Against this program there were, indeed, hostile reactions, especially among the poets and musicians of the early romantic movement, as Mr. HASKELL mentioned; they are fully expressed in the autobiography of BERLIOZ. On the course of physics itself, they had little effect. Most of the evidence for positive contributions

according to regular figures", although they make no use of atoms or central forces. Indeed, NEWTON himself tells us as much in the Preface, where he lists among the "natural powers" he will treat "the resistance of fluids and forces of that sort, whether attractive or repulsive ...," and he mentions not only Book I but also Book II as pertaining to his general program. The forces of coherence and resistance in space-filling bodies do satisfy the principle of "action and reaction", although in general they are not central forces.

by non-mathematical theorists of physics rests on misconceptions about the nature of mathematics. FARADAY's disclaimer is often parroted unhistorically, without a reminder that for him, who, like a modern mathematician, dealt directly with precise and abstract concepts and operations, the word "mathematics" referred only to the ritual algebra of paedagogues in the barrenest period of British mathematics, when the blind use of x, y, z and a, b, c connected by plus and minus signs was allowed to usurp the word that before then did signify and now again does signify the precise theory of relations and sets.

Since the literature of the history and philosophy of science, despite the good intentions and efforts of the best students in those fields, remains jejune in its credulity as to the nature and function of mathematics, I append here some remarks of MAXWELL in the preface to the first edition of his *Treatise:* "As I proceeded with the study of FARADAY, I perceived that his method of conceiving the phenomena was also a mathematical one, though not exhibited in the conventional form of mathematical symbols... FARADAY's methods resembled those in which we begin with the whole and arrive at the parts by analysis.... I also found that several of the most fertile methods of research discovered by the mathematicians could be expressed much better in terms of ideas derived from FARADAY than in their original form." A modern student of natural philosophy might summarize the difference thus: FARADAY preferred global statements of physical laws, but since the mathematical tools necessary for exploiting them locally (GREEN's transformation, *etc.*) did not yet exist, FARADAY's mathematics seemed to be distinct from that already symbolized and commonly taught; MAXWELL, two generations later, easily constructed methods for interconverting local and global statements; the full power of the global viewpoint began to be appreciated in HILBERT's time; the modern creative theorist prefers global principles whenever he can find them. Thus the *real* history of science, the science that is alive and growing, upholds MAXWELL's judgment that FARADAY as a theorist was certainly a mathematician, and a great one. (Not every great mathematician is a master of *all* aspects of mathematics cultivated in his day. Some look more to the future than to the past.) Those historians who adduce FARADAY as an example of a non-mathematical theorist and hence a rebel to NEWTON's program not only fail to understand what mathematics is but also do FARADAY himself a great injustice. They judge him in terms of the mathematics of his own day, while he himself, like other great theorists, could plead a much better case before the higher tribunal of the mathematicians of *our* day. FARADAY may serve indeed as a proof, if one were needed, that the mathematical high priests of any one period are rarely the best mathematicians in it [9].

While some programs remain programs only, even if they are projected by so great a man as NEWTON, and thus may be trodden as paving stones upon the road of good intentions, others stand erect as summaries of practice. As well as many of the former kind, NEWTON left us one of the latter. In the middle of the *Principia*, as a tailpiece to some of its most brilliant successes, NEWTON wrote, "Eodem sensu generali usurpo vocem impulsus, non species virium & qualitates physicas, sed quantitates & proportiones Mathematicas in hoc Tractatu expendens: ut in Definitionibus explicui. In Mathesi investigandae sunt virium quantitates & rationes illae, quae ex conditionibus quibuscunque positis consequentur: deinde ubi in Physicam descenditur, conferandae sunt hae rationes cum Phaenomenis ut innotescat quaenam virium conditiones singulis corporum attractivorum generibus competant. Et tum demum de virium speciebus, causis, & rationibus physicis tutius disputare licebit." "In the same general sense I take over the word 'impulse', pondering in this treatise not kinds of forces and physical qualities, but quantities and mathematical proportions, as I explained in the definitions. In mathematics are to be investigated the quantities of forces and such rules as may follow from whatever conditions be laid down. Whenever we descend therefrom into physics, these rules are to be compared with the phenomena so as to disclose what conditions of forces answer to the several kinds of attractive bodies." NEWTON closes with a characteristic sneer: "And then, finally, argument about the kinds, causes, and physical rules of forces will be safer." He implies that such argument will get us nowhere, but it does not matter, since the real work, the mathematics, is done, so the rest is what you will. Whatever his theology and philosophy, this is NEWTON's grand lesson for us: Disregard the "causes" — whatever that strange word may mean — and cleave to relations among the phenomena. Indeed, "cause" is only a verbal crutch for beginners who do not grasp the concept of function in mathematics. In laying out this program, NEWTON was not alone. In the course of a concrete, specific research on elasticity, LEIBNIZ wrote, "his paucis consideratis, tota haec materia redacta sit ad puram Geometriam, quod in physicis & mechanicis unice desideratum." "These few

[9] *Note added in proof.* There may be some hope that scientists will refuse to let pass by default the misconceptions about the histories of their own fields that the new professionalized history of science is busy to create by its refusal to recognize that science is still alive. The general periodical *Endeavour* in its issue for September, 1967, printed an unsigned commemorative editorial which in one page gives a sounder character of FARADAY than an entire recent biography, *e.g.*

things having been considered, the whole matter is reduced to pure geometry, which is the one aim of physics and mechanics."

8. Our Legacy from Newton

Those who regard the eighteenth century as a period when Newtonian science stood still make the error of equating "Newtonian" to "British". Those who find it a barren period in physics fail to learn a lesson from NEWTON and LEIBNIZ. While first physics relieves philosophy of some of its burdens, at the second stage mathematics reduces the physics of a given class of phenomena to a bare minimum and thereby frees physicists to turn their speculations elsewhere. In brief, the aim is to clear out such physics as is unnecessary, which is most of it. No man before or after ever brought so great a range of physics under the control of specific, explicit, final mathematical theory as EULER did. He was so successful that most of his discoveries are not even credited to him, being taught to all and dismissed by the physicists today as obvious or mathematical and hence, presumably, God-given to physicists at birth. EULER's work lies entirely within the scientific programs of NEWTON and LEIBNIZ, and his specific principles of mechanics unite theirs in their common features, discard them in their differences, and merge them into new, embracing concepts.

Our second great legacy from NEWTON, the specific one, is the concept of force given *a priori*. To the insight that place and time and body and motion and mass are not enough, that decent correlation of the phenomena demands another primitive element, NEWTON won through alone and imperfectly. As Mr. COHEN has remarked, NEWTON's "vires" in the statements of his laws are granules[10] of impulse, not forces as we conceive them today. As further remarked by Mr. COHEN, if in somewhat different words, NEWTON's own laws, if taken strictly, are insufficient even for what he himself claimed to prove from them, let alone to serve as basis for "classical mechanics". It took half of the eighteenth century to render LEIBNIZ's calculus the lingua franca of physical science and hence clear the way for EULER to teach us to walk upright without the crutches of tiny impulses

---

"It is a remarkable fact that although Faraday knew little beyond simple arithmetic, and affected to see little need of mathematics, he clearly had a kind of mathematical intuition that guided him with scarcely a fault through the new fields he opened up."

[10] Those who appeal triumphantly to these granules as evidence that NEWTON conceived all of nature as corpuscular may set themselves the task of explaining why NEWTON did not similarly chop up into gobbets his absolute space and equable time.

and other shadowy, imprecise granules. The concept of force resisted longer. Clear and sufficient mathematical axioms for systems of forces, at the level of HILBERT's axioms for points and lines and CARATHÉODORY's axioms for distributions of mass, were first proposed by NOLL, no earlier than 1960, and are still under revision by him and others.

In surveying reactions to the *Principia*, I did not idly adduce work done only recently. I hope it is not out of place to mention to distinguished philosophers and historians of science that classical mechanics, the mechanics loosely called "Newtonian", is not the corpse it is often pictured as being. Mr. STEIN has given me courage by alluding to the higher levels of concept and precision possible for classical space-time. Refinement and extension of the concept of force is one of the central areas of research in the rational mechanics of deformable bodies today. We stand here upon the shoulders of more giants than one. It is from NEWTON's *Principia* that our forebears learned how to use the concept of force given *a priori*. The structure which that concept binds together we owe to many great mathematicians; among the founders, NEWTON is joined by HUYGENS, LEIBNIZ, and JAMES BERNOULLI, and the great architect is EULER.

References

Most, though not all, of the works mentioned in this lecture are analysed and referenced in my three introductions to volumes in L. EULERI *Opera Omnia*. For the center of oscillation, among other things, see

1. JOUGUET, E., *Lectures de Mécanique*, Paris: Gauthier-Villars, 2 vols., 1908—1909.
2. DUGAS, R., *La Mécanique au XVIIe Siècle*, Neuchatel, Griffon, 1954.

For the principle of moment of momentum, see the fifth essay in this volume, and for the concept of stress, the fourth. While it is customary to refer to MACH's book, for all matters treated in this lecture the account given there is totally unreliable.

Bibliographical Note

This lecture was read at the Newton Tricentennial Celebration, held at the University of Texas, Austin, in 1966, and has been printed in its proceedings in *The Texas Quarterly*, 1967. For permission to print it here also, I am grateful to Professor R. PALTER.

The second half of the lecture was composed while the sessions were in progress, as may be inferred from the references to several other lecturers: Professors I. BERNARD COHEN, FRANCIS HASKELL, CHRISTOPHER HILL, and FRANK E. MANUEL, and Dr. HOWARD STEIN.

IV. The Creation and Unfolding of the Concept of Stress

Fig. 58. Cauchy's first announcement of the stress principle, in the *Bulletin de la Société Philomathique*, 1823

libre du plan élastique, avait considéré deux espèces de forces produites, les unes par la dilatation ou la contraction, les autres par la flexion de ce même plan. De plus, il avait supposé, dans ses calculs, les unes et les autres perpendiculaires aux lignes ou aux faces contre lesquelles elles s'exercent. Il me parut que ces deux espèces de forces pouvaient être réduites à une seule, qui devait constamment s'appeler tension ou pression, et qui était de la même nature que la pression hydrostatique exercée par un fluide en repos contre la surface d'un corps solide. Seulement la nouvelle pression ne demeurait pas toujours perpendiculaire aux faces qui lui étaient soumises, ni la même dans tous les sens en un point donné. En développant cette idée, j'arrivai bientôt aux conclusions suivantes.

Si dans un corps solide élastique ou non élastique on vient à rendre rigide et invariable un petit élément du volume terminé par des faces quelconques, ce petit élément éprouvera sur ses différentes faces, et en chaque point de chacune d'elles, une pression ou tension determinée. Cette pression ou tension sera semblable à la pression qu'un fluide exerce contre un élément de l'enveloppe d'un corps solide, avec cette seule différence, que la pression exercée par un fluide en repos contre la surface d'un corps solide, est dirigée perpendiculairement à cette surface de dehors en dedans, et indépendante en chaque point de l'inclinaison de la surface par rapport aux plans coordonnés, tandis que la pression ou tension exercée en un point donné d'un corps solide contre un très-petit élément de surface passant par ce point, peut être dirigée perpendiculairement ou obliquement à cette surface, tantôt de dehors en dedans, s'il y a condensation, tantôt de dedans en dehors, s'il y a dilatation, et peut dépendre de l'inclinaison de la surface par rapport aux plans dont il s'agit. De plus, la pression ou tension exercée contre un plan quelconque se déduit très-facilement, tant en grandeur qu'en direction, des pressions ou tensions exercées contre trois plans rectangulaires donnés. J'en étais à ce point, lorsque M. Fresnel, venant à me parler des travaux auxquels il se livrait sur la lumière, et dont il n'avait encore présenté qu'une partie à l'Institut, m'apprit que, de son côté, il avait obtenu sur les lois, suivant lesquelles l'élasticité varie dans les diverses directions qui émanent d'un point unique, un théorème analogue au mien. Toutefois le théorème dont il s'agit était loin de me suffire pour l'objet que je me proposais, dès cette époque, de former les équations générales de l'équilibre et du mouvement intérieur d'un corps; et c'est uniquement dans ces derniers temps que je suis parvenu à établir de nouveaux principes propres à me conduire à ce résultat, et que je vais faire connaître.

Du théorème énoncé plus haut, il résulte que la pression ou tension en chaque point est équivalente à l'unité divisée par le rayon vecteur d'un ellipsoïde. Aux trois axes de cet ellipsoïde correspondent trois pressions ou tensions que nous nommerons *principales*, et l'on peut

1. Cauchy's Laws of Motion (1822)

On the 30th of September 1822 CAUCHY announced the *stress principle*, which has ever since then been the foundation of the rational mechanics of continua (Fig. 58). This principle asserts that upon any smooth closed orientable surface $\partial\mathscr{V}$, be it an imagined surface within the body or the bounding surface of the body itself, there exists an integrable field of *traction vectors* $\boldsymbol{t}_{\partial\mathscr{V}}$, equipollent to the action exerted by the matter exterior to $\partial\mathscr{V}$ and contiguous to it on that interior to $\partial\mathscr{V}$ (see Fig. 59). Thus the *total force* $\boldsymbol{F}(\mathscr{V})$ and the *total torque* $\boldsymbol{L}_{\boldsymbol{x}_0}(\mathscr{V})$ acting upon the matter in the region $\mathscr{V}$ included by $\partial\mathscr{V}$, are given by

$$\boldsymbol{F}(\mathscr{V})=\int_{\partial\mathscr{V}}\boldsymbol{t}_{\partial\mathscr{V}}\,ds+\int_{\mathscr{V}}\boldsymbol{b}\,dm, \tag{1}$$

$$\boldsymbol{L}_{\boldsymbol{x}_0}(\mathscr{V})=\int_{\partial\mathscr{V}}(\boldsymbol{x}-\boldsymbol{x}_0)\wedge\boldsymbol{t}_{\partial\mathscr{V}}\,ds+\int_{\mathscr{V}}(\boldsymbol{x}-\boldsymbol{x}_0)\wedge\boldsymbol{b}\,dm, \tag{2}$$

where ds and dm indicate integration with respect to surface area and mass, and where $\boldsymbol{b}$ is the body force per unit mass. The forces and torques acting upon a body are thus of two kinds: those that are absolutely continuous functions of mass, and those that are absolutely continuous functions of boundary surface area. The term "body" is applied to the former, and "contact" or "stress" to the latter. The rational mechanics of materials is the theory of contact forces.

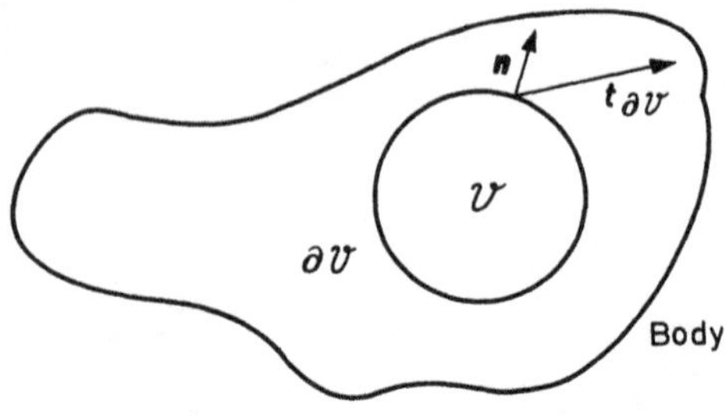

Fig. 59. The traction $\boldsymbol{t}_{\partial\mathscr{V}}$ acting upon the boundary $\partial\mathscr{V}$ of a region $\mathscr{V}$ occupied by a part of a body

CAUCHY assumes that the traction $\boldsymbol{t}_{\partial\mathscr{V}}$ depends upon the surface $\partial\mathscr{V}$ only through the normal $\boldsymbol{n}$:

$$\boldsymbol{t}_{\partial\mathscr{V}}=\boldsymbol{t}_{\boldsymbol{n}}, \tag{3}$$

so that all bodies whose bounding surfaces have a common normal $\boldsymbol{n}$ at a point suffer the same traction there. By appeal to the principle of linear momentum he

then proves that $\boldsymbol{t}_{\boldsymbol{n}}$ changes sign when the orientation of $\partial\mathscr{V}$ is changed:

$$\boldsymbol{t}_{\boldsymbol{n}} = -\boldsymbol{t}_{-\boldsymbol{n}}, \tag{4}$$

or, the traction exerted by the inside on the outside is equal and opposite to that exerted by the outside on the inside. Using this result, which is called *Cauchy's lemma*, by a second appeal to the principle of linear momentum he proves that if $\boldsymbol{t}_{\boldsymbol{n}}$ is assumed to be a continuous function of $\boldsymbol{n}$, it is then a homogeneous linear function of $\boldsymbol{n}$:

$$\boldsymbol{t}_{\boldsymbol{n}} = \boldsymbol{T}\boldsymbol{n}. \tag{5}$$

That is, there exists over the space of vectors a linear transformation $\boldsymbol{T}$, called the *stress tensor*, whose operation upon the unit outer normal to $\partial\mathscr{V}$ at a point yields the traction vector at that point. In particular, in the terms used by CAUCHY himself, the tractions on three mutually perpendicular planes at a point determine uniquely the tractions on all planes there. The existence of the stress tensor is often called *Cauchy's fundamental theorem*. Using it in (1), CAUCHY then proves that at a point interior to a region where at a fixed time the tensor field $\boldsymbol{T}$ is continuously differentiable, where $\boldsymbol{b}$ is continuous, and where the acceleration $\ddot{\boldsymbol{x}}$ exists and is continuous, the principle of linear momentum in an inertial frame is equivalent to the partial differential equation

$$\operatorname{div}\boldsymbol{T} + \varrho\boldsymbol{b} = \varrho\ddot{\boldsymbol{x}}, \tag{6}$$

ϱ being the mass density, also assumed continuous. He then proves that under the same assumptions, when (6) is satisfied, the principle of moment of momentum is equivalent to

$$\boldsymbol{T} = \boldsymbol{T}^{T}. \tag{7}$$

Thus the stress tensor is symmetric. These two equations, (6) and (7), are *Cauchy's first and second laws of motion*. If we grant the principles of linear momentum and moment of momentum, CAUCHY's first law asserts that the resultant contact force per unit volume is $\operatorname{div}\boldsymbol{T}$, while the second is a necessary and sufficient condition that the resultant contact torque be the moment of the resultant contact force.

For modern continuum mechanics these two laws are sometimes insufficiently general. First, that all torques be the moments of forces, as is stated in (2), is not always a fit assumption, since there may be contact couples and body couples, and also in connection with oriented materials it may be necessary to introduce higher-order stresses, spin angular momentum, and so on. Second, the assumptions of smoothness needed to get CAUCHY's laws may be too strong. A long sequence of researches on shock waves and other strong discontinuities, beginning with special cases treated by FOURIER, POISSON, STOKES, RANKINE, HUGO-

NIOT, HADAMARD, and ZEMPLÉN, led finally to a general theorem of KOTCHINE (1926), which when applied to the principle of linear momentum with the resultant force given by (1) delivers the jump condition

$$[\boldsymbol{t}_{\partial\mathscr{V}} + \varrho U \dot{\boldsymbol{x}}] = \mathbf{0}. \qquad (8)$$

Here the bracket denotes a jump across a singular surface at the particular place considered, U is the local speed of propagation[1] of the surface, and $\dot{\boldsymbol{x}}$ is the velocity field. The condition (8) is necessary and sufficient that the principle of linear momentum, with the resultant force given by CAUCHY's stress principle, hold in every neighborhood of the point to which it refers. Thus this condition does not express a new idea but is rather the result of applying the stress principle mathematically to a case when the usual assumptions of smoothness would be too strong. The principle of moment of momentum, provided the resultant torque be given by (2), is also equivalent at a singular surface to KOTCHINE's condition.

Here I shall not discuss these generalizations and extensions; neither shall I present CAUCHY's proofs, which have been reproduced many times, in effect if not always in detail, in the textbooks, nor shall I quote his words in support of his thesis. CAUCHY's papers on this subject are so clear, direct, and elegant that were a historian to describe them, he might do the reader a disservice in deflecting him from reading them for himself (Figs. 60, 61). Clearly this work of CAUCHY's marks one of the great turning points of mechanics and mathematical physics, even though few writers on the history of that subject seem to know it[2], a turning point that could well stand comparison with HUYGENS' theory of the pendulum, NEWTON's theory of the solar system, EULER's theory of the perfect fluid, and MAXWELL's theories of the monatomic gas and the electromagnetic field.

That the stress principle is not an obvious one, any teacher may convince himself by trying to teach it to students, and now that continuum mechanics is no

[1] More precisely, the values U^+ and U^- to be used when the condition (8) is made explicit are the limiting values of the local speed of propagation as the singular surface is approached from the sides labelled + and —.

[2] In a recent and attractive volume setting forth the progress of physics in the nineteenth century, the author, a professional historian of science, has not seen fit to mention the names of CAUCHY or STOKES. Neither did MACH.

Fig. 60. CAUCHY's fundamental theorem on the stress tensor, from his *Exercices de Mathématiques*, Volume 4, 1829

$$(6)\quad\left\{\begin{aligned}
\mathcal{E} &= p_3(\cos\lambda_3\cos\alpha_1 + \cos\mu_3\cos\beta_1 + \cos\nu_3\cos\gamma_1)\,,\\
\mathcal{D} &= p_3(\cos\lambda_3\cos\alpha_2 + \cos\mu_3\cos\beta_2 + \cos\nu_3\cos\gamma_2)\,,\\
\mathcal{C} &= p_3(\cos\lambda_3\cos\alpha_3 + \cos\mu_3\cos\beta_3 + \cos\nu_3\cos\gamma_3)\,.
\end{aligned}\right.$$

Si maintenant on substitue, dans les équations (4), (5), (6), les valeurs de $p_1\cos\lambda_1$, $p_1\cos\mu_1$, etc..., tirées des équations (1), (2), (3), on obtiendra les six formules

$$(7)\quad\left\{\begin{aligned}
\mathcal{A} &= A\cos^2\alpha_1 + B\cos^2\beta_1 + C\cos^2\gamma_1 + 2D\cos\beta_1\cos\gamma_1 + 2E\cos\gamma_1\cos\alpha_1 + 2F\cos\alpha_1\cos\beta_1\,,\\
\mathcal{B} &= A\cos^2\alpha_2 + B\cos^2\beta_2 + C\cos^2\gamma_2 + 2D\cos\beta_2\cos\gamma_2 + 2E\cos\gamma_2\cos\alpha_2 + 2F\cos\alpha_2\cos\beta_2\,,\\
\mathcal{C} &= A\cos^2\alpha_3 + B\cos^2\beta_3 + C\cos^2\gamma_3 + 2D\cos\beta_3\cos\gamma_3 + 2E\cos\gamma_3\cos\alpha_3 + 2F\cos\alpha_3\cos\beta_3\,;
\end{aligned}\right.$$

$$(8)\quad\left\{\begin{aligned}
\mathcal{D} &= A\cos\alpha_2\cos\alpha_3 + B\cos\beta_2\cos\beta_3 + C\cos\gamma_2\cos\gamma_3\\
&\quad + D(\cos\beta_2\cos\gamma_3 + \cos\beta_3\cos\gamma_2) + E(\cos\gamma_2\cos\alpha_3 + \cos\gamma_3\cos\alpha_2) + F(\cos\alpha_2\cos\beta_3 + \cos\alpha_3\cos\beta_2)\,,\\
\mathcal{E} &= A\cos\alpha_3\cos\alpha_1 + B\cos\beta_3\cos\beta_1 + C\cos\gamma_3\cos\gamma_1\\
&\quad + D(\cos\beta_3\cos\gamma_1 + \cos\beta_1\cos\gamma_3) + E(\cos\gamma_3\cos\alpha_1 + \cos\gamma_1\cos\alpha_3) + F(\cos\alpha_3\cos\beta_1 + \cos\alpha_1\cos\beta_3)\,,\\
\mathcal{F} &= A\cos\alpha_1\cos\alpha_2 + B\cos\beta_1\cos\beta_2 + C\cos\gamma_1\cos\gamma_2\\
&\quad + D(\cos\beta_1\cos\gamma_2 + \cos\beta_2\cos\gamma_1) + E(\cos\gamma_1\cos\alpha_2 + \cos\gamma_2\cos\alpha_1) + F(\cos\alpha_1\cos\beta_2 + \cos\alpha_2\cos\beta_1)\,.
\end{aligned}\right.$$

Ces six formules fournissent le moyen de calculer les projections algébriques des pressions p_1, p_2, p_3, supportées par trois plans perpendiculaires entre eux, sur trois demi-axes perpendiculaires à ces mêmes plans, quand on connaît les projections algébriques des pressions supportées par trois plans parallèles aux plans coordonnés sur les axes des x, y, z.

Observons encore que, les demi-axes OL, OM, ON étant perpendiculaires l'un à l'autre, on aura, en vertu de formules connues,

$$(9)\quad\left\{\begin{aligned}
&\cos^2\alpha_1 + \cos^2\alpha_2 + \cos^2\alpha_3 = 1\,, &\quad &\cos\beta_1\cos\gamma_1 + \cos\beta_2\cos\gamma_2 + \cos\beta_3\cos\gamma_3 = 0\,,\\
&\cos^2\beta_1 + \cos^2\beta_2 + \cos^2\beta_3 = 1\,, &\quad &\cos\gamma_1\cos\alpha_1 + \cos\gamma_2\cos\alpha_2 + \cos\gamma_3\cos\alpha_3 = 0\,,\\
&\cos^2\gamma_1 + \cos^2\gamma_2 + \cos^2\gamma_3 = 1\,, &\quad &\cos\alpha_1\cos\beta_1 + \cos\alpha_2\cos\beta_2 + \cos\alpha_3\cos\beta_3 = 0\,.
\end{aligned}\right.$$

Par suite on tirera des équations (4), (5), (6)

signant par $\mathfrak{X}$, $\mathfrak{Y}$, $\mathfrak{Z}$ les projections algébriques de la force accélératrice qui serait capable de produire à elle seule le mouvement effectif d'une particule, et prenant x, y, z, t pour variables indépendantes, on obtiendra, à la place des équations (1), celles qui suivent

$$(2)\quad \left\{ \begin{aligned} &\frac{dA}{dx}+\frac{dF}{dy}+\frac{dE}{dz}+\rho X=\rho\mathfrak{X}, \\ &\frac{dF}{dx}+\frac{dB}{dy}+\frac{dD}{dz}+\rho Y=\rho\mathfrak{Y}, \\ &\frac{dE}{dx}+\frac{dD}{dy}+\frac{dC}{dz}+\rho Z=\rho\mathfrak{Z}. \end{aligned} \right.$$

Enfin, si l'on nomme ξ, η, ζ les déplacements de la particule qui, au bout d'un temps t, coïncide avec le point (x, y, z), mesurés parallèlement aux axes coordonnés, on trouvera, en supposant ces déplacements très-petits,

$$\mathfrak{X}=\frac{d^2\xi}{dt^2},\quad \mathfrak{Y}=\frac{d^2\eta}{dt^2},\quad \mathfrak{Z}=\frac{d^2\zeta}{dt^2},$$

et par conséquent les équations (2) deviendront

$$(3)\quad \left\{ \begin{aligned} &\frac{dA}{dx}+\frac{dF}{dy}+\frac{dE}{dz}+\rho X=\rho\frac{d^2\xi}{dt^2}, \\ &\frac{dF}{dx}+\frac{dB}{dy}+\frac{dD}{dz}+\rho Y=\rho\frac{d^2\eta}{dt^2}, \\ &\frac{dE}{dx}+\frac{dD}{dy}+\frac{dC}{dz}+\rho Z=\rho\frac{d^2\zeta}{dt^2}. \end{aligned} \right.$$

Les formules (1), (2), (3) sont les véritables équations d'équilibre ou de mouvement intérieur des corps considérés comme des masses continues; et pour en déduire, par exemple, les lois de l'équilibre ou du mouvement des corps solides élastiques, il suffit de chercher comment, dans ces derniers, les pressions ou tensions A, B, C, D, E, F doivent s'exprimer à l'aide des déplacements ξ, η, ζ. Nous ferons, à ce sujet les remarques suivantes.

Soient, au bout du temps t, α, β, γ les angles que forme avec les demi-axes des coordonnées positives une droite menée par le point (x, y, z), et représentons par ε la dilatation ou condensation linéaire ε, mesurée suivant cette droite. On aura, en supposant que les déplacements ξ, η, ζ soient très-petits,

longer included in the training of physicists, increasing numbers of younger specialists in such fields as solid-state physics and the statistical mechanics of fluids find themselves at a loss in following even classical ideas on the mechanics of solids and fluids, mainly from failure to master the apparently simple concept of stress. In this lecture I wish to show how one main aspect of mechanical intuition, the subtle idea of stress, grew through the centuries.

2. Subject of this Lecture

My purpose here is to mark the conceptual stages we can now recognize as predecessors of CAUCHY's great and unifying idea. Few of these were cited by CAUCHY himself, and perhaps he did not know of them. Indeed, he mentioned only the work of FRESNEL toward a theory of crystalline elasticity designed to explain the reflection and refraction of light, and a theory of NAVIER in which internal forces in a bent plate were regarded as due to change of volume and curvatures. These were the recent researches of colleagues which CAUCHY regarded it necessary to mention having seen. Naturally he did not feel obliged to attribute to any particular person such material as he chose to use from the common background of the geometers of his day. In the history of ideas, the progress of the uncited central or common domain is just as important as the cited points of departure from it, or perhaps more so. It is mainly this progress I wish to trace.

The general theory of continua grew out of special cases, each of which was a theory of a special kind of continua, special in respect to material properties or to geometry. The geometric possibilities differ in dimension and curvature:

1. One-dimensional
   a) straight
   b) curved
2. Two-dimensional
   a) flat
   b) curved
3. Three-dimensional

Fig. 61. CAUCHY's first law of motion for a continuum, as stated in his *Exercices de Mathématiques*, Volume 4, 1829

While material properties are infinitely various, only four kinds were sufficiently considered before CAUCHY's day to be given concrete mathematical forms:

0. Rigid
1. Fluid
2. Flexible (if the dimension is 1 or 2)
3. Elastic

The first of these is of no use for the concept of stress. Counting the remaining combinations yields fourteen possible special kinds of theories, but not all these were achieved before the general theory of stress. In particular, the three-dimensional theory of linear elasticity was first obtained at just about the same time, to within minor differences of approach and extent and clarity, by FRESNEL, NAVIER, CAUCHY himself, and POISSON. The earlier success in the other possibilities ranged from complete in the case of one-dimensional tubes of fluid and flexible lines and elastic bands to none in the case of curved elastic plates. It is the successes I wish to trace here, but we must bear in mind that some successes, particularly those in too special circumstances, deflect the student from the true general theory rather than lead him toward it.

3. The Central Role of Euler's Hydrodynamics (1750—1755). What it Achieved, and what it Obscured

Plainly a major building block in CAUCHY's storehouse of ideas ready to hand was EULER's hydrodynamics, some seventy-five years old. This was the first correct three-dimensional theory of a deformable material. To obtain it from CAUCHY's general theory, we need only assume the traction $\boldsymbol{t}_{\boldsymbol{n}}$ to be normal to the surface on which it acts:

$$\boldsymbol{t}_{\boldsymbol{n}} = -p\boldsymbol{n}, \tag{9}$$

where the scalar field p is called the *pressure*. Obviously (5) requires then that

$$\boldsymbol{T} = -p\boldsymbol{1}, \tag{10}$$

where $\boldsymbol{1}$ is the unit tensor. CAUCHY's first law (6) reduces to *Euler's dynamical equation*:

$$-\operatorname{grad} p + \varrho\boldsymbol{b} = \varrho\ddot{\boldsymbol{x}}, \tag{11}$$

and of course CAUCHY's second law (7) is satisfied automatically. I mention hydrodynamics now because it serves as a midpoint in our study, and much of the earlier work we shall discuss may be grasped more easily if we look on it as preliminary to EULER's theory of "perfect fluids", which was to render it in the end obsolete or at least subservient.

EULER's dynamical equation for fluids as a mere string of symbols is unlikely to suggest the greatest lesson CAUCHY and others learned from it. This is, that the way to discuss the motion of a body is to cut it in imagination into two parts, an inside and an outside, and to represent the entire action of the exterior on the interior by fields defined on the boundary. This is *Euler's cut principle*. Its importance may be gathered from a famous passage in OSGOOD's *Mechanics*, p. 102. "What comes first is *a material system*..., and then this system is acted on by forces.

ISOLATE THE SYSTEM

"The man who first uttered these words deserves a *monumentum aere [perennius]*."

I cannot say who first uttered these words. As with most lapidary phrases fit to be set over the monumenta of science, the concept it expresses was groped for and glimpsed and fumbled and rightly applied long before its distillation into clear words. Surely EULER was not the first even to state it, but, as we shall observe, he used it again and again and with ever increasing ease, generality, and directness, to the point that his hydrodynamical researches at mid-century laid it bare for everyone to see, once and for all. This CAUCHY inherited and took for granted. Indeed, he wrote in introducing the traction $\boldsymbol{t}_n$ that it is "of the same nature as the hydrostatic pressure exerted by a fluid at rest against the exterior surface of a body. Only, the new pressure does not always remain perpendicular to the planes subjected to it, nor the same in all directions at a given point."

The halting, even faltering steps toward the cut principle constitute a great deal of the history we are to trace here, but far from all of it. Hydrodynamics, while it rests upon the fully general cut principle, is still an old-fashioned, primitive theory in that it confounds material properties with the general laws of continuum mechanics. In modern terms, constitutive equations and the principles of momentum and moment of momentum are stated all together, undistinguished. The confusion is subtle in that it rests upon a purely geometric statement, namely, that the traction is normal to the surface on which it acts, as asserted in (9). Moreover, the confusion is deep, for, first, it obscures the tensorial character of the stress, as expressed by CAUCHY's fundamental theorem (5), and, second, it renders the principles of linear momentum and moment of momentum equivalent, as they are for many degenerate systems[3].

Thus even after the creation of hydrodynamics there remained much left to be done. First, the concept of a component of traction parallel to the surface on which it acts, now called *shear stress*, had to be seen and stated. Second, it was necessary to free the concept of stress from *any* geometric restriction laid down *a priori*. In modern terms, the concept of constitutive equation as the definition of an ideal material had to be separated and set aside as belonging to a different part of the subject, not restricting the general principles of mechanics. Third, the *principle of moment of momentum* had to come into its own place, side by side with the principle of linear momentum, and equal in status.

The growth of these three ideas had begun before EULER's birth, and since each of them, like all great ideas of mathematical physics, was discovered by necessity in the solution of special problems, their histories are intertwined with each other and with that of the cut principle. We shall trace and unravel them now, though not quite in chronological order, through the dark gropings from case to case in the earlier researches which gained, one by one, those individual intuitive elements that CAUCHY had only to cleanse of superfluous special properties and weld into one conceptual frame.

4. Stevin's Principle of Solidification (1586)

That liquids exert forces upon walls confining them is common experience, but it was only from centuries of thought about this and other facts of mechanics, phrased in terms of the elementary mathematical theories of force and equilibrium, that any definite concept of *internal* pressure was evolved.

[3] *E.g.*, for a punctual system subject to pairwise equilibrated central forces. This fact has given rise to a confusion, sedulously nourished by physics courses and textbooks, that the principle of moment of momentum is only a consequence of the principle of linear momentum or an "integral of motion".

The truth is quite the reverse, as may be seen from an argument due to NOLL. As basic principle of mechanics we may lay down the demand that the *mechanical power be frame-indifferent*; *i.e.*, invariant under time-dependent rigid transformation of Euclidean (or a more general "neo-classical") space-time. Any such transformation may be decomposed into a translation and a rotation. Considering translations yields the principle of linear momentum; considering rotations yields the principle of moment of momentum. Thus the independence of the two principles reflects a basic property of space-times in which (1) there is an absolute concept of simultaneity and (2) the section of the universe at a fixed time is a Euclidean space.

The earliest reference to the force exerted by fluid on fluid occurs in ARCHIMEDES' theory of bodies floating in a spherical ocean covering a spherical earth (c. 250 A.C.). His Postulate I is[4]: "Let the fluid be granted to have such a nature that of the parts of it which are at the same height and side by side, what is pressed the less is pushed away by what is pressed the more, and that each of its parts is pressed by the fluid plumb above it, provided the fluid be not shut up in anything nor pressed by anything else." The meaning is disclosed by the propositions which follow. First, the condition at the end serves no purpose but to leave ARCHIMEDES uncommitted as to fluids subjected to other circumstances, such as confinement in a vessel subject to outside pressure. The "pushing away" refers to what happens before equilibrium is established. In equilibrium, ARCHIMEDES asserts, the surfaces bearing constant weight per unit area[5] are spheres about the

---

When these ideas are applied to punctual systems, it turns out that under reasonable assumptions of smoothness the principle of moment of momentum is *equivalent* to the requirement that the mutual forces be central and pairwise equilibrated. The argument of CAUCHY mentioned in § 1 may be interpreted, likewise, as showing that for any three-dimensional material whose constitutive equation forces the stress tensor to be symmetric, the principle of linear momentum implies as a corollary the principle of moment of momentum, provided the moment of momentum be defined as usual and the torque be given by (2).

[4] This postulate has given a lot of trouble. HEATH's English translation, the only one available until recently, is from a faulty text which omits the essential word "not" (μή) in the disclaimer at the end. DIJKSTERHUIS' translation is essentially correct but obscure from its awkward syntax and its failure to bring out the concept of plumb line (κάθετος).

[5] Of course, ARCHIMEDES nowhere says just this, but I can find no other specific meaning for "pressed by the fluid which lies plumb above it." Moreover, for the postulate to be true, we must regard the part pressed as extending to the center of the earth, or through it, so that the plumb line (κάθετος) sweeps out *two* conical parts above the given one. See Fig. 62, in which the heavily shaded part is that pressed, while the lightly shaded ones are those "plumb above" it. Otherwise the postulate does not provide all the forces necessary to maintain equilibrium.

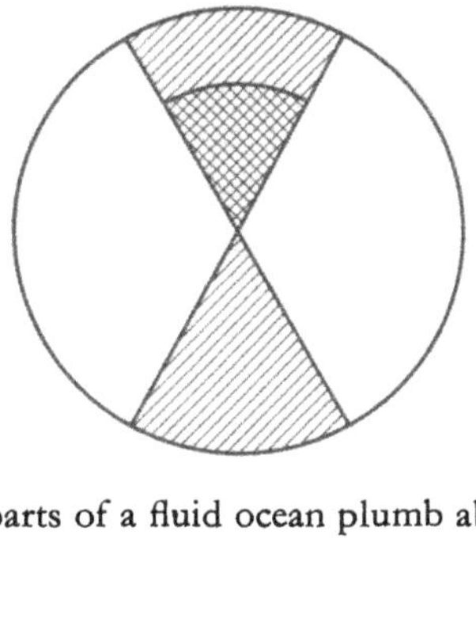

Fig. 62. Sketch to illustrate the parts of a fluid ocean plumb above a given part

center of the earth. In the application to the equilibrium of floating bodies, ARCHIMEDES assumes that this state of affairs still prevails, undisturbed by the immersed portions of the rigid solids. He seems not to recognize that a new axiom would be needed here in a strict theory, since the fluid is certainly pressed by "something else" than itself.

While ARCHIMEDES makes it plain that he has some idea about the pressure of fluid on fluid, he does not make any such concept explicit, let alone general, and indeed he pushes it aside as soon as it has served his purpose and given him a tool for the analysis of the position of a floating body.

STEVIN's *principle of solidification* is often stated nowadays as follows: In any fluid at rest, if any portion be replaced by a rigid solid, the forces exerted by the remainder will not be altered. No statement so embracing or explicit, however, is to be found before the work of CLAIRAUT (1743). Perhaps the nearest approach of STEVIN himself (1586) is his Theorem IV: "A solid body of equal specific gravity to water keeps any place given to it in water." His theory is restricted to incompressible heavy fluids on the surface of a flat earth. He derives what amounts to a correct formula for the weight sustained by a flat horizontal bottom and shows that it is independent of the shape of the vessel, if there is one, confining the fluid above this bottom, or even below it (Fig. 64). We can express his result as

$$\text{Total weight} \propto \sigma h A \tag{12}$$

where σ is the specific gravity, h is the depth, and A is the area of the bottom. Again some concept of internal pressure must stand behind his inference that adding or taking away solid boundaries does not alter the force on the bottom, but he does not venture to apply this result to walls that are not flat and horizontal.

Indeed, when STEVIN comes to calculate the "weight" of water on an inclined wall, he resorts to a proof by exhaustion in ARCHIMEDES' style. Of interest to us here is not the calculation of the limit but the mechanical axiom that enables it

---

Fig. 63 (p. 197). SIMON STEVIN (1548—1620), portrait by an unknown artist, not earlier than 1620, reproduced by courtesy of the Library, University of Leiden

Fig. 64 (p. 198). STEVIN's argument for the principle of solidification, from *De Beghinselen des Waterwichts*, 1586, as republished in French in 1634. (The bottom EF should have been shown as the same in all figures.)

Fig. 65 (p. 199). STEVIN's argument for calculating the weight of water on an inclined surface, from *De Beghinselen des Waterwichts*, 1586, as republished in French in 1634

NATUS BRUGIS Anno 1548
DENATUS HAGÆ CO MITIS Anno 1620
SIMON STEVIN,
MATHEMATICUS INSIGNIS
CONSILIAR. CELSISS. MAURITII PRINCIPIS
ARAUSIO NENSIUM.

la colomne soit celle qui est comprise entre le fond pour sa base E F, & G E hauteur, assavoir la colomne G E F H.

Le requis. Il faut demonstrer que sur le fond E F repose un poids egal à la pesanteur de l'eau de la colomne G E F H.

Demonstration.

Si sur le fond E F repose un poids plus grand que G E F H, cela viendra à cause de l'eau prochaine: Soit, s'il est possible, de l'eau A D E G, & H F C B; & de mesme pourra-on dire que

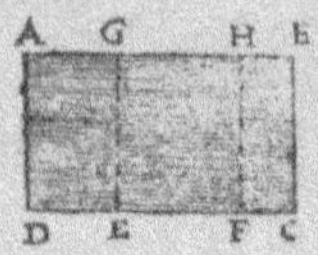

sur le fond D E repose plus que l'eau A D E G, & sur F C plus que l'eau H F C B; tellement que sur D C reposera plus que l'eau ADCB; ce qui est absurde, estãt iceluy un parallelipipede rectangle. Semblablement on demonstrera que sur E F ne repose pas moins que G E F H, & par consequent sur E F reposera precisement un poids egal à la colomne d'eau G E F H.

Corollaire I.

En l'eau A B C D de la 10 proposition, mettons I L un corps solide minugrave à l'eau, c'est à dire flottant sur l'eau, N L dedans, & N K dehors, comme cy-joignant; alors le solide I L, sera equiponderant à l'eau N O L M, par la 5 proposi-

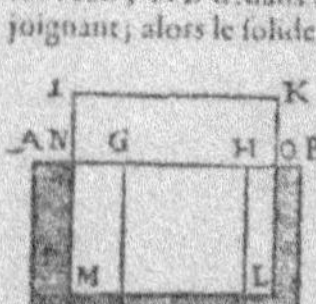

tion, d'où s'ensuit que le corps I L, avec le reste de l'eau d'alentour, est equiponderant à un corps d'eau, egal à A B C D; parquoy nous dirons encor selon la proposition, que sur le fond E F, repose un poids egal à la pesanteur de l'eau d'une telle colomne que G E F H, de laquelle E F est la base, & G E perpendicle, entre le fond & la fleur de l'eau pour la hauteur d'icelle: D'où l'on peut conclurre que quand quelque matiere flotte sur l'eau, qu'elle n'apporte aucun changement au poids que le fond soustient lors que l'eau demeure en la mesme hauteur.

Corollaire II.

Soit encore en l'eau A B C D, un ou plusieurs corps solides parigraves à l'eau, tellement qu'il n'y aye place que pour l'eau I K F E L M, alors ce corps ne surcharge,

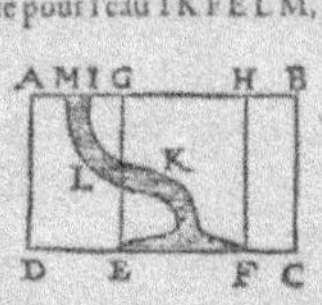

ny n'allege le fond E F pas plus que devant: Et partant selon la proposition, sur le fond E F, repose un poids egal à la pesanteur de l'eau, laquelle est egale à la colomne, ayant ledit fond pour base, & la hauteur egale à la perpendicle sur l'horizon, qui est entre le fond & fleur d'eau.

Corollaire III.

Soit derechef A B C D entierement eau, EF un fond en icelle parallele à l'horizon; alors l'eau poussera autant par dessous, en elevant que par dessus en abaissant; Autrement le plus fort emporteroit le plus foible: ce qui n'arrive pas ainsi, d'autant que tout tient la disposi-

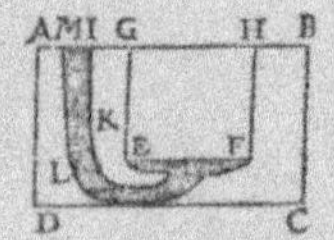

tion qu'on luy donne, par la 1 proposition. Soyent maintenant dedans l'eau quelques corps solides parigraves à icelle, & tellement disposez que l'eau I K E F L M pousse par dessous E F, assavoir contre le corps solide, comme devant contre l'eau, mais l'un contrepoussoit egalement l'autre: donc contre E F y a un effort qui le pousse enhant, de mesme que la colomne d'eau G E F H pousse le mesme fond E F embas, selon la proposition; car la hauteur G E est la perpendicle interceptée entre la fleur de l'eau & le fond E F.

Corollaire IV.

Que si on mettoit les corps solides des deuxiesme & troisiesme corollaires en leurs lieux, & l'eau vuidée, y ayant une place vuide I K F E L M; alors le fond E F ne portera aucune pesanteur; que si on remply le lieu vuidé, avec de l'eau, le fond patira autant d'effort, que lors que le vaisseau estoit entierement plein d'eau (ayant osté les corps solides.)

Corollaire V.

Mais si on ostoit la matiere inutile des corps solides, & s'il n'y en demeuroit pas d'avantage que pour tenir l'eau en ceste figure; alors le fond E F patira autant d'effort,

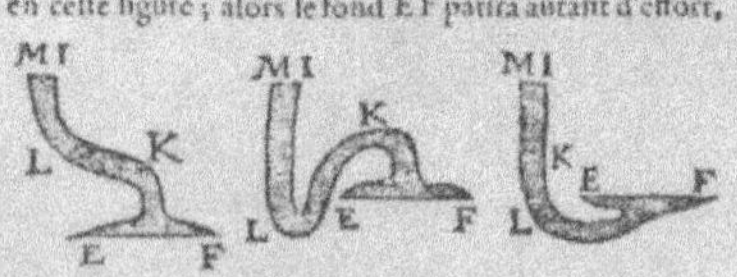

que s'il y avoit dessus une colomne d'eau, comprise sur le mesme fond comme base, dont la hauteur soit la perpendicle entre ledit fond & la fleur d'eau.

Conclusion. Sur le fond de l'eau parallele à l'horizon, repose donc un poids, &c.

Lisez les espreuves deduites en l'Appendice de la practique de l'Hydrostatique.

Notez.

Nous pourrions aussi proposer la 10 proposition comme s'ensuit:

SUr quelconque fond d'eau, de superfice parallele à la fleur d'icelle, repose un poids egal à la pesanteur de l'eau, comprise dans un secteur spherique, tronqué par une superfice spherique parallele, ou homocentrique à la superfice spherique de la terre.

Nous eussions fait aussi les demonstrations comme dessus, mais nous l'avons delaissé pour les raisons descrites en la septiesme petition.

Theoreme IX. Proposition XI.

SUr un fond convenant, duquel le plus haut poinct est à fleur d'eau, repose un poids egal à la demi-colomne d'eau, de laquelle la base est pareille audit fond, & sa hauteur egale à la perpendicle comprise entre les niveaux qui passent par le plus haut & plus bas poinct dudit fond.

1 *Exemple.*

Le donné. Soit A B un vaisseau plein d'eau, & le fond A C D E soit premierement un parallelogramme, non parallele à l'horizon, comme devant, mais perpendiculaire à iceluy; duquel le plus haut costé A C est à fleur d'eau,

d'eau, & A E soit la perpendicle comprise entre les deux niveaux qui passent par les extremitez plus haute & plus basse dudit fond : & soit marquée H dans D B, parallele à l'horizon, tellement que D H soit egale à D C, & menée C H, tellement que par A C H D E soit denotée la demi-colomne, ayant A C D E pour base, & D H hauteur egale à A E perpendicle.

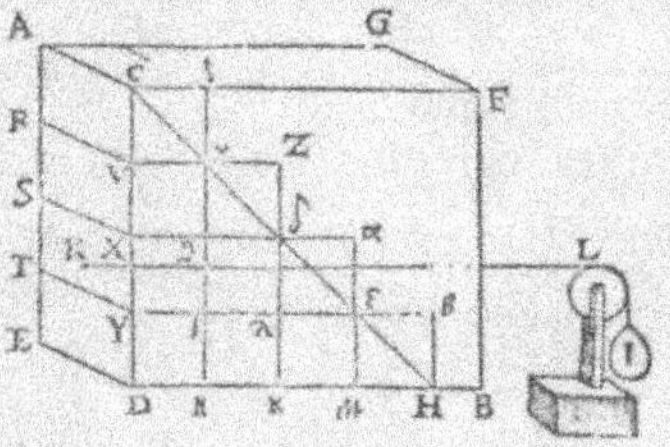

Le requis. Il faut demonstrer que le poids qui repose contre le fond A C D E est egal à la demi-colomne A C H D E : C'est à dire (pour declarer le tout plus amplement) que si I estoit un poids oblique, equipondérant A C H D E, & K L corde soit parallele à D H, & K centre de gravité de l'effort du poussement congregé à l'encontre du fond (lequel centre de gravité se trouvera par la 18 proposition suivante) le poids I equipondérant à l'effort de l'eau tient le fond (posant qu'il soit mobile) en ceste disposition.

Ou bien pour donner à entendre ce que dessus autrement, Soit M N O P un fond, pareil à A C D E, assavoir M P homologue à A C ; M N à A E, sur lequel fond M N O P repose un corps solide M N O P Q, pareil & equipondérant à ACHDE, & Q O (egale à D H) soit perpendicle à l'horizon; je dis que le pressement que MNOPQ fait sur le fod MNOP, (assavoir pressant plus vers NO, que vers MP, pource qu'il est là plus espais & pesant qu'icy)

est le mesme que l'eau contre le fond ACDE, (assavoir poussant plus vers E D, que vers A C.)

Preparation. Soit divisée A E en 4 parties egales par R, S, T, & menées R V, S X, T Y paralleles à A C; & V Z, X α, Y β paralleles à D H, coupans C H en γ, δ, ι & tellement que chacune des lignes γ Z, δ α, ι β, soyent egales à V γ. Soit aussi menée par γ la ligne ζκ, parallele à C D, coupant X α en ϑ, & Y β en ι ; de mesme la ligne Z κ par δ, coupant Y β en λ, & ainsi la ligne α μ par ι, & finalement β H.

DEMONSTRATION.

Contre le fond A C V R repose plus que rien, car si ledit fond estoit tout à fleur d'eau, rien ne reposeroit à l'encontre, mais il vient plus bas, donc il repose plus que rien à l'encontre d'iceluy : D'avantage je dis qu'à l'encontre d'iceluy repose moins que le corps d'eau A C ζ γ V R, car s'il estoit parallele à l'horizon par R V, alors ledit corps y reposeroit, par la 10 proposition; mais il est plus haut, & par consequent il y repose moins à l'encontre. Je dis semblablement que contre le fond R V X S repose plus que le corps d'eau A C ζ γ V R, veu qu'il vient plus bas que V γ, mais ledit corps d'eau est egal à R V γ ϑ X S : D'avantage je dis que contre iceluy repose moins que le corps d'eau A C ζ ϑ X S, veu que le fond R X est plus haut que le fond S X ϑ ; or ledit corps d'eau A C ζ ϑ X S est egal à R V Z δ X S; parquoy contre le fond R X repose moins que R V Z δ X S. De mesme contre le fond S Y, repose plus que A C ζ ϑ X S, car il vient plus bas que le fond S X ϑ, lequel corps est egal à S X δ λ Y T ; parquoy contre S Y repose plus que S X δ λ Y T. D'avantage je dis qu'il y repose moins que A C ζ ι Y T, car il est plus haut que le fond T Y ι ; or ledit corps A C ζ ι Y T est egal à S X α ι Y T, donc contre le fond S Y repose moins que S X α ι Y T. Pareillement contre le fond T D repose plus que A C ζ ι Y T, car il vient plus bas que le fond T Y ι ; iceluy corps est egal à T Y ι μ D E; parquoy contre le fond T D repose plus que T Y ι μ D E. D'avantage il y repose moins que le corps A κ, car il vient plus haut que le fond E D κ ; mais ledit corps A κ est egal à T Y β H D E ; donc contre le fond T D, repose moins à l'encontre que T Y β H D E. Or veu que contre

Le fond	repose plus que	& moins que
A V	rien	A C ζ γ V R.
R X	R V γ ϑ X S	R δ Z.
S Y	S X δ λ Y T	S ι α.
T D	T Y ι μ D E	T H β.

Il s'ensuit que contre le fond entier A C D E reposera plus que les corps susdits ensemble, qui sont le corps inscrit R V γ ϑ δ λ ι μ D E dans la demi-colomne A C H D E, & que contre ledit fond A C D E entier reposeront moins que les corps circonscrits ensemble, qui sont le corps A C ζ γ Z δ α ι β H D E, lequel est circonscrit à la demi-colomne A C H D E.

Que si on divisoit le fond A C D E en plus de 4 parties egales, soit en 8, il appert que les corps inscrits & circonscrits ne differeroyent que de la moitié de la difference precedente ; & est manifeste qu'on pourroit partir le fond en tant de parties egales que la difference des corps inscrits & circonscrits à la demi-colomne, differeroyent moins qu'aucun corps donné, si petit puisse-il estre, d'où j'argumente ainsi.

A. *Toute pesanteur qui differe moins du poids reposant contre le fond A C D E, qu'on ne sçauroit donner, est egale au poids reposant contre le fond A C D E.*

I. *Le poids de la demi-colomne A C H D E, est une pesanteur differant moins du poids, qui repose contre le fond A C D E, qu'on sçauroit donner.*

I. *Le poids donc de la demi-colomne A C H D E est egal au poids qui repose contre le fond A C D E.*

2. Exemple.

Le donné. Soit encore A B un vaisseau plein d'eau, & le fond A C D E d'iceluy soit un parallelogramme oblique à l'horizon, AC la plus haute extremité à fleur d'eau A C F G; lequel soit divisé comme au premier exemple, & soit A ν perpendiculaire à l'horizon, jusques au niveau par E D.

Le requis. Il faut demonstrer que le poids d'eau qui repose contre le fond A C D E, est egal à la moitié de la colomne qui a pour base le fond A C D E, & hauteur A ν.

Preparation. Soit divisée A ν en 4 parties egales par les poincts ο, π, ρ.

DEMONSTRATION.

Contre le fond A V repose un poids plus que rien, d'autant qu'il est plus bas que le fond A ζ, contre lequel rien ne repose : D'avantage il repose contre ledit fond

to be applied to hydrostatics. STEVIN rules off the inclined plane into strips by horizontal lines. If the strips be rotated about their upper edges until they are horizontal, they will sustain less weight, says STEVIN, because there is less fluid above them; likewise, if they are turned level along their lower edges, they will sustain greater weight because there is more fluid above them. Thus the true weight is bounded above and below by weights on horizontal bottoms, which can be calculated by (12). Passage to the limit as the breadth of the strips is diminished indefinitely yields the same result for both the lower and upper sums, and hence the true weight (Fig. 65).

While the modern reader, to whom integral calculus is second nature, sees that STEVIN's hypothesis is neither more nor less than the statement that the traction on a boundary with outer unit normal $\boldsymbol{n}$ is of the form

$$\boldsymbol{t}_{\boldsymbol{n}} = -k\sigma h\boldsymbol{n}, \tag{13}$$

STEVIN cannot have had this clear and explicit idea, for if he had, his limit process would have been unnecessary for the string of theorems[6] he proved by its means.

The work of PASCAL (1663) seems to be entirely derivative as far as concerns incompressible fluids; from his book it became known that air in equilibrium follows much the same laws as does water.

5. Galileo's Rupture Stress (1638)

In GALILEO's *Two New Sciences* (1638), Salviati undertakes to refute the common error that a long rope cannot hold up so great a weight as a short one (Fig. 67).

[6] *E.g.*, the splendid Theorem X: "Given a regular bottom whose highest point is below the water's upper surface: the weight resting against it is equal to the prism of water whose base is equal to that bottom and whose height is the vertical from the plane through the water's upper surface to the highest point of the bottom, and in addition thereto the half of the vertical from the highest point of the bottom to the plane parallel to the horizon through the lowest point of the bottom."

Fig. 66 (p. 201). GALILEO GALILEI (1564—1642), after an engraving by OTTAVIO LEONI, before 1631

Fig. 67 (p. 202). GALILEO's argument to prove that a long rope is no weaker than a short one (1638)

Fig. 68 (p. 203). Part of GALILEO's discussion of the breaking strength of a cantilever (1638)

Galileus Galileus Florentinus
Superior licentia
16 24
Eques Octauius Leonus Roman. pictor fecit

allungando, e non ſolo nell' vſo traſuerſale, mà ancora per lo lungo, in quel modo appunto che veggiamo vna corda lunghiſſima eſſer molto meno atta à reggere vn gran peſo, che ſe fuſſe corta: onde io credo che vna verga di legno, ò di ferro più peſo aſſai potrà reggere ſe ſarà corta, che ſe ſarà molto lunga; intendendo ſempre vſata per lo lungo, e non in trauerſo; & anco meſſo in conto il ſuo proprio peſo, che nella più lunga è maggiore.

Salu. *Dubito S. Simp., che in queſto punto voi con molti altri v' inganniate, ſe però hò ben compreſo il voſtro concetto, ſi che voi vogliate dire, che vna corda lunga, v. gr., quaranta braccia non poſſa ſoſtenere tanto peſo, quanto ſe fuſſe vn braccio, ò due della medeſima corda.*

Simp. *Coteſto hò voluto dire, e ſin quì mi par propoſizione aſſai probabile.*

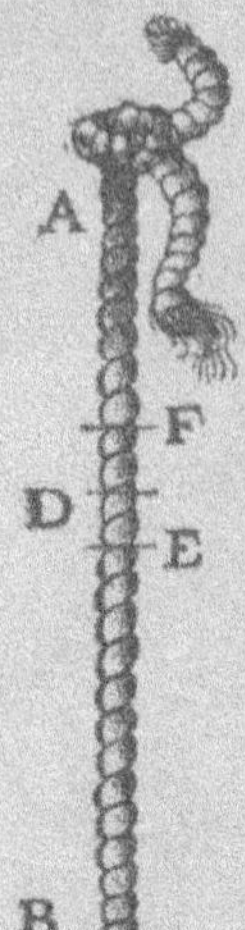

Salu. *Mà io l'hò per falſa, non che per improbabile; e credo di poterui aſſai agevolmente cauar d'errore. Però ponghiamo queſta corda* A B *fermata di ſopra dal capo* A, *e dall' altro ſia il Peſo* C, *dalla cui forza debba eſſa corda eſſere rotta. Aſſegnatemi voi S. Simp. il luogo particolare doue debba ſeguir la rottura.*

Simp. *Sia nel luogo* D.

Salu. *Vi domando qual ſia la cagione dello ſtrapparſi in* D.

Simp. *E' la cauſa di ciò, perche la corda in quella parte non era potente a reggere, v. gr., cento libbre di peſo, quanto è la parte* D B *con la pietra* C.

Salu. *Adunque tutta volta che tal corda nella parte* D *veniſſe violentata dalle medeſime cento libbre di peſo, ella lì ſi ſtrapperebbe.*

Simp. *Così credo.*

Salu. *Mà ditemi ora; chi attaccaſſe il medeſimo peſo non al fine della corda* B, *mà vicino al punto* D, *come ſarebbe*

in E,

ſin quì dichiarate, non ſarà difficile l'intender la ragione, onde auuenga, che vn Priſma, ò Cilindro ſolido di vetro, acciaio, legno, ò altra materia frangibile, che ſoſpeſo per lungo ſoſterrà graviſſimo peſo, che gli ſia attaccato, mà in trauerſo (come poco fà diceuamo) da
Prop. I *minor peſo aſſai potrà tal'volta eſſere ſpezzato, ſecondo che la ſua lunghezza eccederà la ſua groſſezza. Imperò che figuriamoci il Priſma ſolido* AB, CD *fitto in vn muro dalla parte* AB, *e nell'altra eſtremità s'intenda la forza del Peſo* E, *(intendendo ſempre il muro eſſer eretto all'Orizonte, & il Priſma, ò Cilindro fitto nel muro ad angoli retti) è manifeſto che douendoſi ſpezzare ſi romperà nel luogo* B, *doue il taglio del muro ſerue per ſoſtegno, e la* BC *per la parte della Leua, doue ſi pone la forza, e la groſſezza del ſolido* BA *è l'altra parte della Leua, nella quale è poſta la reſiſtenza, che conſiſte nello ſtaccamento, che s'hà da fare della parte del ſolido* BD, *che è fuor del muro, da quella che è dentro; e per le coſe dichiarate il momento della forza poſta in* C *al momento della reſiſtenza che ſtà nella*

Given a particular rope, he considers attached at *C* a weight just sufficient to break it, and he asks Simplicio where the break will occur. Simplicio chooses a point *D*, "because at this point the rope is not strong enough to support, say, 100 lbs." Salviati then imagines the rope fixed at *F*, just above *D*, and at *E*, just below *D*, he proposes to attach the weight. He says the rope is still subject to the same pull at *D*, and thus the short segment *FE* will break again at *D*, by Simplicio's admission. This *reductio ad absurdum* is a typical example of GALILEO's method of concealing his assumptions in rhetoric. In fact, he has assumed that the whole effect on the rope above *D* of the rope and weight *DC* is to exert at *D* a certain force in the direction *DE*. That is, the action of the system below *D* on that above *D* is equipollent to a force at *D*.

Of course the argument itself is ridiculous, since it proves that a rope either never breaks at all or else breaks everywhere at once. The concept of critical load just does not explain rupture of a uniform rope, and in common experience, as everyone knows, a long section from a spool of cable will generally, though not always, break under a lesser load than a short one taken at random from the same spool. GALILEO has introduced a good concept here, but in a case where it does not belong.

In GALILEO's consideration of the rupture of cylindrical beams in tension, he asserts that the force required is as the base area, "since so many more are the fibres, the filaments, or the tenacious parts that hold together the parts of solids." That is, he assumes that rupture in tension occurs when a certain force per unit area is attained. *Cf.* Fig.68.

6. Pardies' and James Bernoulli's Concept of Tension in a Flexible Line (1673, 1691—1704)

In his study of the catenary curve and the suspension bridge, PARDIES assumed that the form of a flexible line remains unchanged if we solidify any part, or, further, if we replace the parts above two points *A* and *a* by *suitable forces acting along the tangents at A and a* (Fig. 69). This principle was used in all the later analyses of the catenary; in particular, by JOHN BERNOULLI and LEIBNIZ in 1690.

In a long series of researches on flexible lines of arbitrary thickness and subject to arbitrary distributed loading (1691—1704), JAMES BERNOULLI adopted PARDIES' principle without any change. He introduced the tension (firmitas) explicitly;

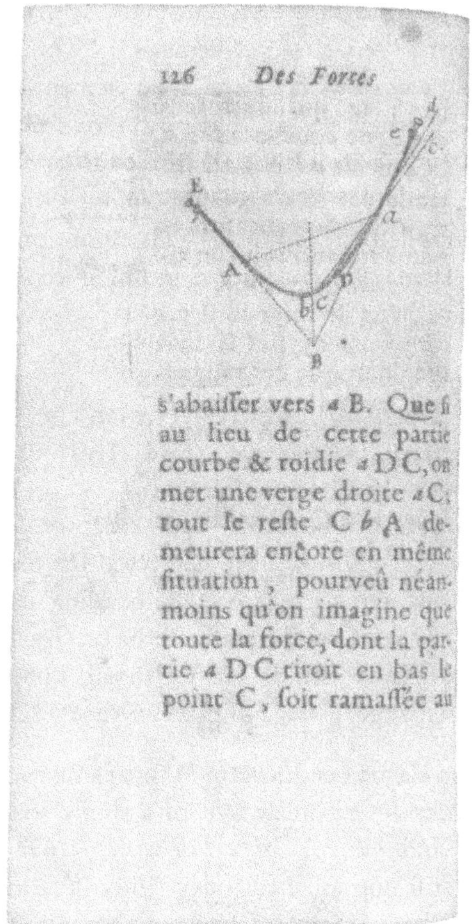

126 Des Forces

ſ'abaiſſer vers *a* B. Que ſi au lieu de cette partie courbe & roidie *a* D C, on met une verge droite *a* C, tout le reſte C *b* A demeurera encôre en même ſituation, pourveû néanmoins qu'on imagine que toute la force, dont la partie *a* D C tiroit en bas le point C, ſoit ramaſſée au

Fig. 69. PARDIES' principle (1673)

denoting it by T, we may write as follows one of the forms in which he gave the general equations of equilibrium for a plane flexible line:

$$T\frac{dx}{ds} = T_0 - \int_0^s B_x\, ds, \qquad T\frac{dy}{ds} = -\int_0^s B_y\, ds, \tag{14}$$

where B_x and B_y are the components of body force per unit length. While these results were not published until 1744, meanwhile (1716) HERMANN, who had been a pupil of the great JAMES BERNOULLI, had evolved them and had published an explicit definition of the "tenacitas vel firmitas": "The tension or compression of a thread or body at any of its points or at an element of the curve is that force of the thread or body which resists that power or force growing from all the applied powers and tending by pulling the thread in opposite directions to tear it apart. This tension exactly equals or is equipollent to that tearing force resulting from all the powers applied to the body." This definition is not as inclusive as its words suggest, for HERMANN considered nothing more general than a plane perfectly flexible curve.

7. Leibniz, James Bernoulli, and Parent on the Variation of Tension in the Fibres of a Loaded Beam (1684, 1691, 1713)

In an analysis of rupture of a beam by transverse force, LEIBNIZ (1684) *allowed the tensions of the fibres to vary arbitrarily*. He neglected the extensions of the fibres in response to the tensions, and he regarded the bending itself small enough as not to alter the strength. In effect, then, he treated the beam as rigid (Fig. 71). His ideas were expressed more clearly and in greater detail, but without any substantial improvement, by VARIGNON (1702) and BÜLFFINGER (1729).

In his first work on elastic bending (1691), JAMES BERNOULLI replaced the action of all the fibres of a cross-section by that of a single spring on the convex side. In a series of deep and difficult later researches (1694—1705) on the neutral fibre, most of them still unpublished today (Figs. 72 and 73), he sought to take into account the variation of tension simultaneously with bending, but he adopted a wrong hypothesis and went astray. Here he tried in vain to achieve what he had

---

Fig. 70 (p. 207). GOTTFRIED WILHELM LEIBNIZ (1646—1716), after an engraving of 1775 by JOHANN FRIEDRICH BAUSE, following a painting by ANDREAS SCHEITS

Fig. 71 (p. 208). LEIBNIZ' consideration of the tensions in a loaded but undeformed cantilever, from the *Acta Eruditorum* for July, 1684

Fig. 72 (p. 209). Page 308 of JAMES BERNOULLI's notebook, showing his second attempt to determine the neutral fibre (about 1687, not previously published, reproduced here by courtesy of the Handschriften-Abteilung der Universitätsbibliothek Basel)

Fig. 73 (p. 210). Page 352 of JAMES BERNOULLI's notebook, showing his figures for his last attempt to determine the neutral fibre (about 1704, not previously published, reproduced here by courtesy of the Handschriften-Abteilung der Universitätsbibliothek Basel)

G. W. v. LEIBNITZ

bentur, & quidem fibris 1 H 2 H minus resistet trahenti, quam fibra 1 B 2 B, idque in duplicata ratione distantiæ ab A, seu ex duplici capite a distantia sumto. Nam *primo* pondus in C, quo opus esset ad tendendam fibram 1 H 2 H tantundem, quantum fibram 1 B 2 B, foret minus pondere requisito ad tendendam fibram 1 B 2 B, in ratione A H ad A B: verbi gratia si A H sit tertia pars ipsius A B, tunc & pondus in C, quod solam fibram 1 H 2 H ita extendere potest, ut fiat æqualis ipsi 1 B 2 *B*, erit tertia pars ponderis tendentis solam fibram 1 *B* 2 B. Verum nunc *secundo* cum ambæ simul tenduntur a pondere appenso in C, utiq; fibra 1 H 2 H non est tantum tensa quantum fibra 1 B 2 *B*, sed multo minus, idque rursus in ratione AH ad AB. nam si AH sit tertia pars ipsius AB, erit 1 H 2 H tertia pars ipsius 1 B 2 B. Itaque (ex hypothesi alibi confirmata, quod extensiones sint viribus tendentibus proportionales) ad eam ita tendendam tertia tantum ponderis parte opus erit, qua ad eam, tantundem quantum 1 B 2 B, tendendam opus fuisset; id est tertia parte tertiæ partis ponderis ipsam 1 B 2 B tendentis seu parte ejus nona. Itaque generaliter in hac simultanea tensione omnium fibrarum ad quævis puncta existentium, resistentiæ in quolibet puncto erunt in duplicata ratione distantiarum a fulcro imo, seu centro vel axe librationis, sumtarum; id est resistentia in H erit ad resistentiam in B, ut quadratum ipsius AH ad quadratum ipsius AB. Itaque si jam pondus F in *fig 3.* sit corpus parabolicum NRSQN, libere suspensum ex C, in quo altitudo NR sit æqualis basi RS (uti AB æqualis est ipsi AC) & sint ordinatim applicatæ quadratis altitudinum proportionales, seu PQ ad RS, ut quadratum N P ad quadratum N R: tunc posito basin RS repræsentare resistentiam in B, ordinata PQ repræsentabit resistentiam in H; si scilicet altitudines NP, NR, sint altitudinibus respondentibus AH, AB, proportionales; totum vero trilineum parabolicum concavum NRSQN repræsentabit resistentiam totius lineæ AB; si scilicet trabs ABC transversim seu per modum vectis a pondere appenso F deprimatur. At quadratum RNTS huic trilineo parabolico circumscriptum, repræsentaret resistentiam ejusdem lineæ AB directam, si scilicet trabs directe ex pariete esset evellenda, ut in *fig. 2.* Nam quia AB & AC æquales, resistentia puncti B transversa eadem erit quæ directa, nempe repræsentata per RS in *fig. 3.*: jam si directe evellatur trabs (ut in fig. 2.) resistentia omnium punctorum eadem est, ergo resistentia directa

308.

CCL.

Curvatura laminae Elasticae ubi supponitur latus [illegible] partem externam extendi et internam contrahi. vid. Act. Lips. a. 1695. p. 539. vid. constructionem ejus datam infra § CCLXXX.

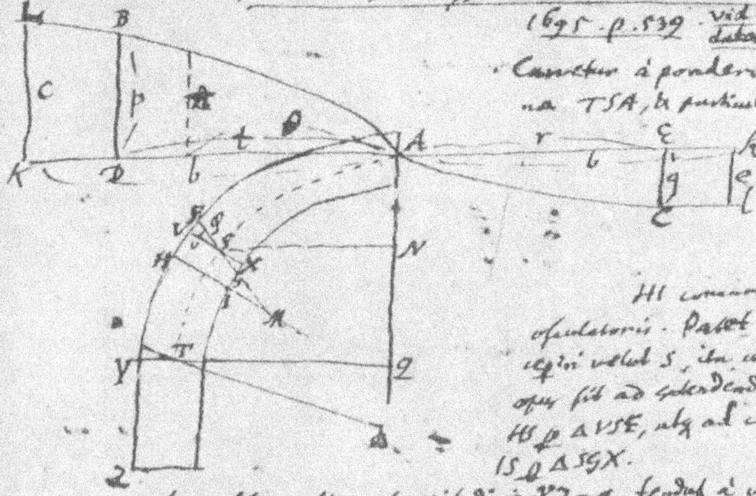

Curvetur a pondere appenso in $\mathfrak{A}$ lamina TSA, et particula convexa HV extendatur usque in F, dum concava IX contrahitur ad G, sint VSX fibrae aegrae FSG, quae productae HI conveniant in M centro circuli osculatoris. Patet in recta VSX punctum aliquod esse velut S, ita comparatum, ut eadem opus sit ad extendendam partem laminae HS per Δ VSF, atque ad comprimendam alteram IS per Δ SGX.

Jam fibra aliqua longitudinis $YZ = a$, tendatur a vi $AK = b$, per rectam $\ddagger$ & a vi θ $KL = c$; & a vi $AD = t$ per rectam $DB = p$. $\ddagger$ eadem vero comprimatur a vi $\ddagger$ per rectam π; $AK = b$ per rectam $kl = e$; & a vi $A\varepsilon = r$ per rectam $\varepsilon C = q$. Ergo fibra HV vel $IX = ds$, tendatur vi $AD = t$, d. per $\frac{pds}{a}$, & comprimatur vi $A\varepsilon = r$ per $\frac{qds}{a}$; unde erit $VF = \frac{pds}{a}$ & $GX = \frac{qds}{a}$ sit $VX = a$, erit $VS = \frac{ap}{p+q}$ & $SX = \frac{aq}{p+q} = g$: Ergo fibra vS $\frac{\pi ds}{a}$: erit $VF . VS :: v\varphi . vS$ unde $\frac{pds}{a} . f :: \frac{\pi ds}{a} . \frac{f\pi}{p}$ ejusque differentialis $\frac{fd\pi}{p}$, quae ducta in vim in distant. $S = \frac{f\pi}{p}$ ductum facit momentum tensionis, respectu puncti S $= \frac{f\pi \theta d\pi}{fp}$; quare vis qua tenduntur omnes fibrae $= \frac{ff}{pp}\int \theta \pi d\pi = \frac{aa}{app+q}\int \theta \pi d\pi = \frac{aa}{ap+q}\int tpdp$. Similiter vis qua comprimuntur omnes est $\frac{aa}{ap+q}\int rqdq$, quare

1\. $\int tpdp = \int rqdq$; unde cum t detur per p, & r per q ob datas curvas AB, AC, dabitur $\int tpdp$ per p, & $\int rqdq$ per q; adeoque etiam relatio ipsius q ad p innotescit.

2\. $\frac{aa}{ap+q}\int tpdp + \frac{aa}{ap+q}\int rqdq$ seu $\frac{2aa}{ap+q}\int tpdp$ (vis omnis qua tendit & comprimit particulae laminae HK) $=$ ponderi appenso $\mathfrak{A}$ in distantiam NS seu $x = a\mathfrak{A}x$ adeoque $\frac{2a}{\mathfrak{A}p+q}\int tpdp = x$

3\. $VF . vS :: HV . SM$ unde $\frac{pds}{a} . \frac{ap}{p+q} :: ds . \frac{aa}{p+q}$ ... unde $\overline{p+q} . SM = aa$ ... $\frac{\overline{p+q}\, dpds}{ddy} = aa$ ... $\overline{p+q}\, dpds = aaddy$ ... differentiale $dy = \frac{dx \int \overline{p+q}\, dx}{aa}$

NB. Convenit haec solutio cum illa in Actis Erud. ..., ubi curvae AB, AC ponuntur similes, adeo ut sit $at = p^{m}$ & $br = q^{m}$.

CCLXXX.

Emendatio Constructionis Curvae Elasticae supra § CCL. allatae p. & schediasmatis Leibnitiani de resist. solidorum correctio.

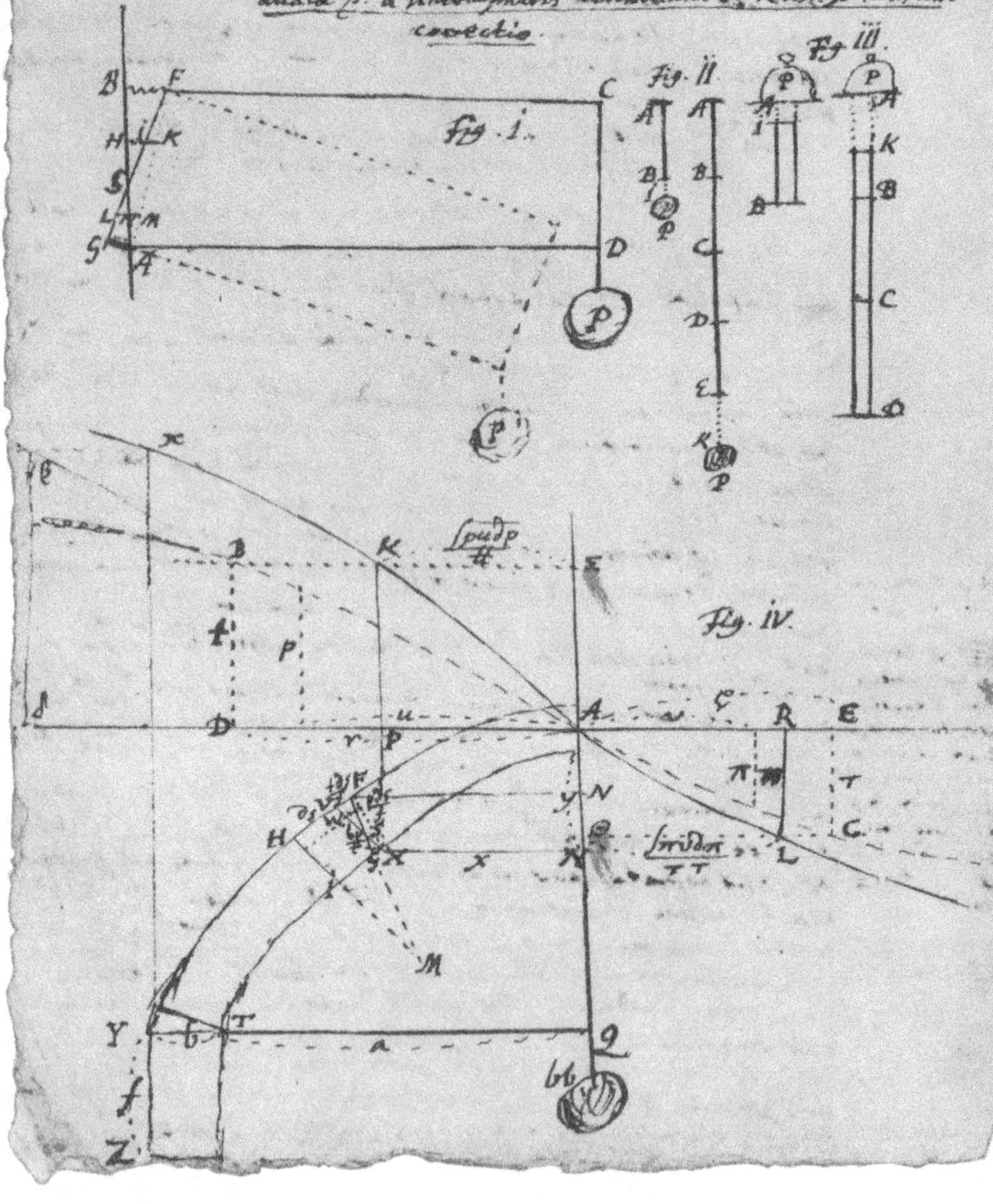

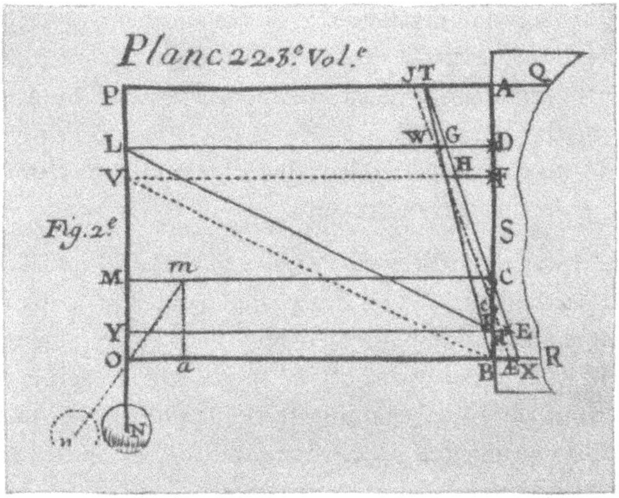

Fig. 74. PARENT's diagram indicating shear force in a loaded beam

done with entire success for perfectly flexible lines, namely, to express the principles of mechanics in general, without use of any constitutive equation defining a particular kind of material.

The work of JAMES BERNOULLI was taken up by PARENT (1713), who was the first to consider the equilibrium of interior forces in addition to the equilibrium of moments in the analysis of a loaded beam. Allowing different moduli for linearly varying normal forces, and representing them on a diagram, he showed that the area of the triangle of compression must equal the area of the triangle of tension (Fig. 74). In a single sentence, he remarked also that forces tangent to the cross-section, *i.e. shear stresses,* are necessary for equilibrium. PARENT's excellent work, which really foreshadows the stress principle, was published obscurely, drew no interest, and had no influence.

8. Mathematical Generalizations of Archimedes' and Stevin's Hydrostatics before the General Equations (1687—1749)

Brilliant as were its successes, the early mathematical hydrostatics developed without a clear or central concept of internal pressure. The modern student easily sees how such a development was possible, unnatural as it seems to him, since he thinks at once of the partial differential equations of hydrostatics and their general solution, so he can understand that one or another property of that general solution can have been painfully discovered, guessed, or even proved by early theorists before the differential equations themselves had been established and made the basis of the whole science. Here, not attempting to summarize the acievements, I do no more than collect the hints and conjectures that to some extent foreshadow the concept of pressure.

Newton (1687) was the first to *isolate in imagination a part of the interior fluid* and to assert that it is "equally pressed on every side", and that "any internal part of a fluid is in the same state as a submerged body." However, he referred only to fluid inclosed within an unmoved spherical vessel which itself is pressed equally on every side, and he published nothing that comes any closer to the fact that in *any* fluid in equilibrium subject to *any* field of force, the pressures on all elements at a point are normal and equal. Newton's treatments of special problems suggest rather that he did not see this generality, since they rest on ingenious conjectures making no use of the concept of pressure. For example, to determine the figure of a spinning fluid earth, he considered two perpendicular canals, one axial and the other in the equatorial plane, bored straight down to the earth's center, and he asserted that the columnar masses of fluid in these two canals had the same weight (Fig. 75). Likewise, when Newton turned to the motion of fluids, where he was indeed the first person ever to present a mathematical proof of anything, so as to obtain Torricelli's rule for the speed at which water spurts from a hole at the bottom of a vessel he had recourse to a model in which a cataract of ice, slipping past an unmoved funnel of water, melts as it comes to the bottom (Fig. 44, p. 148). This original and false conjecture about the motion of water, which seems to extrapolate Stevin's principle of solidification, makes no use of the concept of pressure.

Let it not be thought that Newton, like some modern physicists, preferred vague formalism and disconnected guessing to a sound mathematical theory. Centuries had still to pass before "physical intuition" was invented[7]. Newton

[7] It is not mentioned in the article "Intuition" in the *Oxford Dictionary* (1900) or in the supplement (1933). The nearest approach is sense 6, "direct or immediate insight", introduced by a quotation of 1762 in regard to a painter and supported by Froude on Caesar: "Rashness if it fails is madness, and if it succeeds is the intuition of genius." This cannot be the modern sense, since physicists who fail, however rashly, are not commonly regarded as mad.

Fig. 75. Newton's treatment of the figure of the earth in the *Principia*

paulò altior esset sub æquatore quàm ad polos, Maria ad polos subsiderent, & juxta æquatorem ascendendo, ibi omnia inundarent.

Prop. XIX. Prob. II.

Invenire proportionem axis Planetæ ad diametros eidem perpendiculares.

Ad hujus Problematis solutionem requiritur computatio multiplex, quæ facilius exemplis quàm præceptis addiscitur. Inito igitur calculo invenio, per Prop. IV. Lib. I. quod vis centrifuga partium Terræ sub æquatore, ex motu diurno oriunda, sit ad vim gravitatis ut 1 ad $290\frac{4}{5}$. Unde si *APBQ* figuram Terræ designet revolutione Ellipseos circa axem minorem *PQ* genitam; sitque *ACQqca* canalis aquæ plena, à polo *Qq* ad centrum *Cc*, & inde ad æquatorem *Aa* pergens: debebit pondus aquæ in canalis crure *ACca* esse ad pondus aquæ in crure altero *QCcq* ut 291 ad 290, eò quòd vis centrifuga ex circulari motu orta partem unam è ponderis partibus 291 sustinebit & detrahet, & pondus 290 in altero crure sustinebit partes reliquas. Porrò (ex Propositionis XCI. Corollario secundo, Lib. I.) computationem ineundo, invenio quod si Terra constaret ex uniformi materia, motuque omni privaretur, & esset ejus axis *PQ* ad diametrum *AB* ut 100 ad 101: gravitas in loco *Q* in Terram, foret ad gravitatem in eodem loco *Q* in sphæram centro *C* radio *PC* vel *QC* descriptam, ut $126\frac{2}{15}$ ad $125\frac{2}{5}$. Et eodem argumento gravitas in loco *A* in Sphæroidem, convolutione Ellipseos *APBQ* circa axem *AB* descriptam, est ad gravitatem in eodem loco *A* in Sphæram centro *C* radio *AC* descriptam, ut $125\frac{2}{15}$ ad $126\frac{2}{15}$. Est autem gravitas in loco *A* in Terram, media proportionalis inter gravitates in dictam Sphæroidem & Sphæram, propterea quod

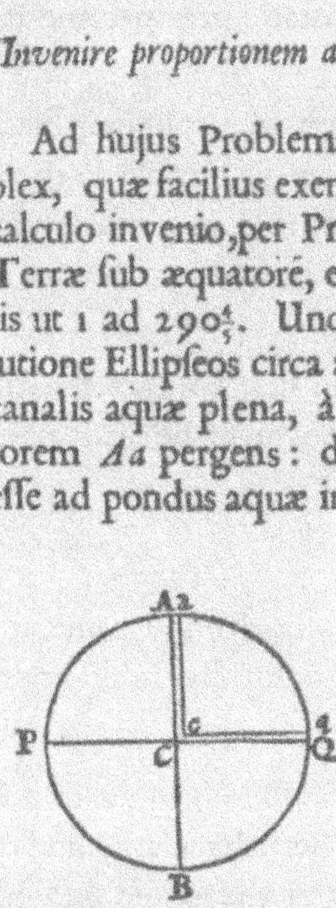

Sphæ-

is the rational physicist *par excellence*. When he turns to guessing and fumbling, it is only from necessity, for want of concepts and principles on which to base a rational theory. Indeed, in the preface to the *Principia*, after vaunting that he had found the system of the world by mathematical proof from a set of mechanical axioms, he wrote, "would that it were possible to derive the rest of the phenomena of nature from mechanical principles by the same kind of argument." Thus NEWTON's failure to obtain clear and correct solutions to the problems of fluid mechanics he brilliantly attacked, shows with certainty that he was *not* the master of concepts and principles of hydrostatics and hydrodynamics — in particular, the two essentials: the concept of hydrostatic pressure and the general principle of linear momentum.

In an attempt to describe the relative equilibrium of a spinning mass of fluid, HUYGENS (1690) asserted that the plumb line was always normal to the surface of the sea. When an early author established a proposition in general hydrostatics, he did so by showing it to follow from NEWTON's principle of columns, HUYGENS' principle of the plumb line, or both.

TAYLOR, who was NEWTON's disciple, was the first to give a mathematical theory concerning compressible fluids (1715). The postulate he really used, though he did not state it, is equivalent to

$$dF \propto \sigma A\, dh, \tag{15}$$

where F is the force which would be exerted on a horizontal plane at height h. Thus TAYLOR regarded STEVIN's formula (12) as true locally, in differential form, even for fluids of varying specific gravity σ.

EULER's *Scientia Navalis* (completed 1738, published 1749) concerns only incompressible fluids near the surface of a flat earth. With his usual clarity and directness, EULER states on the first page the principle that forms the axiom on which all his hydrostatic theorems rest. Namely, *the pressures in all directions at a given point are equal and are normal to the surface elements on which they act.* Lemma 1 is equivalent to

$$\boldsymbol{F} = -\sigma \int_S h\boldsymbol{n}\, dS, \tag{16}$$

where $\boldsymbol{F}$ is the resultant force on the surface S and where h is the depth at the point where the unit outer normal to S is $\boldsymbol{n}$. Thus, as always, EULER goes straight to the heart of his subject and states openly the assumption he makes. This is just enough for the class of problems he wishes to treat. In a single equation, he exhausts STEVIN's hydrostatics but does not go beyond it. In particular, EULER's postulate here is not strong enough to cover the problems attacked by NEWTON, HUYGENS, and TAYLOR.

In 1740 MACLAURIN proved his splendid hydrostatic theorem: Every oblate spheroid is a figure of equilibrium. We now know that this theorem, along with its supplement asserting that no prolate spheroid is a figure of equilibrium, is true according to the general equations, but these lay more than ten years in the future when MACLAURIN wrote.

CLAIRAUT's brilliant *Figure de la Terre* (1743) rests on the axiom that the external and mutual forces acting along any canal within a figure of equilibrium and terminating in the surface must equilibrate each other (Fig. 76). This axiom includes as special cases NEWTON's principle of columns and HUYGENS' principle of the plumb-line. While its consideration of imagined divisions of the fluid into parts suggests the cut principle, CLAIRAUT does not make any hypothesis about the action of fluid on fluid. His interest is in conditions on the body force, and he shows that in order for equilibrium to be possible, the differential of the body force must be exact. CLAIRAUT's book, like NEWTON's, is a great triumph of mechanics; in particular, it presents for the first time a general vector field, a careful distinction between force and weight, the concept of potential function, and the general theory of level surfaces, but, again like NEWTON's, it does not attain the concept of internal pressure; *a fortiori*, it does not contain the general differential equations of hydrostatics.

9. Influences of Constitutive Properties on the General Concept of Stress in the Work of James Bernoulli and Euler on the Bending of Bars (1705, 1727, 1779)

The traction $\boldsymbol{t}_{\partial \mathscr{V}}$ or $\boldsymbol{t}_{\boldsymbol{n}}$ in Equations (1)—(5) is not a force but rather a surface density of force, *i.e.* a force per unit area. For a proper general theory of stress, the distinction is of the essence, but in primitive special cases it did not matter. As we have seen in § 5, GALILEO was the first to state a property of a material in terms of force per unit area rather than force. In most of the other early work, this distinction was not seen. For example, in HOOKE's crude experiments on elasticity, the constant

Fig. 76a, b (pp. 216 and 217). CLAIRAUT's principle of equilibrated canals (1743)

THEORIE DE LA FIGURE DE LA TERRE.

PREMIERE PARTIE.

Principes généraux pour trouver les hypotheses dans lesquelles les Fluides peuvent être en équilibre, & pour déterminer la figure de la Terre & des autres Planetes, lorsque la Loi de la pesanteur est donnée.

CHAPITRE I.

Exposition d'un Principe général dont l'observation est nécessaire pour l'équilibre des Fluides, avec les propositions préliminaires pour faire usage de ce principe.

§. I.

UNE *masse de Fluide ne sçauroit être en équilibre, que les efforts de toutes les parties qui sont comprises dans un canal de figure quelconque qu'on imagine traverser la masse entiere, ne se détruisent mutuellement.*

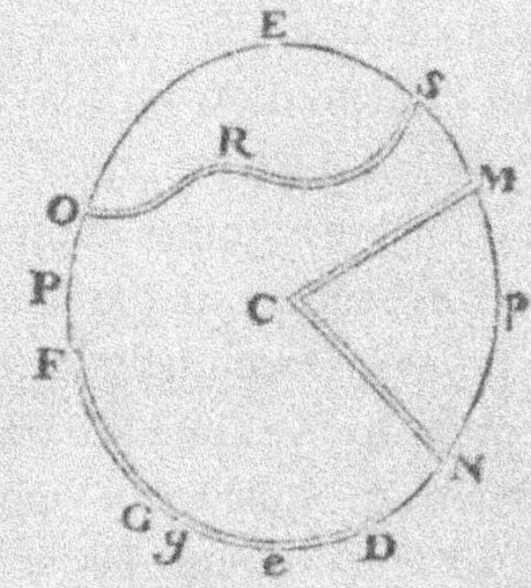

Puisque la masse entiere *PEpe* est supposée en équilibre, une partie quelconque du Fluide pourroit devenir solide, sans que le reste changeât de situation. Supposons que toute la masse se durcisse, excepté ce qu'il faut de Fluide pour former le Canal *ORS*, ce Canal sera donc en équilibre ; or cela ne peut arriver que les efforts de *OR* pour sortir vers *S*, ne soient égaux à ceux de *SR* pour sortir vers *O*.

of proportionality in "HOOKE's law" is a force, not a stress, and is of course different for two bars of different shape cut from the same piece of iron. The next after GALILEO to see the light here was JAMES BERNOULLI, who in his great, final paper of 1705 asserted that the elastic law for the fibres of a beam should state the fractional extension as a function of the force divided by the area. Perhaps since it contained errors on other matters, the ideas in this paper were not mastered by those who followed, even PARENT. Since, as in the work of LEIBNIZ on inextensible and inflexible beams, the shape of the cross-section has an explicit influence on the modulus, no definite answer can come out of the analysis. The first point where it becomes *mathematically necessary* to use a general equation for the *material* rather than equations for different bodies occurs when we derive the BERNOULLI law of bending from a law of extension for the fibres so as to obtain also the now well known result $\mathcal{B}=EI$, where E is "YOUNG's modulus" of the material and I is the moment of inertia of the cross-section about the axis normal to the plane of bending at the neutral line. The quantity I had occurred in the work of LEIBNIZ on essentially inextensible and inflexible bars in 1687, mentioned above. In order to integrate the tractions over the cross-section, as LEIBNIZ did, one has to have an explicit rule for the tension per unit area, not just the resultant tension in a bar. Although JAMES BERNOULLI tried his life long to get a theory of bending based on a law of extension for the fibres, he never succeeded, despite his having gathered sufficient concepts and principles, and it was left to EULER to achieve the proof in the course of his first attempt on the theory of elasticity, done in his twentieth year, when he was a student in Basel, but not published until 1862. This paper uses explicitly a one-dimensional stress-strain relation of the type proposed twenty years earlier by JAMES BERNOULLI, but linear, so what is now called "YOUNG's modulus" must be introduced. EULER did not come back to the matter again until 1779, at the very end of his life. In his last papers on the vibrations of bars, published in 1782, he presented and explained the definition of "YOUNG's modulus[8]" and the scaling laws for rods based upon its use.

[8] It should be unnecessary to remark, although unfortunately it is, that YOUNG himself did not introduce this modulus or recognize a law of elasticity as expressing a property of a material rather than of a body of that material. The passages concerning elasticity and vibration in YOUNG's lectures of 1807 consist in repetition of the easier parts of researches by the BERNOULLIS and EULER, with equations translated into words, and with proofs, often inadequate, phrased in terms of fluxions and without references.

In all this work, of course, the vectorial character of the traction, while obvious enough, is not brought out into the open. In the mean time, moreover, researches on fluids, which we shall describe next, had been forced to take account of the distinction between a force and a surface density of force and had also made the direction of that force density explicit if unduly special.

10. The Hydraulics of Daniel Bernoulli, John Bernoulli, and Euler (1730–1752)

In 1730 DANIEL BERNOULLI succeeded in relating the speed of a steady flow of incompressible fluid in a tube to the pressure it exerts upon its walls. While today we know his result in the form

$$p+\tfrac{1}{2}\varrho v^2+\sigma h=\text{const.}, \tag{17}$$

he does not so express it. Indeed, assuming that $h=$ const., he writes it in the form (Figs. 77, 78)

$$v\frac{dv}{dx}=\frac{a-v^2}{2c}, \tag{18}$$

where the units are so chosen that the acceleration of gravity is $\frac{1}{2}$, where $\sqrt{a}$ is a reference velocity, and where dv is the impulsive increment of speed that water flowing with the speed v would experience in traversing the distance dx if the tube, supposed horizontal, were suddenly to dissolve into thin air at the point x. To obtain this result, BERNOULLI appeals to ideas connected with the conservation of energy. Thus (18) is not a differential equation; rather, $cv\frac{dv}{dx}$ is itself the accelerating force, or, as we should say today, $cv\frac{dv}{dx}$ is the internal pressure per unit density. BERNOULLI, however, does not say this at all, and it is only by intro-

---

Fig. 77 (pp. 220 and 221). The last two pages of DANIEL BERNOULLI's letter of 17 July 1730 to GOLDBACH, giving the earliest statement of "Bernoulli's Theorem" (from the Central State Archives of Ancient Acts, Moscow, by courtesy of Dr. G. K. MIKHAILOV). Some letters are lost in the binding of p. 94 at the left and p. 94v. at the right. The passage begins on the previous page: "Pour moi je me suis entierement plongé dans les eaux, qui font mon unique occupation, et renonce depuis quelque tems à tout ce qui n'est pas hydrostatique ou hydraulique . . . Pour en faire le"

Fig. 78 (p. 222). "BERNOULLI's Theorem" as published in the *Hydrodynamica*, 1738

Fig. 79 (p. 223). Plate XI from DANIEL BERNOULLI's *Hydrodynamica*

ommencement, j'aurai l'honneur de vous dire, que j'ai fait ces jours
assés quelque nouvelle decouverte, qui peut etre d'un grand usage
our la structure des conduits des eaux, mais qui sur tout aportera
n nouveau jour à la physiologie; c'est d'avoir trouvé la statique
s eaux courantes, science que personne n'a considerée avant moi
utant que je sache et que, quand meme on l'auroit voulu entreprendre
roit sans doute demeurée fort imparfaite, puisqu'on ne s'étoit pas encor
i de ces principes dont je me sers et qu'un grand nombre d'experiences
a fait trouver tres justes. Il s'agit ici de trouver l'effort des eaux,
i sont poussées avec une force quelconque par un tuyau quelconque.
ici un cas le plus simple.

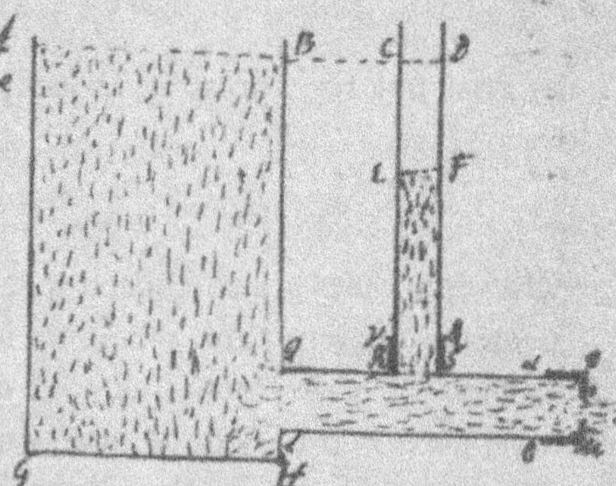

t ACHD un vaisseau avec un tuyau cylindrique
izontal LMPQ, dont le fond PM a une ouverture
; à ce tuyau est ajouté un autre tuyau
ical SDCR: si on bouche l'ouverture on et
remplisse le tout d'eau jusqu'en AB, on
, que l'eau se mettra de niveau dans le
au vertical, jusqu'en CD: mais si l'on
bouche l'ouverture on, de sorte que
commence à couler à travers le tuyau
izontal d'une vitesse plus ou moins grande selon que l'ouverture on est
s ou moins grande, je dis que la surface d'eau dans le tuyau vertical
scendra comme jusqu'en EF, quoique la surface AB ne descende pas,
it que le vaisseau soit fort grand, soit qu'on y verse continuellement autant
eau qu'il en ecoule. Cela etant, il est clair que l'effort d'une eau
urante est moindre, que de l'eau dormante, et que ces efforts sont
recisement comme SF à SD: Il etoit donc necessaire, que l'on determinat
u juste la hauteur de SF; et voici ce que j'ai trouvé par ma nouvelle
heorie. Soit le cercle dont le rayon est QL ou PM au cercle de l'ouverture
nt le rayon est on comme n à 1. Soit la hauteur SD = a, je dis, qu'on
trouvera toujours $SF = \frac{nna - a}{nn}$; de sorte que s'il n'y avoit point de fond
c'est à dire si tout le tuyau etoit percé, toute la surface descendroit jusqu'en RS.

946.

marque que dans ce cas le tuyau horizontal ne souffre aucun effort ce qui donne de nouvelles regles pour la structure des tuyaux qui conduisent l'eau, et ce que Mr. Bulffinger ne pouvoit s'imaginer jusques à ce que je lui en fis l'experience en presence des autres Academiciens. Je faisois les experiences moyennant un cylindre tres poli de fer que je me fis faire avec differens couvercles, qui avoit des trous de differente grandeur tel qu'etant d… au milieu du cylindre etoit soudé un petit bout de tuyau VRSQ propre à embrasser un tuyau de verre CRSD: toutes les experiences ont parfaitement reussi.

Mais je vois qu'il m'arrive la meme chose qu'à ceux qui ont eu … et qui veulent le conter à tout le monde. Pardonnez moi, Monsieur, cette digression ou plutot cette indiscretion. Quant au probleme de changer $\frac{dz}{1+zz}$ en $\frac{n\,dv}{1+vv}$, je vous prie de ne les pas croire ces circonstances, qui m'ont voulu vous persuader, qu'il m'auroit eté difficile de l'eau que pendant mes … itations: comme tout le mystere ne depend, que des logarithmes, dont je savois parfaitement bien le calcul alors, je crois qu'il m'auroit aussi eté egalement facile mais je comprens bien quelles sont ces circonstances; c'est que vous croyez, que j'aye employé les series recurrentes et veritablement on peut s'en servir, mais on peut aussi s'en passer: marque de cela c'est que pour transformer par ex. $\frac{dz}{1-zz}$ en $\frac{n\,dx}{1-xx}$, il faut poser $z=\frac{(1+x)^n-(1-x)^n}{(1+x)^n+(1-x)^n}$, qui n'a point de ressemblance avec les termes generaux des suites recurrentes. Quant au probleme de changer $\frac{dz}{1+zz}$ en $\frac{nv^p\,dv}{1+vv}$, on ne sauroit donner des substitutions generales, mais on peut bien donner des substitutions generales pour chaque valeur particuliere de p. ainsi si $p=1$, on changera $\frac{dz}{1+zz}$ en $\frac{nv\,dv}{1+vv}$ par la substitution de $z=\frac{1-(1+vv)^n\sqrt{-1}}{(1+vv)^n-\sqrt{-1}}$. pour votre cas particulier de $n=-\sqrt{-1}$ et $p=-1$, il faut poser $z=\frac{\sqrt{-1}}{1+2vv}$. On parvient à ces substitutions sans aucune inconnue et en ne travaillant que synthetiquement, et chaque substitution à trouver ne coute qu'une minute de tems. J'aurai l'honneur de vous entretenir sur ces sortes de problemes, quand j'aurai le plaisir de vous revoir. j'espere que ce sera dans peu. Je ne manqueray pas d'aporter tout le soin à fin que votre dissertation soit bien imprimée. Je suis avec toute sorte de consideration et d'estime

Monsieur

Votre tres humble et
tres obeissant serviteur
Daniel Bernoulli.

R. à Moscou le 23. Juill. 1730.

laterum proportionalem esse accelerationi seu incremento velocitatis, quod aqua sit acceptura, si in instanti omne obstaculum motus evanescat, sic ut immediate in aërem ejiciatur.

Res igitur jam eo perducta est, ut si durante fluxu aquæ per o, tubus E D in temporis puncto abrumpatur in cd, quæratur quantam accelerationem guttula $acbd$ inde sit perceptura : tantam enim pressionem sentiet particula ac in lateribus tubi sumta à præterfluente aqua : Hunc in finem considerandum est vas ABEcdC, atque pro eo invenienda acceleratio particulæ aqueæ effluxui proximæ, si hæc habuerit velocitatem $\frac{\sqrt{a}}{n}$: Istud negotium fecimus generalissime in *paragrapho tertio sect. V.* Attamen quia in hoc casu particulari brevis est calculus, motum in vase decurtato ABEcdC hic iterum calculo subducemus.

Sit velocitas aquæ in tubo Ed, quæ nunc ut variabilis consideranda est, $= v$: amplitudo tubi ut antea $= n$, longitudo E$c = c$: indicetur longitudo ac particulæ aqueæ infinite parvæ & effluxui proxime per dx : Erit guttula æqualis in E tubum ingressura eodem temporis puncto quo altera $acdb$ ejicitur : dum autem guttula in E, cujus massa $= ndx$, tubum ingreditur acquirit velocitatem v, atque *vim vivam* $nvvdx$, quæ *vis viva* tota fuit de novo generata ; nullum enim, ob amplitudinem vasis A E infinitam, motum guttula in E habuit tubum nondum ingressa : huic *vi vivæ* $nvvdx$ addendum est incrementum *vis vivæ*, quod aqua in Eb accipit, dum guttula ad effluit, nempe $2ncvdv$: aggregatum debetur *descensui actuali* guttulæ ndx per altitudinem BE seu a: habetur igitur $nvvdx + 2ncvdv = nadx$ sive $\frac{vdv}{dx} = \frac{a-vv}{2c}$.

In omni autem motu est incrementum velocitatis dv proportionale pressioni ductæ in tempusculum quod hic est $\frac{dx}{v}$: igitur in nostro casu est pressio, quam guttula ad patitur, proportionalis quantitati $\frac{vdv}{dx}$, id est, quantitati $\frac{a-vv}{2c}$.

Est vero in eo temporis puncto, quo tubus abrumpitur, $v = \frac{\sqrt{a}}{n}$ vel $vv = \frac{a}{nn}$, hic igitur valor substituendus est in expressione $\frac{a-vv}{2c}$, quæ sic

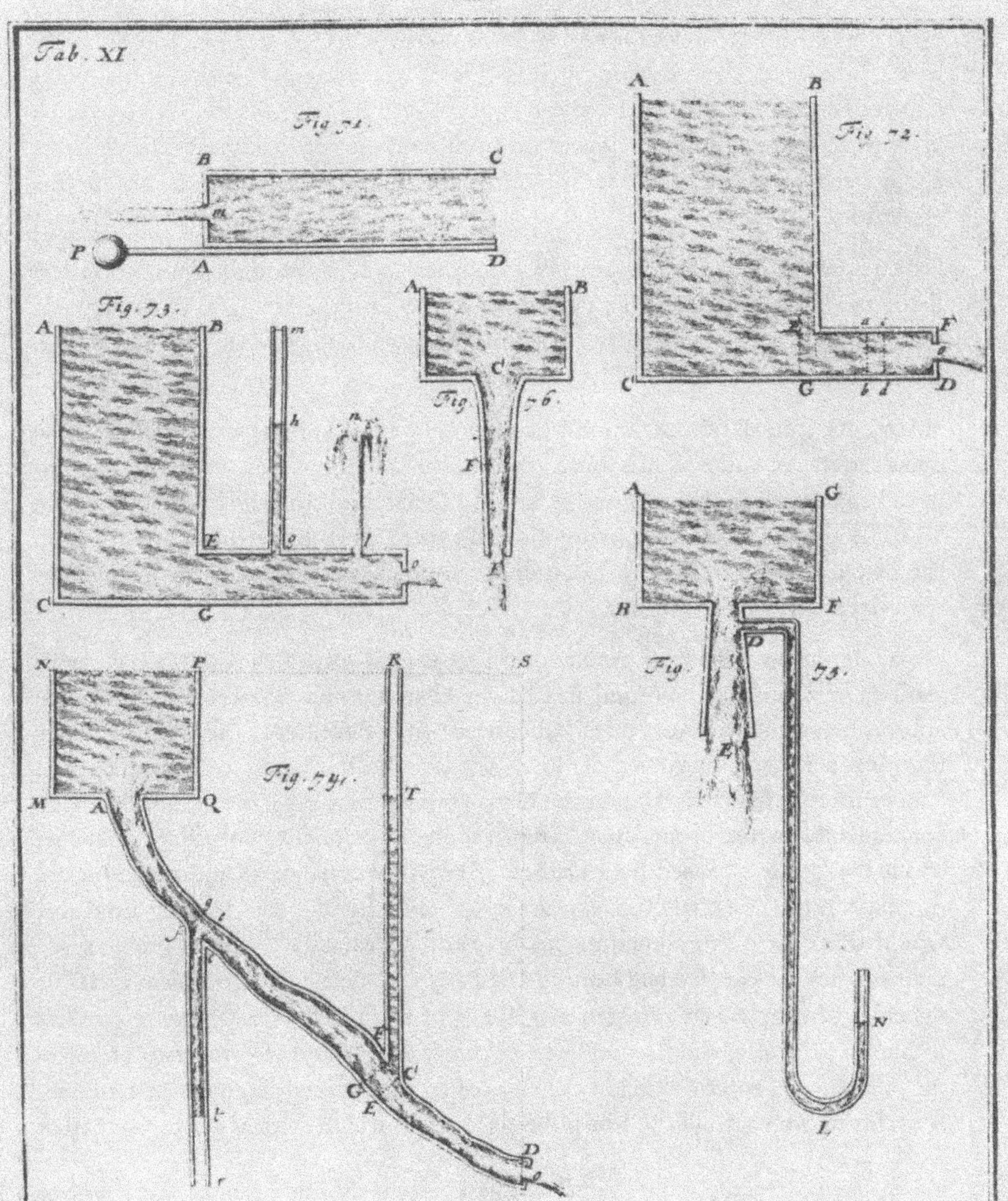
Tab. XI
Fig. 71
Fig. 72
Fig. 73
Fig. 74
Fig. 75
Fig. 76

ducing ideas not used by him that we can couple his result to what is today called "BERNOULLI's theorem". Indeed, BERNOULLI by a further argument identifies $v\frac{dv}{dx}$ with the pressure on the wall of the tube when the tube exerts sufficient force to keep the water from bursting it. The line of reasoning here makes it plain that DANIEL BERNOULLI did not have the concept of internal pressure. Of course, he knew that water exerted pressures on itself, and he made some vague remarks about such pressures, but they do not enter his argument, his result, or his applications of it.

The success of the *Hydrodynamica*, published in 1738, brought upon its author the envy of two of the leading geometers of the day: the young D'ALEMBERT, of whom we shall hear more presently, and DANIEL BERNOULLI's own old father, the mighty and pugnacious JOHN BERNOULLI.

In 1743 JOHN BERNOULLI published his "Hydraulica, nunc primum detecta ac demonstrata ex fundamentis pure mechanicis". Although he is known not to have begun writing the manuscript before 1738 and not to have finished it before 1740, he attempted to appropriate his son's main work by setting upon his own the date 1732, a year before DANIEL BERNOULLI had deposited his manuscript with the Petersburg Academy.

While the problems and results of the treatise of JOHN BERNOULLI are generalizations of what may be found in Chapter 12 of DANIEL BERNOULLI's book, the approach and method are entirely different. JOHN BERNOULLI carefully separates the kinematics and dynamics of flow. To describe the latter, he introduces the "inner force", which acts upon the cross-sections of fluid in the vessel. While he concealed the nature of this inner force in some intentionally confusing statements about the "eddy" caused by a change of cross-sectional area, and while he did not explain the essentials for fear, as he wrote to EULER, the "scurvy English" would think he owed something to NEWTON's "cataract", the progress of his mathematics is clear. He had been the first hero of the catenary problem in 1690; his elder brother's later researches, while kept from him in detail, were familiar to him in principle, and one thing he certainly knew how to do was to calculate the difference of forces acting at the ends of an infinitesimal segment of a curve[9]. With his lightning ability to grasp the right trick, which even as a septuagenarian

[9] In another late work, published at the same time, he was the first author to obtain a set of differential equations as the complete statement of the motion of a deformable system, namely, the compound pendulum.

he had not lost, he at once applied to fluid in a tube the simplest method of treating the catenary, now making the assumption that there is a hydraulic pressure playing the part of the tension in the flexible line. Thus old JOHN BERNOULLI was the first to give the "Bernoulli equation" in essentially its modern form (Fig. 80):

$$p = \frac{\alpha^2 h^2 - \beta^2 \omega^2}{2\alpha^2 h} v^2 + \frac{h \omega v \, dv}{dx} \int \frac{ds^2}{y \, dt} . \tag{19}$$

It is not necessary to explain all the symbols. We see at once that the pressure is there, along with adjustments for unsteadiness of flow and for change of cross-section and height.

EULER enthusiastically took up JOHN BERNOULLI's idea and made it the basis of his clear and general presentation of hydraulics in a sequence of works on engineering problems about pumps, water wheels, thermal convection, *etc.* (1749 to 1752). Particularly worth notice here is EULER's project and design of a turbine, which did not attract any attention from the "practical men" of his own day or later until it was finally built by ACKERET in 1944 and found to perform nearly as well as the best modern machines of similar capacity and head. In this work, not only did the "Coriolis force" make its first appearance, but also the true *hydraulic pressure p* assumed from this moment its central position in fluid mechanics, but still only for incompressible fluids in tubes, leading to a crude one-dimensional theory.

11. The first Attempts at Hydrodynamics by Euler and D'Alembert (1745—1749)

While there are a few words about a velocity field at one point in DANIEL BERNOULLI's book, all the specific early work was limited by the "hypothesis of parallel sections", according to which all fluid particles on a cross-section of a tube have the same velocity, so that there is a single independent variable of position. The first to divide up the flow past an obstacle into stream fillets was EULER in his translation (1745) of ROBINS' *New Principles of Gunnery* (Fig. 81). By an appeal to the balance of momentum at infinity, EULER gave a correct and fairly general proof that a steady flow of fluid past an obstacle exerts no force on that obstacle. Since this somewhat startling fact was rediscovered or appropriated some years later by D'ALEMBERT, who proved it by a misleading and unnecessary appeal to symmetry, it is called "the D'ALEMBERT paradox". That this early work of

V I.

Hæ duæ vires, *hydrostatica* & *hydraulica*, componunt vim totalem, quæ nimirum generatur ab actione vis primitivæ p, quæ inventa est Art. III $= gha$. Æquando itaque hanc, cum aggregato illarum duarum Art. IV & V inventarum, obtinebimus æquationem generalissimam, pro determinatione velocitatis quacum liquor quovis momento effluit, quæ æquatio hæc est $\frac{vv(\alpha\alpha hh - \beta\beta\omega\omega)}{2\alpha\alpha h} + \frac{h\omega v dv}{\alpha dx}\int\frac{ds^2}{ydt}, = gha$. Ubi notandum per $\int\frac{ds^2}{ydt}$ intelligi summam omnium $\frac{ds^2}{ydt}$ quæ continentur non tantum inter Cc & Ff, sed omnino inter extremas ab una ad alteram omnes comprehendendo.

V I I.

Quod si lubeat exprimere æquationem per z, seu altitudinem unde corpus naturali gravitate g præditum delabendo acquirat velocitatem quæsitam v; scribendum tantum est, per principia dynamica, $2gz$ pro vv & gdz pro vdv; id quod dabit hanc æquationem $\frac{gz(\alpha\alpha hh - \beta\beta\omega\omega)}{\alpha\alpha h} + \frac{gh\omega dz}{\alpha dx}\int\frac{ds^2}{ydt} = gha$, vel reductione peracta hanc $(\alpha\alpha hh - \beta\beta\omega\omega)zdx + \alpha hh\omega dz \times \int\frac{ds^2}{ydt} = \alpha\alpha hhadx$, vel quia $\int\frac{ds^2}{ydt}$, sumendum per totam axis longitudinem, ut constans, adeoque ut datum supponi potest, saltem per quadraturas, nominetur illud M; eritque æquatio ad hanc formam reducta $(\alpha\alpha hh - \beta\beta\omega\omega)zdx + \alpha Mhh\omega dz = \alpha\alpha hhadx$. Ex resolutione hujus æquationis, invenietur z in quantitatibus datis per x, & constantes $M, a, h, \omega, \alpha, \beta$.

COROLLARIUM I.

Existente effluxus velocitate uniformi, ad quam sensibiliter

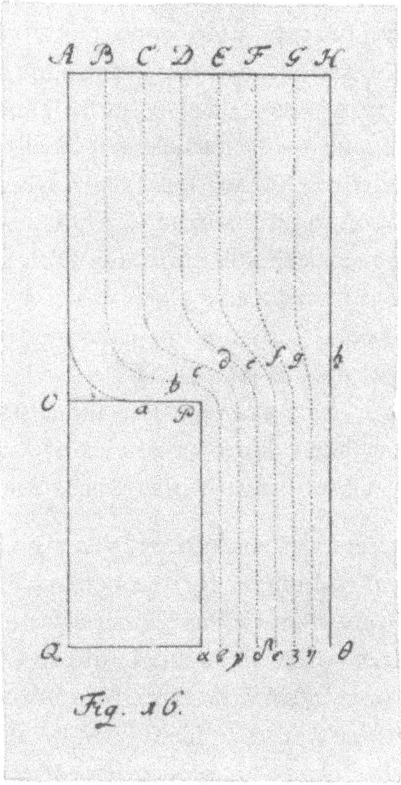

Fig. 81. The earliest drawing of the stream fillets of a steady flow (from EULER's *Artillerie*, 1745)

EULER did not introduce the internal pressure is no surprise, since it precedes his hydraulic researches, described just above.

While D'ALEMBERT's *Traité des Fluides* of 1744 added nothing to the subject, it is a different matter with his *Essay on the general cause of winds*, which won the Berlin prize of 1746. Since D'ALEMBERT assumed that the winds arise from the attractions of the sun and moon, we cannot protest at DANIEL BERNOULLI's disgust and contempt: "When one has read it all, one knows as much about the

---

Fig. 80. JOHN BERNOULLI's generalization of "BERNOULLI's theorem", employing the internal pressure (manuscript completed in 1740, alleged to be a work of 1732, printed in 1743 in Volume 4 of his *Opera*, dated 1742)

winds as before." Tortuous, obscure, ugly, pretentious, wearisome with detours, useless and complicating restrictions, and irrelevant numerical examples to six or eight figures as it creeps from one special or even inconsistent case to an infinitesimally more general one — nevertheless it finally reaches some partial differential equations asserted to govern the axially symmetric flow of a compressible fluid in small motion upon a sphere[10]. After several attempts over the past fifteen years, I have never been able to follow D'Alembert's profuse notations, devious concepts, and interminable algebra sufficiently to decide whether or not his results[11] are correct. Certainly, his contemporaries could not follow him either[12]. The importance of D'Alembert's essay lies in its *attempt* — the first — to get general equations of a class of two-dimensional motions of a continuum. While D'Alembert had a kinematical idea of "accelerating force", he certainly wrote nothing that could suggest the concept of fluid pressure[13].

A great and definite success was attained by D'Alembert in his *Essay on a new theory of the resistance of fluids*, submitted for the Berlin prize of 1750. While this work did as little to explain resistance as his former essay had done to explain the winds, it did achieve correct partial differential equations for axially symmetric and plane flows of the type now called irrotational (Fig. 82), though the hypotheses, both physical and kinematical, from which D'Alembert inferred them are obscure. While D'Alembert claims to reduce hydrodynamics to hydrostatics, which he says has been reduced in turn to "the single principle of experience, the equality of pressure in all directions", yet he nowhere introduces the pressure into the mathematics. In arriving at his equations, he refers to "forces" in the interior but takes these as being reversed accelerations. His dynamical equations

[10] This work seems never to have been analyzed in detail by a recent student. My summary of it on p. LVII of *Euleri Opera Omnia* (II) **12** (1954) is insufficient and partly unjust.

[11] The defining equations of his theory seem to be those numbered (M) through (R) in § 77.

It is worth noting that D'Alembert was the first author ever to derive a partial differential equation as the statement of an entire physical theory, namely, the heavy hanging rope (1743).

[12] *Cf.* Euler's respectful and cautious evaluation, "Recensio dissertationis de ventis," *Opera postuma* **2**, 793—797 (1862).

[13] Of course, the word "pression" occurs frequently, but it means the weight of a portion of fluid.

Fig. 82. The general equations of isochoric axisymmetric irrotational flow, as published in 1752 in D'Alembert's prize essay of 1749

Propositions suivantes, nous ne considérons que de pareils solides engendrés par la révolution de la figure *ADC* (Fig. 13) autour de l'Axe *AC*, & nous n'aurons égard qu'à une seule Section *ADC* par l'Axe, parce que le calcul doit être le même pour toutes les autres.

PROPOS. VIII. THEORÊME.

45. *Les mêmes choses étant posées que dans l'art.* 43, *je dis que* $B' = -A - \frac{p}{z}$ *&* $A' = B$.

Soit *KQM'* (Fig. 17) la Courbe que décrivent les particules du Fluide infiniment proches de la surface *AMN*, & soient menées les ordonnées infiniment proches *PNM'*, *pnm'*; *NR* perpendiculaire à *pm'*, & *NQ* à *AN* ou *QMm'*; il est évident

1°. Que la vitesse du Fluide en *N* suivant *Nm*, est en raison renversée de la surface conique décrite par la révolution de *NQ* autour de *FP*. Donc la vitesse en *N* est comme $\frac{1}{NQ \times PN}$. Donc si on appelle *U* la vitesse suivant *Nm*, & qu'on fasse $PN = z$ ou y, (*) on aura $NQ = \frac{\alpha^2 a}{Uz}$, *a* étant la vitesse en *g*, & α une constante, pour garder la loi des homogenes.

(*) Le point *N* peut être regardé, ou comme étant sur la surface du corps, ou comme appartenant en général à une ligne

are the appropriate special cases of the condition

$$\boldsymbol{v}(\boldsymbol{x}, t) \cdot d\boldsymbol{x} \quad \text{is exact} \tag{20}$$

where $\boldsymbol{v}$ is the velocity field. Thus the pressure does not appear in his results.

12. Euler's Hydrodynamics (1750—1766)

EULER as director of the Berlin Academy had to read both of D'ALEMBERT's prize essays. We may well conjecture that they gave him the final impulse to the creation of his general hydrostatics and hydrodynamics. A modern reader sees at once that only two elements, beyond the calculus of partial derivatives, were still lacking in order to get at one blow general dynamical equations of the type for which D'ALEMBERT struggled in vain over many pages. These were, the general principle of linear momentum, and the general concept of internal pressure.

Both these elements EULER quickly perfected. For many years past, he had been approaching the general principle of linear momentum, which we now call *Euler's first law of motion*: For any portion $\mathscr{V}$ of any body, the resultant force $\boldsymbol{F}(\mathscr{V})$ equals the time-rate of change of the linear momentum $\boldsymbol{M}(\mathscr{V})$

$$\boldsymbol{F}(\mathscr{V}) = \dot{\boldsymbol{M}}(\mathscr{V}). \tag{21}$$

This principle, in differential form, he finally stated in 1750 as being the one and sufficient general law of mechanics. He was led to it through special cases in celestial mechanics, in the theory of flexible lines and linked bars, in rigid dynamics, and in the theory of sound, as well as well as in hydraulics.

The concept of internal pressure he already had in hydraulics. All that was left was to see it in space and to combine it with the momentum principle. This EULER did, almost at the first attempt. To explain the internal pressure, EULER in 1750 speaks of an element of water as being "subject to the pressures of the particles of water surrounding it"; in his next treatment (1752), of "the pressures with which the particles of water everywhere actually act upon each other." The result, of course, is EULER's dynamical equation (11), in which, for the first time, the pressure is a field varying in space and time, and the only way in which it falls short of CAUCHY's traction is that the force of the pressure is assumed to

---

Fig. 83. EULER's fundamental equations of hydrodynamics as first published, from the *Mémoires de l'Académie des Sciences*, Berlin, for 1755

or il eſt aiſé de montrer que $\left(\frac{dds}{dx\,dy}\right) = \left(\frac{dds}{dy\,dx}\right)$, d'où la vérité de cette propriété devient évidente.

XV. Maintenant je poſe pour abréger :

$$X = \left(\frac{du}{dt}\right) + u\left(\frac{du}{dx}\right) + v\left(\frac{du}{dy}\right) + w\left(\frac{dw}{dz}\right)$$

$$Y = \left(\frac{dv}{dt}\right) + u\left(\frac{dv}{dx}\right) + v\left(\frac{dv}{dy}\right) + w\left(\frac{dw}{dz}\right)$$

$$Z = \left(\frac{dw}{dt}\right) + u\left(\frac{dw}{dx}\right) + v\left(\frac{dw}{dy}\right) + w\left(\frac{dw}{dz}\right),$$

& l'équation différentielle qui détermine la preſſion p eſt :

$$\frac{dp}{q} = P\,dx + Q\,dy + R\,dz - X\,dx - Y\,dy - Z\,dz,$$

dans laquelle le tems t eſt ſuppoſé conſtant. Or l'autre équation tirée de la continuité du fluide eſt :

$$\left(\frac{dq}{dt}\right) + \left(\frac{d.qu}{dx}\right) + \left(\frac{d.qv}{dy}\right) + \left(\frac{d.qw}{dz}\right) = 0,$$

& ce ſont les deux équations qui contiennent toute la Théorie tant de l'équilibre que du mouvement des fluides, dans la plus grande univerſalité qu'on puiſſe imaginer.

XVI. Lorsqu'il eſt queſtion de l'équilibre, on n'a qu'à faire évanouïr les trois viteſſes u, v, & w, & puisque alors les quantités X, Y, & Z, évanouïſſent auſſi, toute la Théorie de l'équilibre des fluides eſt contenuë dans ces deux équations :

$$\frac{dp}{q} = P\,dx + Q\,dy + R\,dz,$$ le tems t étant conſtant,

& $$\left(\frac{dq}{dt}\right) = 0,$$

 de

be normal to the surface on which it acts. Although EULER never gives any fuller or clearer explanation in words except in the case of equilibrium, the mathematical treatment of pressure is entirely explicit and clear (Fig. 83). Indeed, the concept is so natural that once it is achieved, one has only to look upon the equations, which speak for themselves. EULER's own earlier writings on hydrostatics and hydraulics, where only special cases occur, include more careful physical motivation. This is but one of many examples in the history of science indicating that as soon as a good mathematical theory has been stated explicitly, "physical intuition" begins to develop to such a point that finally what has been gotten in fact as end result of long struggles of thought becomes "obvious".

13. Resultant Shear Force, and Euler's General Theory of Deformable Lines (1771—1774)

As we all know, and as he himself found in time, EULER was too hasty in claiming that the principle of linear momentum "includes all the laws of mechanics." It was the theory of the elastica, and, finally, of the rigid body, that revealed to him the principle of moment of momentum, *viz*,

$$\boldsymbol{L}_{\boldsymbol{x}_0}(\mathscr{V}) = \dot{\boldsymbol{H}}_{\boldsymbol{x}_0}(\mathscr{V}), \qquad (22)$$

where $\boldsymbol{L}_{\boldsymbol{x}_0}(\mathscr{V})$ and $\boldsymbol{H}_{\boldsymbol{x}_0}(\mathscr{V})$ are, respectively, the resultant torque and the moment of momentum, with respect to $\boldsymbol{x}_0$, of the matter in $\mathscr{V}$, as an independent and equally general law. Here[14] I shall describe only the second great fruit of EULER's reconsideration, late in life, of the theory of deformable lines. While in his earliest research on elasticity (1727), he had succeeded in taking into account a linear variation of tension over the cross-section of a bent beam so as to derive JAMES BERNOULLI's law of bending from HOOKE's law of extension, he was unsatisfied with the statical foundation of the theory of the elastica and sought for most of his lifetime, in repeated trials, to establish the equations of the elastica, like those of the general catenary, directly from the balance of forces acting on a portion of the curve. That this is impossible, is one of the major pieces of evidence that led him, finally, to perceive that the balance of moments is a mechanical law

[14] For the principle of moment of momentum, see the next essay in this volume.

Fig. 84. EULER's general equations of the statics of a plane deformable line, from the *Novi Commentarii Academiae Scientiarum Petropolitanae* for 1770

que elasticum, vel etiam vbi in statu suo naturali non situm rectum tenet, sed secundum curuam quamcunque datam sit formatum, quocirca adhuc duos casus sequentes adiungamus.

Casus Tertius pro filis inaequaliter elasticis.

XV. Talis inaequalitas scilicet locum habere potest, si vel ipsum filum non vbique sit aeque crassum etiamsi ex eadem constet materia, vel si adeo ex diuersis materiis fuerit compositum, hoc igitur casu elasticitas in singulis punctis, non simpliciter formulae $\frac{d\Phi}{ds}$ erit proportionalis, sed praeterea a functione quadam pendebit, ad punctum quoduis M pertinente, vnde manifestum est hanc functionem per ipsam portionem fili $AM = s$, determinari debere, sit igitur S ista functio elasticitatem absolutam definiens, atque loco constantis illius A, casu praecedente hic scribi oportebit S, sicque tota solutio sequenti modo se habebit: Ante omnia debet esse $Vv = \frac{S\,d\Phi}{ds}$, cui insuper vt ante adiungi conuenit has tres:

I. $dT + Vd\Phi = pds$; II. $dV - Td\Phi = qds$; III. $d.Vv = Vds$

ex vltima aequatione statim concluditur

$$V = \frac{S\,dd\Phi + dS\,d\Phi}{ds^2},$$

posito elemento ds constante, hincque vicissim

$$v = \frac{S\,d\Phi\,ds}{S\,dd\Phi + dS.d\Phi},$$

qui

independent, in general, of the balance of forces. Before achieving this degree of clarity, however, he drew profit from his studies of systems of forces. In a passage written about 1742 in one of his unpublished notebooks, EULER calculated the force which has to be supplied at a point A on the elastica in order that the band at one side of A retain its shape when all the material to the other side of A is cut away; he found that it is not tangential; *i.e.*, *shear force* is generally necessary in addition to the tension that suffices in the case of a perfectly flexible line. In 1771 he finally succeeded in separating the statical equations from the constitutive equations. Introducing the stress resultant acting upon a plane curve as a vector having tangential and normal components T and V, EULER obtained the fully general dynamical equations when there are no distributed couples, *viz* (Fig. 84),

$$\begin{aligned} \frac{dT}{ds} - V\varkappa &= -B_t + \sigma \ddot{x}_t, \\ \frac{dV}{ds} + T\varkappa &= -B_n + \sigma \ddot{x}_n, \\ \frac{dM}{ds} - V &= 0, \end{aligned} \tag{23}$$

where $\varkappa$ is the curvature and M is the bending moment, where s is arc length, and where t and n denote tangential and normal components. Later (1774) he announced the principle

$$\boldsymbol{S}_+ = -\boldsymbol{S}_-, \tag{24}$$

where $\boldsymbol{S} = T\boldsymbol{t} + V\boldsymbol{n}$ and where the $+$ and $-$ signs refer to the effects of the material on the opposite sides of the point in question. This last result we recognize as being the one-dimensional counterpart of CAUCHY's lemma (4). The first two lines in (23), of course, can be written in the form

$$\frac{d\boldsymbol{S}}{ds} + \boldsymbol{B} = \sigma \ddot{\boldsymbol{x}}, \tag{25}$$

which is the one-dimensional counterpart of CAUCHY's first law of motion (6), though hardly typical of it.

Fig. 85. COULOMB's figures for discussion of the tensions and shear forces in an undeformed cantilever

que elaſticum, vel etiam vbi in ſtatu ſuo naturali non ſitum rectum tenet, ſed ſecundum curuam quamcunque datam ſit formatum, quocirca adhuc duos caſus ſequentes adiungamus.

Caſus Tertius pro filis inaequaliter elaſticis.

XV. Talis inaequalitas ſcilicet locum habere poteſt, ſi vel ipſum filum non vbique ſit aeque craſſum etiamſi ex eadem conſtet materia, vel ſi adeo ex diuerſis materiis fuerit compoſitum, hoc igitur caſu elaſticitas in ſingulis punctis, non ſimpliciter formulae $\frac{d\Phi}{ds}$ erit proportionalis, ſed praeterea a functione quadam pendebit, ad punctum quoduis M pertinente, vnde manifeſtum eſt hanc functionem per ipſam portionem fili A M $= s$, determinari debere, ſit igitur S iſta functio elaſticitatem abſolutam definiens, atque loco conſtantis illius A, caſu praecedente hic ſcribi oportebit S, ſicque tota ſolutio ſequenti modo ſe habebit: Ante omnia debet eſſe $Vv = \frac{S\,d\Phi}{ds}$, cui inſuper vt ante adiungi conuenit has tres:

I. $dT + Vd\Phi = pds$; II. $dV - Td\Phi = qds$; III. $d.\,Vv = Vds$

ex vltima aequatione ſtatim concluditur

$$V = \frac{S\,dd\Phi + dS\,d\Phi}{ds^2},$$

poſito elemento ds conſtante, hincque viciſſim

$$v = \frac{S\,d\Phi\,ds}{S\,dd\Phi + dS.\,d\Phi},$$

qui

independent, in general, of the balance of forces. Before achieving this degree of clarity, however, he drew profit from his studies of systems of forces. In a passage written about 1742 in one of his unpublished notebooks, EULER calculated the force which has to be supplied at a point A on the elastica in order that the band at one side of A retain its shape when all the material to the other side of A is cut away; he found that it is not tangential; *i.e.*, *shear force* is generally necessary in addition to the tension that suffices in the case of a perfectly flexible line. In 1771 he finally succeeded in separating the statical equations from the constitutive equations. Introducing the stress resultant acting upon a plane curve as a vector having tangential and normal components T and V, EULER obtained the fully general dynamical equations when there are no distributed couples, *viz* (Fig. 84),

$$\begin{aligned} \frac{dT}{ds} - V\varkappa &= -B_t + \sigma \ddot{x}_t, \\ \frac{dV}{ds} + T\varkappa &= -B_n + \sigma \ddot{x}_n, \\ \frac{dM}{ds} - V &= 0, \end{aligned} \tag{23}$$

where $\varkappa$ is the curvature and M is the bending moment, where s is arc length, and where t and n denote tangential and normal components. Later (1774) he announced the principle

$$\boldsymbol{S}_+ = -\boldsymbol{S}_-, \tag{24}$$

where $\boldsymbol{S} = T\boldsymbol{t} + V\boldsymbol{n}$ and where the $+$ and $-$ signs refer to the effects of the material on the opposite sides of the point in question. This last result we recognize as being the one-dimensional counterpart of CAUCHY's lemma (4). The first two lines in (23), of course, can be written in the form

$$\frac{d\boldsymbol{S}}{ds} + \boldsymbol{B} = \sigma \ddot{\boldsymbol{x}}, \tag{25}$$

which is the one-dimensional counterpart of CAUCHY's first law of motion (6), though hardly typical of it.

Fig. 85. COULOMB's figures for discussion of the tensions and shear forces in an undeformed cantilever

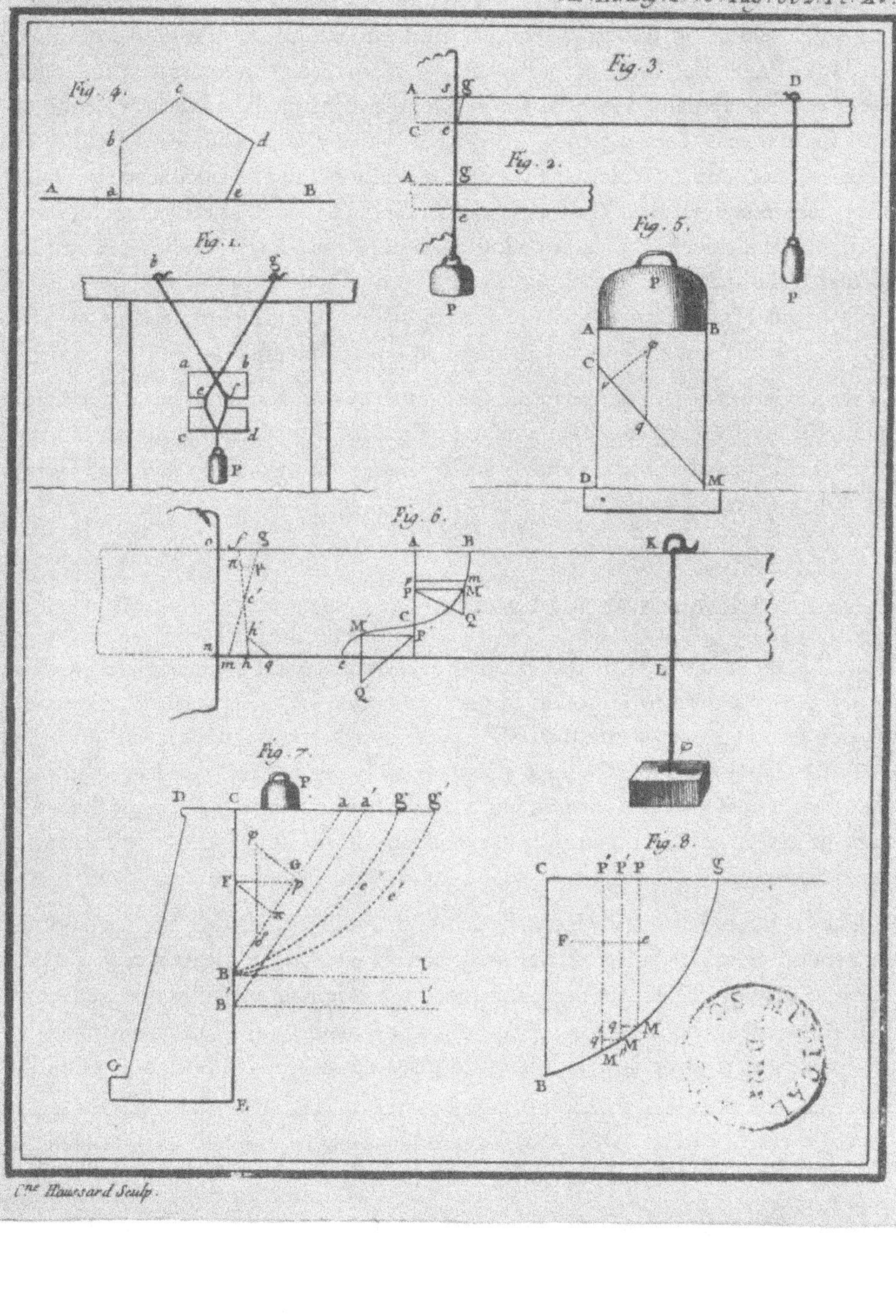

Cne Houssard Sculp.

14. Explicit Recognition of Shear Stress by Coulomb (1773)

In a celebrated paper, COULOMB revived and made more explicit the considerations of PARENT. He envisaged arbitrary variation of the forces, both normal and tangential, acting on the cross-section of a beam subject to transverse terminal load. Like PARENT's, these forces may be regarded as interior stresses in the plane case. COULOMB actually wrote down, in integrated form, all equations of equilibrium for the cross-section; thus he concluded that the area under the curve of pressures must equal that under the curve of tensions, and he calculated the resultant shear stress in terms of the load (Fig. 85). Going beyond PARENT, COULOMB was the first to consider the stresses acting upon different planes through a point; he calculated the shear stress upon a plane arbitrarily inclined to the direction of the applied force, in a beam supposed to be in a state of pure compression, and he showed that the shear stress was greatest on a plane inclined at 45°.

While COULOMB's ideas were clear, his achievements should not be exaggerated. Failing by far to reach the level attained by EULER in hydrodynamics and in the theory of deformable lines, COULOMB did not obtain any differential equations, nor did he foreshadow the stress tensor.

15. A Final Remark about Cauchy

From the above account, it is clear that every conceptual element in CAUCHY's theory of stress was to be found in one or another of the special theories constructed in the previous century. Moreover, in researches of FRESNEL done in 1821—1822, with which CAUCHY must certainly have been familiar, many of CAUCHY's results are more or less implied, although in FRESNEL's work the concepts of stress and strain are always connected through a presumed linearly elastic response.

Thus it might seem that CAUCHY's achievement in formulating and developing the general theory of stress was an easy one. It was not. CAUCHY's concept has the simplicity of genius. Its deep and thorough originality is fully outlined only against the background of the century of achievement by the brilliant geometers who preceded, treating special kinds and cases of deformable bodies by compli-

Fig. 86. AUGUSTIN CAUCHY (1789—1857) after a lithograph by JEAN BELLIARD, based on a drawing by JEAN ROLLER

cated and sometimes incorrect ways without ever hitting upon this basic idea, which immediately became and has remained the foundation of the mechanics of gross bodies.

Nothing is harder to surmount than a corpus of true but too special knowledge; to reforge the tradition of his forebears is the greatest originality a man can have.

Bibliographical Note

The lecture here printed was delivered at the Center for the Philosophy of Science, University of Minnesota, in 1967. It was based on and supersedes two earlier articles:

Zur Geschichte des Begriffes "innerer Druck", *Physikalische Blätter* **12**, 315—326 (1956).

Stages in the development of the concept of stress, *Problems of Continuum Mechanics* (Muskhelisvili Anniversary Volume), pp. 556—564 (pp. 439—447 of the Russian edition), 1961.

The first of these articles, in turn, grew out of a lecture given at the Institute of Mathematical Sciences, New York University, in 1954.

V. Whence the Law of Moment of Momentum?

1. The Standard Law of Moment of Momentum for Physicists

While mechanics is vulgarly thought a part of physics, it is not through the efforts of physicists that the last two decades have seen a new flourishing of classical mechanics, much in applications, indeed, but even more in basic theory, presented in half a dozen expressly founded journals and a floodlet of books. To the typical physicist, in natural philosophy classical mechanics seems a finished chapter, yet nearly nothing of what he learned about it in the hasty introduction included in his training in theoretical physics is considered generally correct, let alone well done, by the modern theorist of pure mechanics.

The principle of moment of momentum illuminates a fork in method and standpoint as blunt as any professional demarcation today. Let us recall how the principle is presented in a typical textbook of physics. For definiteness, we choose

Joos's *Theoretical Physics*[1], but nearly any other would serve as well, except that the more recent are often less careful to state explicitly what is done. First, in a system of mass-points the equation of motion for the kth particle is

$$m_k \ddot{\boldsymbol{r}}_k = \boldsymbol{F}_k + \sum_i \boldsymbol{F}_{ik}, \tag{1}$$

in standard notation. "NEWTON's third law of the equality of action and reaction" is laid down in the form $\boldsymbol{F}_{ik} = -\boldsymbol{F}_{ki}$.

Forming $\dot{\boldsymbol{H}}$, where the moment of momentum $\boldsymbol{H}$ is defined as

$$\boldsymbol{H} = \sum_k \boldsymbol{r}_k \times m_k \dot{\boldsymbol{r}}_k, \tag{2}$$

shows that

$$\dot{\boldsymbol{H}} = \boldsymbol{L} + \tfrac{1}{2} \sum_{k,j} (\boldsymbol{r}_k - \boldsymbol{r}_j) \times \boldsymbol{F}_{jk}, \tag{3}$$

where $\boldsymbol{L}$ is the total torque exerted by the external forces:

$$\boldsymbol{L} \equiv \sum_k \boldsymbol{r}_k \times \boldsymbol{F}_k. \tag{4}$$

If the mutual forces $\boldsymbol{F}_{jk}$ are central, so that $(\boldsymbol{r}_k - \boldsymbol{r}_j) \times \boldsymbol{F}_{jk} = \boldsymbol{0}$, then (3) reduces to

$$\dot{\boldsymbol{H}} = \boldsymbol{L}, \tag{5}$$

the desired law. Joos states the result as follows: "*For a system of particles in which the forces between any two particles are in the direction of the line joining these particles, the rate of change of the total angular momentum is equal to the sum of the moments of the applied forces.*" No one can criticize this statement or its proof. However, Joos is uneasy about the *relevance* of the result, for he adds in fine print:

"The limitation made above is actually of little importance. From considerations of symmetry, it is difficult to imagine a force acting between two points which does not coincide in direction with the line joining them, for there is no other preeminent direction. If the Law of *Biot-Savart* (p. 290) seems an exception, it must be remembered that this law deals with the force between a magnet-pole and an elementary *segment* (*i.e.*, not a point or particle) of an electrical conductor."

---

[1] G. Joos, *Theoretical Physics*, New York, Stechert, 1934. The passages to which we refer are on pp. 101, 103—104, 135—145, 158—160.

Despite this reservation, Joos applies (5) explicitly to a rigid body, beginning with simple cases and ending with the Euler equations of motion. He does not state anything about the forces among or between the particles, presumably finite in number since (5) was derived by him only for finite systems, that are presumed to make up the rigid body. Then he treats continuous media, with particular reference to elastic bodies in equilibrium. He states that there are "two kinds of force operating". In addition to "forces like that of gravity", there are "forces operating on the surface particles which are due to the presence of the contiguous particles." These are "not proportional to the volume of the element, but to the area of its surface." Denoting forces of the first kind by $\boldsymbol{g}$, those of the second kind by $\boldsymbol{P}$, Joos claims to apply "the fundamental equations of the mechanics of systems of particles":

$$\begin{aligned} \int \boldsymbol{g}\, d\tau + \oint \boldsymbol{P}\, dS &= \boldsymbol{0}, \\ \int \boldsymbol{r}\times\boldsymbol{g}\, d\tau + \oint \boldsymbol{r}\times\boldsymbol{P}\, dS &= \boldsymbol{0}. \end{aligned} \tag{6}$$

From this start he derives various equations of continuum mechanics, such as that asserting the symmetry of the stress tensor, in the usual way. He never mentions angular momentum again, and for continuous media he never writes an equation of the form (5). However, he assumes obliquely that the stress tensor remains symmetric in kinetic as well as in static problems of elasticity and linear viscosity[2].

I repeat that Joos' book was chosen only to give a typical example of development and use of moment of momentum for physicists. The textbook literature is so vast that some variations must exist, but the approach I have met commonly in conversation with practising physicists suggests that their training went along the above line.

Few if any specialists in mechanics think of their subject in this way. By them, classical mechanics is based on three fundamental laws, asserting the conservation or balance of *force*, *torque*, and *work*, or, in other terms, of *linear momentum*, *moment*

---

[2] The assumption is buried well. On p. 169 he derives the kinetic equations of linear elasticity from the static ones by adding to the body forces "the inertial forces given by D'Alembert's principle". On p. 205 he obtains the formula of linear viscosity by saying: "In order to calculate the stresses on a volume element of a viscous fluid we can avail ourselves of the formulae of the theory of elasticity ..." The dynamical principle he uses in elasticity is insufficient to obtain the equations of linear viscosity, which he in fact asserts on the basis of a (fanciful) *analogy* to elasticity, not using any mechanical principle whatever.

of momentum, and *energy*[3]. They regard Equation (5) itself as the statement of the second basic law. They do not regard the principle of moment of momentum as a consequence of the principle of linear momentum or of "NEWTON's Laws" or of any other more familiar principle[4].

Those who cultivate mechanics easily find half a dozen errors of logic and method in the usual approach of the physicists; physicists, in turn, rarely consent to regard (5) as a fundamental law but insist that it is derived, or at least ought to be derived, from NEWTON's laws or the "Newtonian" equations[5]. Their extreme reverence for what they consider NEWTON's approach to have been seems to

---

[3] It is largely due to the intensive and widespread study of gas dynamics in the 1940's that the principle of energy came to be understood to some extent by a large body of scientists, although it had been used now and then in mechanical problems for a century.

A general and detailed presentation of the three principles is given by C. TRUESDELL & R. A. TOUPIN, *The Classical Field Theories*, Handbuch der Physik III/1 (1960). Because of the recent tendency to write specialized monographs, even as textbooks for beginners, it is difficult to cite a general introduction to modern mechanics on an elementary level comparable to Joos' book for physicists. Works on elasticity stress the independence of the first two principles but do little justice, usually, to the third. Works on gas dynamics emphasize the independence of the third principle but often slight the second.

[4] Of course, formal rearrangements are possible. *E.g.*, a sufficiently general "variational" principle (*i.e.*, a formal expression in variations, not a true minimum principle) can be made equivalent to the first two laws, or even to all three.

[5] In this essay I use the term NEWTON's *Laws* to denote the laws actually stated by NEWTON; the "*Newtonian*" *equations*, to denote the various equations, of which (1) is representative, alleged in physics books as stating NEWTON's Laws. Examples of the varying usage of physical writers are given in Footnote 8 of the paper by J. C. C. MCKINSEY, A. C. SUGAR, & P. SUPPES, "Axiomatic foundations of classical particle mechanics," *J. Rational Mech. Anal.* **2**, 253—272 (1953). I add further examples. W. V. HOUSTON, *Principles of Mathematical Physics*, New York, McGraw-Hill, 2nd ed., 1948, in § 3 of Ch. 2 calls $\boldsymbol{F}_{ij} = -\boldsymbol{F}_{ji}$ "NEWTON's third law" and infers that it is "a statement of the conservation of [linear] momentum"; in § 5 he states that the additional condition $(\boldsymbol{r}_i - \boldsymbol{r}_j) \times \boldsymbol{F}_{ij} = \boldsymbol{0}$ "is satisfied in a great many cases." In § 2 of Ch. IX, HOUSTON makes it clear he does *not* regard the latter condition as general; see Footnote 26, below. In § 1. 4 of their *Principles of Mechanics*, New York, McGraw-Hill, 2nd ed., 1949, among the laws they assert to be "equivalent to those used by NEWTON", J. L. SYNGE & B. A. GRIFFITH state $(\boldsymbol{r}_j - \boldsymbol{r}_k) \times \boldsymbol{F}_{jk} = \boldsymbol{0}$, "summed up by saying: action and reaction are equal and opposite." They do not, however, state that all mutual forces are binary. But these examples are too clear to be typical. Turn now to A. SOMMERFELD's *Vorlesungen über theoretische Physik*, **1**, *Mechanik*, Leipzig, Akad. Verlagsg., 5th ed., 1955. There in § 1 NEWTON's own laws are quoted verbatim and taken as axioms; in § 13 $\boldsymbol{F}_{jk} = -\boldsymbol{F}_{kj}$ is inferred from "actio = reactio"; two pages later we are told to conclude that $\boldsymbol{r}_k \times \boldsymbol{F}_{ik} + \boldsymbol{r}_i \times \boldsymbol{F}_{ki} = \boldsymbol{0}$ by looking at a figure in which the forces are drawn as central! Such examples could be multiplied indefinitely.

furnish the motive for their tendency to brush off any criticism of it or major deviation from it as mere mathematical quibbling.

Elsewhere I design to take up the whole question more thoroughly. Here I wish to open a historical analysis. How did the two viewpoints arise and separate? Whence came the principle of moment of momentum? I have not yet uncovered all I seek on either question, but it seems to me that research on the history of rational mechanics is still enough of a novelty that an example of method during the working, before it is yet known which conjectures are right and which leads point to ends, might not be unwelcome in a circle accustomed to studies of an older, less formally mathematical, natural philosophy.

2. The Two Concepts of the Law of Moment of Momentum

First we must decide what the question really is. Turning our backs on physics, metaphysics, and prejudice, if we are to get anything solid in this matter we must begin by seeing *how the principle of moment of momentum is used today*. For punctual systems it serves only as a tool in the integration. This fact supports the viewpoint of the physicists that it is a subsidiary, derived principle. For rigid or deformable bodies, it has little use in solution of special problems but instead is one of the two basic principles from which the dynamical equations are commonly derived. For rigid bodies, the principle of moment of momentum yields the EULER equations; for deformable continua, the symmetry of the stress tensor. If we try to follow the physicists here, we see that for them the principle of moment of momentum serves *to obviate the need for specifying the mutual forces among the particles of a rigid or deformable body*. It is enough that the mutual forces contribute nothing to the resultant torque. In this sense we shall understand the problem. For us, it is not the trivial result of taking the cross-product of (1) by $\boldsymbol{r}_k$ and summing over k that will be seen as the principle of moment of momentum, but rather the equation $\dot{\boldsymbol{H}}=\boldsymbol{L}$ where $\boldsymbol{L}$ is *the torque of the applied forces only*, whether in the form (4) for admittedly punctual systems, or in more general form for space-filling bodies. Therefore, we shall search the origins of two statements:

A. *The resultant torque of a system of mutual forces is zero, whether as a consequence of Newton's Laws, "Newtonian equations", or otherwise.*

B. *The equation $\dot{\boldsymbol{H}}=\boldsymbol{L}$ is a fundamental, independent law of mechanics.*

3. Newton, Huygens, and Lagrange Published Nothing on the Two Concepts

For Statement A a beginner might appeal to NEWTON's *Principia*, but he would soon note

Datum 1. In NEWTON's *Principia* there is neither Statement A nor Statement B. Nor does NEWTON anywhere assert that $(\boldsymbol{r}_k - \boldsymbol{r}_j) \times \boldsymbol{F}_{jk} = \boldsymbol{0}$.

Often I have seen physicists surprised by this conclusion, sometimes to the point of refusing to believe it, but truly it should conform to widespread faiths in the empirical nature of science. I take it as the first principle of historical research in mechanics that the meaning is to be inferred from the use, since successful application has always preceded statement of the principle being applied. There being in NEWTON's book no theory of general dynamical systems, of rigid bodies[6], or of the stress in a continuum, no empiricist ought to be surprised to learn that NEWTON's system of mechanics was not general enough to yield such theories. Since HUYGENS was the first to solve a non-trivial problem about rotation, namely, the problem of the center of oscillation, the beginner might try to read the *Horologium Oscillatorium* and other works of HUYGENS, but he would add

---

[6] This fact does not imply that NEWTON had never thought about these matters, as is shown by the famous comment following Law I: "A top, whose parts by their cohesion constantly draw each other back from rectilinear motions, does not stop spinning except in so far as it is slowed by the air." It would be interesting to see if among NEWTON's papers there is anything mathematical concerning the motion of rigid bodies.

Note added for the reprinting, 1967. Such a paper has been found by J. W. HERIVEL and printed as Manuscript V in his *The Background to Newton's Principia*, Oxford, Clarendon Press, 1965. While on p.82 HERIVEL concludes, "It is clear that [NEWTON's] unerring intuition had led him to an almost perfect physical appreciation of the problem," I do not find in § 8 of Manuscript V good grounds for such sweeping praise. NEWTON does indeed perceive that free rotation is possible only about certain axes, not all, but to discuss motion about other axes he seems to confuse, as other early writers sometimes did in regard to the pendulum and the vibrating string, effects of friction with those of inertia. For example, NEWTON virtually asserts that rotation about a principal axis is the only periodic motion a free rigid body can have: "And as the axis is continually moved in the body, so it continually moves in the space too with some kind or other of spirall motion, always drawing nearer and nearer to a center or parallelisme with it selfe, but never attaining to it. Nay tis so far from ever keeping parallel to it selfe, that it shall never be twice in the same position."

Of course, that NEWTON in this short note did not come any nearer to the true dynamical laws for rigid bodies, is no reproach to him and should be no surprise to us. The note is remarkable enough in itself, without the exaggerated interpretation HERIVEL gives it.

Datum 2. In HUYGENS' solution of the problem of the physical pendulum there is nothing whatever regarding Statement A or Statement B.

Neither of these data will surprise a historian of science, for, on the basis of his general knowledge, he will not expect any development of mechanics as a whole along "Newtonian" lines by NEWTON or anyone else before the middle of the nineteenth century, nor will he expect the *general* principle of moment of momentum to be stated before its first great application, namely, the theory of general motion of a rigid body, was discovered. He will more likely start by checking the material as presented in the first systematic treatise on analytical dynamics, the *Méchanique Analitique* of LAGRANGE, published in 1788 and thus bisecting the period separating HUYGENS and NEWTON from the textbook writers at the end of the last century[7]. In the *Méchanique Analitique*, the basic law of mechanics is the principle of virtual work, and from it LAGRANGE derives easily the general integrals of energy, momentum, and moment of momentum for a system of mass-points. Since any forces that do no work may be left out of the expression for the virtual work, we may infer that in LAGRANGE's equations of the form $\dot{\boldsymbol{H}} = \boldsymbol{L}$, the mutual forces do not contribute to $\boldsymbol{L}$, but *Lagrange* does not say so. When we search for explanation of a concept in LAGRANGE's writings, usually we search in vain; in this case all we find is that the forces are "those that at the same instant act upon each point of the mass m along some given directions, that is, the velocities that each of these forces would impress upon the mass m if they acted separately and equally during the time taken as a unit. However variable may be the action of these forces, nevertheless one can regard it as constant during an instant." Also he speaks of the accelerating forces as "tending to given centers". Beside the looseness of statement, recalling D'ALEMBERT's mode of expression, we see here the typical evasive vagueness of LAGRANGE. In the statement of the principle of areas[8] LAGRANGE in fact makes us suspect that he does *not* perceive the

[7] *Méchanique Analitique*, Paris, Veuve Desaint, 1788. The general principle is stated in the Seconde Section of the Seconde Partie; the integrals are obtained in the Troisième Section. The history to which we refer is given in the Section Première. In the second and later editions, as reprinted in LAGRANGE's *Œuvres* **11**, there are considerable changes in the first and second sections, but they do not materially alter the part of the history to which I refer here, and, while the entire development of the principle of virtual work is recast, it becomes no clearer.

In his *Théorie des Fonctions Analytiques*, Paris, Imp. République, An V (1797) = *Œuvres* **9**, LAGRANGE gives what he regards as an improved treatment. In § 205 he asserts that mutual forces are central and are analogous to forces of constraint, and in § 217 he derives the integral of moment of momentum for a system subject to steady holonomic constraints.

[8] Paragraph 7 of the Troisième Section in the first edition.

generality of the integral of moment of momentum, for he adduces only instances where the torque on each several body vanishes: "If the system were not subject to any accelerating force, or if the only forces present all tend to the point we have selected as origin of co-ordinates...." Search of other parts of the book confirms. Datum 3. In the *Méchanique Analitique* there is nothing relevant to Statement B; as for Statement A, from LAGRANGE's treatment it may be inferred that forces which do no virtual work do not contribute to the resultant torque, but LAGRANGE gives no evidence of seeing this fact.

The historian will consult the *Méchanique Analitique* for a second reason, namely, that it includes in sections at the beginnings of the various parts the first history of mechanics. He who reads that history will obviate the need to consult the secondary works[9], because in regard to rational mechanics since GALILEO's time these do little more than quote, paraphrase, extend, or correct in detail the little sketches by LAGRANGE.

LAGRANGE writes that the principle of areas "... seems to have been discovered at the same time by Messrs. EULER, DANIEL BERNOULLI, and the Chevalier D'ARCY, but in different forms. According to the two first, this principle consists in the fact that in the motion of several bodies about a fixed center, the sum of the products of the mass of each body by the velocity of circulation around its center and by the distance from that same center is always independent of the mutual action that the bodies can exert upon each other, and it remains constant so long as there is neither exterior action nor an exterior obstacle ... The principle of Mr. D'ARCY ... is that the sum of the products of the mass of each body by the area that its radius vector describes about a fixed center is always proportional to the time. It is plain that this principle is a generalization of the beautiful theorem of NEWTON on the areas described in virtue of arbitrary centripetal forces ..."

While LAGRANGE's book is a good starting place, experience with it has led me to the following working hypotheses:
1. There was little new in the *Méchanique Analitique*; its contents derive from earlier papers of LAGRANGE himself[10] or from works of EULER and other predecessors.

---

[9] E. MACH, *Die Mechanik in ihrer Entwicklung*, 1883; E. JOUGUET, *Lectures de Mécanique*, 1908—1909; R. DUGAS, *Histoire de la Mécanique*, 1950.

[10] *E.g.*, the famous Lagrangean equations had been derived before, namely in §§ 7—11 of his "Théorie de la libration de la lune, et des autres phénomènes qui dependent de la figure non

2. General principles or concepts of mechanics are misunderstood or neglected by LAGRANGE.
3. LAGRANGE's histories usually give the right references but misrepresent or slight the contents.

When we read LAGRANGE's sarcastic comment about D'ARCY, "... he even made out of it a kind of metaphysical principle, which he calls the *conservation of action*..., as if vague and arbitrary names were the essence of the laws of nature and could by some secret virtue raise to final causes some simple consequences of the known laws of mechanics," Hypotheses 3 and 2 suggest that maybe D'ARCY had something. However, this is a bad lead; we find D'ARCY *assumes* $\dot{\boldsymbol{H}} = \boldsymbol{0}$ for bodies that "act on each other in any way, whether by wires, by inflexibles lines, by laws of attraction, *etc.*[11]." As a reason he says only, "It is known that a body, all of whose parts are connnected together, cannot take on any motion in virtue of their reciprocal actions."

---

sphérique de cette planète," *Nouv. Mém. Acad. Berlin* 1780, 203—309 (1782)= *Œuvres* **5**, 5—122. Here LAGRANGE begins by writing the inertial forces in rectangular Cartesian co-ordinates, while the accelerating forces are expressed in terms of distances from "any centers" (§ 5). He states (§ 13) that the Lagrangean equations hold for "an infinity of particles subject to any forces proportional to functions of the distances"; the meaning of this statement is not certain, but with any meaning I can conjecture it is generally false. In this paper LAGRANGE stops short of deriving the equations of motion of a rigid body by his method.

Likewise, the general integrals derived in the *Méchanique Analitique* are foreshadowed in the paper, "Remarques générales sur le mouvement de plusieurs corps qui s'attirent mutuellement en raison inverse des carrés des distances," *Nouv. Mém. Acad. Berlin* 1777, 155—172 (1779)= *Œuvres* **4**, 401—418; this paper rests on the "Newtonian" equations with a "Newtonian" potential function. While in §§ 4—8 LAGRANGE derives the integrals of momentum, moment of momentum, and energy, use of special properties of the potential function tends to conceal their meanings. For three bodies, the results are given in § II of his "Essai sur le problème des trois corps," *Prix de l'acad. sci. Paris* **9**, 1772= *Œuvres* **6**, 229—324.

Finally, the principle of virtual work for dynamics, on which the entire *Méchanique Analitique* is founded, had been given more than twenty years earlier in § IV of his "Recherches sur la libration de la lune, dans lesquelles on tâche de résoudre la question proposée par l'Académie royale des sciences pour le prix de l'année 1764," *Prix de l'acad. sci. Paris* **9**, 1764= *Œuvres* **6**, 5—61.

[11] See the argument based on Fig. 2 in his "Principe général de dynamique, qui donne la relation entre les espaces parcourus et les temps, quel que soit le système de corps que l'on considère, et quelles que soient leurs actions les unes sur les autres" (read in 1746), *Mém. Acad. Sci. Paris* 1742, 348—356 (1752), and an addition of 1747, pp. 356—361. The paper is not easy to read. Nothing is added in his later "Réflexions sur le principe de la moindre action de M. de Maupertuis," *Mém. Acad. Sci. Paris* 1749, 531—538 (1753).

Turning to Hypothesis 1, we can choose first to follow up LAGRANGE's own earlier work. Moving slowly backward in his *Œuvres* is a tedious process. The task is lightened by use of a fourth working hypothesis:

4. LAGRANGE's best ideas in mechanics derive from his earliest period, when he was studying EULER's papers and had not yet fallen under the personal influence of D'ALEMBERT.

Thus we turn at once to his paper of 1760 on least action[12]. Here we find the area integral of kinetics shown to follow from the principle of least action "if the system is entirely free, or if it is constrained to move about a fixed point, and if all the forces acting on the bodies come together at this point ..." The same references to works of EULER, DANIEL BERNOULLI, and D'ARCY are given here as later in the *Méchanique Analitique*. The law (5) is never mentioned or used[13].

4. The Fundamental Work of James Bernoulli on the Center of Oscillation (1686—1703)

Since search of other works on general dynamical systems[14] fails to reveal anything toward answering our questions, we must turn to researches on space-filling bodies. At the same time, we quickly see that LAGRANGE's references to

[12] § XII of "Application de la méthode exposée précédente à la solution de différens problêmes de dynamique," *Misc. Taur.* **2**$_2$ (1760—1761), 196—298 [1762] = *Œuvres* **1**, 365—468.

[13] *E.g.*, in §§ XXXIV—XXXIX are incredibly elaborate calculations deriving the equations of motion of a rigid body from the principle of least action. LAGRANGE uses unsymmetrical notations which complicate the formulae. He does not state that the body he treats is rigid, but the results are not generally correct otherwise. His quantities $[P]$ $[Q]$ $[R]$ are in fact angular components of $\dot{\boldsymbol{H}}$, but he never says so. There is no trace of the concept of torque.

[14] D'ALEMBERT's *Traité de Dynamique*, 1743, contains no allusion to moment of momentum, torque, or the area integral, although Chapters II and IV of the Seconde Partie concern the integrals of momentum and energy, respectively.

The *general* "Newtonian" equations of motion for a single mass-point make their first appearance in § 18 of EULER's paper E 112 of 1747, "Recherches sur le mouvement des corps célestes en général," *Mém. acad. sci. Berlin* [3] (1747), 83—143 (1749) = *Opera omnia* II **25**, 1—44. In 1744 he had been the first to set up "Newtonian" equations for a system of n mass-points subject to a particular law of force; see § 26 of E 174, "De motu corporum flexibilium," *Opusc.* [*var. arg.*] **3**, 88—165 (1751) = *Opera omnia* II **10**, 177—232. For this system he had derived the integrals of linear momentum and kinetic energy but had not noted the integral of moment of momentum.

EULER and DANIEL BERNOULLI bring us to papers where a rigid body plays an essential part. However, rather than creep backward we now jump to the point where, had we not needed to buck the currently widespread belief that the mass-point is the beginning and end of mechanics, we ought to have started, namely, at JAMES BERNOULLI's great paper of 1703 on the center of oscillation[15] (Fig. 87a, b). As its title suggests, this paper applies *the principle of the lever* to the motion of a pendulum. Now the "principle of the lever" is nothing else than the equilibrium of moments. As is well known, this paper of JAMES BERNOULLI contains the germ of the D'ALEMBERT-EULER principle that reversed accelerations are equipollent to forces per unit mass. Combination of these two principles yields (5) at once. Of course, in JAMES BERNOULLI's paper we cannot find a general statement of (5) or even an explicit statement in the context of the pendulum, but we recognize the *use* of the principle to determine the center of oscillation. The question is now, did BERNOULLI attempt to prove the extended principle of the lever from something else, or did he recognize it as an independent axiom? He inferred it from the "principle of the lever pulled or pushed by powers in motion", which he said was "demonstrated by the late Mr. MARIOTTE in Prop. 13 of the second part of his treatise *On the impact of bodies*, and there is no one who disagrees with it." Now MARIOTTE "demonstrated" this proposition *by experiment*[16]. We may justly doubt that his experiments establish the principle as used by BERNOULLI, but only fantasy could suggest that BERNOULLI thought of his principle as following from NEWTON's laws or from any other then known laws of mechanics. Rather, expressly remarking of the special dynamical postulate used by HUYGENS that "there are plenty of persons to whom this requirement has seemed a little bold and who have never been able to agree that it is obvious," BERNOULLI clearly sought and found in the extended principle of the lever a *new and more general method* in kinetics.

[15] "Démonstration générale du centre de balancement ou d'oscillation, tirée de la nature du levier," (letter of 13 March 1703), *Mém. Acad. Sci. Paris* 1703, 4to ed. (Paris), 78—84 (1705) = 12mo ed. (Amsterdam), 96—104 (1707) = *Opera* **2**, 930—936. In JAMES BERNOULLI's unpublished notebook, cited above, p. 206, entry no. 247 is an important study of the center of oscillation, fuller than anything BERNOULLI published.

[16] *Traité de la percussion ou choc des corps*, Paris, 1679. The passage in question is reprinted on pp. 135—136 of JOUGUET's *Lectures de Mécanique* I, Paris, Gauthier-Villars, 1908.

Fig. 87a (p. 250). JAMES BERNOULLI's final treatment of the center of oscillation as published in the *Mémoires de l'Académie des Sciences*, Paris, for 1703

Fig. 87b (p. 251). The same, continued

nouvelle espece de centre, que j'appelle *centre de tension*, où l'hypothêse de M. Huguens ne sçauroit avoir lieu: j'expliqueray en son tems ce que j'entends par-là. Et comme je n'ay encore rien publié de tout cela, je veux vous l'envoyer par parties, pour pouvoir être presenté à l'Academie, si vous trouvez qu'il le merite. Je commence par la premiere.

Principe du Levier tiré ou poussé par des puissances qui sont en mouvement.

FIGURE I. Soient *AC*, *AC*, *AD*, *AD*, les branches d'un levier mobile autour du point *A*; soient *C*,*C*,*D*,*D*, des poids ou des puissances muës avec des vîtesses *CB*,*CB*, *DE*, *DE*, lesquels fassent impression suivant les directions *CB*, *CB*, *DE*, *DE*, perpendiculaires aux bras de levier *AC*, *AC*, *AD*, *AD*. Je suppose que si tous les produits des puissances *C* par *AC* & *CB*, sont égaux à tous les produits des puissances *D* (qui agissent en sens contraire) par *AD* & *DE*; ou bien si tous les produits de *C* par *AC* & *CB* (entant qu'on conçoit toutes les puissances agir en même sens) sont égaux à rien, le levier doit demeurer en équilibre.

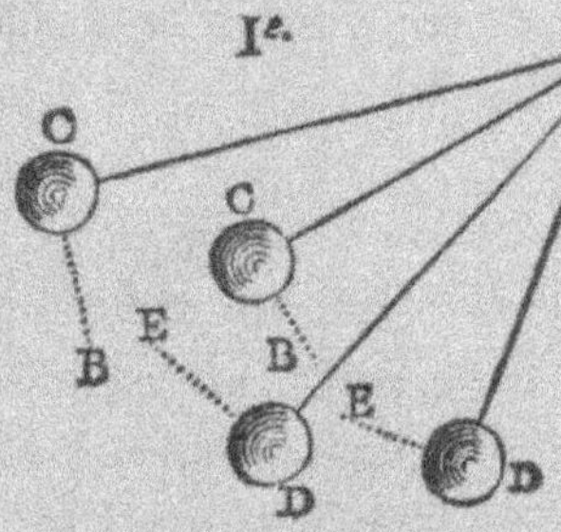

Ce principe a été démontré par feu M. Mariotte dans la Prop. 13. de la seconde Partie de son Traité de la Percussion des corps; & il n'y a personne qui en disconvienne.

SOLUTION.

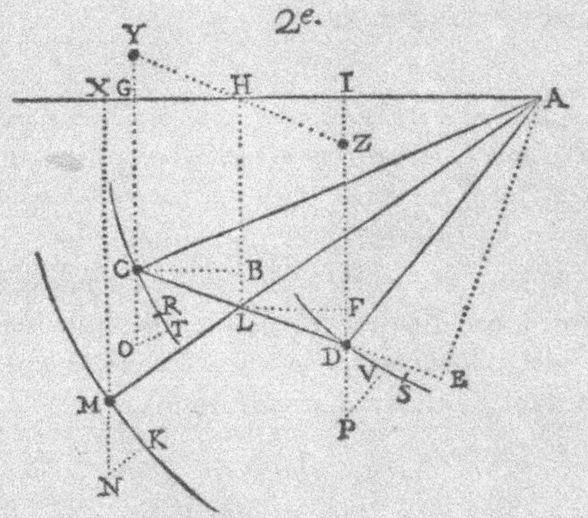

SOLUTION.

Soit maintenant *A* l'axe horizontal du balancement; FIGURE II. *AXM* un plan vertical droit à l'axe; *AM* le diametre de la figure qui balance, auquel on ait appliqué dans le même plan l'ordonnée *CLD* à angle donné *ALD*, laquelle ait *CL* = *LD*. Soient de plus *C* & *D* deux petites parcelles de la figure, lesquelles décrivent dans leur balancement les arcs *CT*, *DS*; soit aussi *AM* la longueur du Pendule simple, qui fait ses vibrations dans le même tems que la figure qui balance.

De ce que le balancement tant de *M*, que de *C* & *D*, s'acheve par l'hypothêse en même tems, il s'ensuit que les vîtesses dont ces poids se meuvent à chaque instant, sont proportionnelles à leurs distances *AM*, *AC*, *AD*, de l'axe *A*; & que par conséquent leur mouvement peut être continué avec ces vîtesses, sans que les poids *C* & *D* agissent en aucune maniére l'un sur l'autre *M*: De sorte qu'il ne faut considerer que la seule impulsion que la pesanteur ajoûte à chaque moment aux vîtesses acquises. Soit donc

This much established, we can follow the idea back a little to JAMES BERNOULLI's first and incorrect attempt[17] (Fig. 88), published in 1686 (be it noted, before he could even have seen NEWTON's *Principia*), which L'HOSPITAL[18] recognized[19] as resting on a principle "which is nothing else than that of the lever." In this earlier attempt, moreover, BERNOULLI does not cite any evidence, either theoretical or experimental, for his principle.

Thus I assert

Conjecture 1. *The principle of moment of momentum, as an independent law of mechanics and as a generalization to kinetics of the principle of equilibrium of moments in statics, is due to* JAMES BERNOULLI (1686, corrected 1703); *in concept, though not in correct statement, it antedates* NEWTON's *laws* (1687).

Of course, BERNOULLI did not state the principle either clearly or in any generality, nor did he claim that it could *not* be derived from the laws later published by NEWTON. We must now follow the way in which the principle was used and developed until it reached the generality in which theorists of mechanics conceive it today.

5. Daniel Bernoulli and Euler on the Dependence or Independence of the Law of Moment of Momentum in 1744

As LAGRANGE noted, in 1745 both DANIEL BERNOULLI and EULER put forward a form of the principle in the course of their solutions of the problem of a mass-

[17] "Narratio controversiae inter Dn. Hugenium et Abbatem Catelanum agitatae de centro oscillationis quae loco animadversionis esse poterit in responsionem Dn. Catelani, num. 27. Ephem. Gallic. anni 1684, insertam," *Acta Erud. Leipzig*, July 1686, 356—360 = *Opera* **1**, 277—281. Here BERNOULLI took the effective forces as being the negatives of the velocities acquired in a finite time rather than the instantaneous accelerations, or, as L'HOSPITAL said in his note of correction, "the beginning velocities".

[18] "Lettre de Mr. le Marquis de l'Hopital à Monsieur Huygens, dans laquelle il prétend démonstrer la régle de cet auteur touchant le centre d'oscillation du pendule composé par sa cause physique, et répondre en même temps à Mr. Bernoulli," *Hist. Ouvrages Sçavans* June-Aug. 1690, 440—449 = *Jac. Bernoulli Opera* **1**, 454—457 = *Œuvres de Huygens* **9**, 403—406.

[19] This was no small feat on the part of L'HOSPITAL. BERNOULLI's first attempts are always extremely hard to follow. He does say at one point, "since the rod, in this case, can be considered after the fashion of a lever ...."

Fig 88. JAMES BERNOULLI's first published attempt to find the center of oscillation by use of the law of the lever (*i.e.*, the earliest statement of the principle of moment of momentum), in the *Acta Eruditorum* for July, 1686

junctim dimissum, ob inflexilem virgam, nequit pertingere nisi ad L, dum pondus B jam est in N, hinc sequitur, gravitatis vim in pondere A non esse exhaustam; adeoque residuum harum virium, ex una parte urgere debere corpus B, ex altera ipsum axem D, eundemque premendo aliquam sui partem ibidem insumere & deperdere; siquidem virga hocce casu instar vectis considerari possit; prout extra dubium est, quod si corpus B infinite tarde moveri, id est, firmum & stabile esse intelligatur, sicut axis D corpus A partem sui ponderis, æque in axem D atque in corpus B transferret. Ex hactenus dictis colligere proclive est, si quis examinare vellet, quantam partem celeritatis suæ pondus A in premendo axe D consumere debeat, eum exinde, imitando Dn. Catelani ratiocinium, veritatem aut falsitatem Hugenianæ Hypotheseos, inque hac fundatæ Propositionis detegere posse. Rogantur hac occasione eruditi, ut examinent, qualem legem communicationis celeritatum observent corpora mota, quæ ex una parte innituntur firmo fulcimento, ex altera alii corpori itidem, sed tardius moto: si namque celeritatis excessus, qui hinc inde communicandus est, in eadem ratione distribueretur, in qua distribuitur onus aliquod, quod vecti duobus sustento fulcris impositum est, nimirum in ratione reciproca distantiarum mobilis a fulcris, tum imitando ratiocinium Dn. Catelani, deprehenderemus, summam altitudinum, ad quas ascendunt separata penduli pondera, vicissim nunc minorem esse summa altitudinum, e quibus antea conjunctim descenderant, quod iterum Hugenianam Propositionem everteret.

En calculum: Esto altitudo AL = 1 ped.

Altitudo BN = 4 ped.

Celeritas ponderis A acquisita in puncto L, ubi descendit separatim = 1

Celeritas ponderis B acquisita in puncto N, quando cadit separatim = 2

Celeritas ponderis A acquisita in puncto L, quando descendit conjunctim = x

Igitur Excessus celeritatis ponderis A, qui tam in axem, quam in pondus B redundat = 1 − x

point constrained to slide within a rigid, rotating tube. The method used by EULER[20] is the same as JAMES BERNOULLI's, but it is clearly explained, and EULER gives (5) explicitly in the special case of plane motion about a fixed axis ($L = I\,\alpha$); his alleged proof is no more than an assertion of the extended law of the lever. While DANIEL BERNOULLI[21] obtained the same result[22], he tried to prove it from other principles:

"... the action exerted mutually by the tube and the ball upon each other does not disturb the said [moment of] momentum." He asserts that the "action" of the tube and ball upon each other is normal to the tube, so he replaces that "action" by a little thread connecting the two. Thus the tube and ball exert

---

[20] E 86, "De motu corporum in superficiebus mobilibus," *Opusc. varii argum.* **1**, 1—386 (1746) = *Opera omnia* (II) **6**, 75—174. See §§ 42, 45.

[21] "Nouveau problème de mécanique résolu," *Hist. Mém. Acad. Sci. Berlin* [**1**] (1745), 54—70 (1746). See §§ IX—XI.

[22] It follows by combining his Lemmas VIII and IX. Then he adds the following characteristic remarks:

"This lemma has been proved in several places: For my part, I proved it in a paper *On the motion of bodies arising from excentric percussion etc.*, §§ 4 and 5, which I sent to the Academy of Petersburg some years ago, at a time when nobody had yet thought about this subject, so far as I know. My reason for saying this is certainly not to get myself any credit for it, but rather so as not to be held plagiarist by those who will find the same propositions in other works since printed; and this justification holds also for all the new propositions in my *Hydrodynamics* and my various memoirs on mechanics; nevertheless I do not claim that those who have solved my problems or proved my theorems after me owe the least thing to me in this respect, and I esteem infinitely the wisdom of their solutions and their proofs.

"This property has been noticed also by Mr. Euler, and he pointed it out to me in one of his letters. For my part I could not show that I found it without him. I had used other terms in my calculations, pointing out that the sum of the live forces resulting from the circulating motion of the system, divided by the angular velocity of the tube, remains always the same. That comes down to the same thing, and to do justice to Mr. Euler, I have preferred his expression to mine."

It will be noted that DANIEL BERNOULLI's generalization is usually false; it is *not* the general principle of moment of momentum.

Fig. 89. From Volume 3 of EULER's *Opusculi*, showing the earliest instance in which the principles of linear momentum and moment of momentum are laid down as independent basic axioms of mechanics

His igitur motus punctorum K, L, M, N, &c. ita afficientur, ut fit:

$$\begin{array}{l|l}
2K\,ddp = \mathfrak{B}\,dt^2 & 2K\,ddx = B\,dt^2 \\
2L\,ddq = \mathfrak{C}\,dt^2 - \mathfrak{B}\,dt^2 & 2L\,ddy = C\,dt^2 - B\,dt^2 \\
2M\,ddr = \mathfrak{D}\,dt^2 - \mathfrak{C}\,dt^2 & 2M\,ddz = D\,dt^2 - C\,dt^2 \\
2N\,dds = \text{- - -} - \mathfrak{D}\,dt^2 & 2N\,ddv = \text{- - -} - D\,dt^2 \\
\quad\&c. & \quad\&c.
\end{array}$$

Tum vero fingulorum articulorum motus rotatorii a momentis harum virium ita accelerabuntur; ut fiat:

$$2K\,kkdd\zeta = (\mathfrak{B}b\cos\zeta - Bb\sin\zeta)\,dt^2$$

$$2L\,lldd\eta = (\mathfrak{C}c\cos\eta - Cc\sin\eta)\,dt^2 + (\mathfrak{B}\beta\cos\eta - B\beta\sin\eta)\,dt^2$$

$$2M\,mmdd\theta = (\mathfrak{D}d\cos\theta - Dd\sin\theta)\,dt^2 + (\mathfrak{C}\gamma\cos\theta - C\gamma\sin\theta)\,dt^2$$

$$2N\,nndd\iota = \text{- - - - - -} + (\mathfrak{D}\delta\cos\iota - D\delta\sin\iota)\,dt^2$$

&c.

Ex prioribus follicitationibus oriuntur istæ binæ æquationes

$$K\,ddp + L\,ddq + M\,ddr + N\,dds + \&c. = 0$$

$$K\,ddx + L\,ddy + M\,ddz + N\,ddv + \&c. = 0$$

quæ bis integratæ dabunt:

$$Kp + Lq + Mr + Ns + \&c. = \mathfrak{F}t + f$$

$$Lx + Ly + Mz + Nv + \&c. = \mathfrak{G}t + \mathfrak{F}$$

quibus

on each other equal and opposite forces. DANIEL BERNOULLI's idea of what he had done is expressed in his letter[23] of February 1744 to EULER:

"... I have derived and proved the principle of conservation of moment of the rotational motion from the usual principles ..."

Hence I assert

Conjecture 2. *The idea that the principle of moment of momentum follows from the principle of linear momentum in the sense that the mutual actions of a system of bodies exert no resultant torque is due to* DANIEL BERNOULLI (1744).

Did this idea influence the discovery of the general equations of motion for rigid bodies? Perhaps so. First I note a step in the opposite direction:

Conjecture 3. *In a work of 1744,* EULER[24] *was the first to use the principles of linear and angular momentum as independent mechanical laws for setting up the equations of motion of a system.*

The system considered was *n* linked bars in a plane (Fig. 89).

---

[23] Published by P.-H. FUSS, *Correspondance mathématique et physique de quelques célèbres géomètres du XVIIIe siècle*, St. Pétersbourg, 1843. The letter of EULER to which BERNOULLI refers above seems to have been lost.

I have searched the remaining correspondence of the BERNOULLIS with EULER as published by ENESTRÖM, *Bibliotheca Math.* (3) **4**, 344—388 (1903); **5**, 249—291 (1904); **6**, 16—89 (1905); **7**, 126—156 (1906—1907). It does not reveal anything further that is conclusive in the present inquiry, though some dozen letters are relevant. It would do the modern Newtonians good to read these great geometers' repeated complaint that *new principles* are needed in order to set up the equations of motion for various simple mechanical systems. Apparently it never entered their heads to apply NEWTON's Laws or "Newtonian equations". On 23 April 1743 DANIEL BERNOULLI wrote to EULER that he wished he could determine the plane motion of three bodies linked by threads. "This problem may not be one of the easiest. I believe, however, that if we knew all the universal laws of motion, this problem would become rather easy and would at the same time give the impulse for solving more general problems of this kind." The remark was prophetic, for EULER shortly did solve this problem, publishing his solution in § 26 of the paper cited in the next footnote. EULER's solution is the earliest example of use of the "Newtonian" equations to set up the differential equations of motion for a dynamical system. For further discussion of the point at issue here, see §§ 26 and 30 of my history of elastic and flexible bodies, *L. Euleri Opera Omnia* II $\mathbf{11}_2$ (1960).

[24] E 174, "De motu corporum flexibilium," *Opusc.* [*var. arg*] **3**, 88—165 (1751) = *Opera omnia* II **10**, 177—232. Presentation date: 5 November 1744. This paper is described and translated in part on pp. 223—229 of my book, cited just above.

6. Euler's General Principle of Linear Momentum (1750) and Discovery of the Equations of Motion of a Rigid Body

His special case of the principle of moment of momentum did not lead DANIEL BERNOULLI to the equations of motion of a rigid body. In a letter of 4 December 1745 to EULER he described the general problem of rotation as "extremely difficult, which will not be solved easily by anybody .... One might ask how to determine the axis of rotation by a summation sign, such that the centrifugal forces destroy each other." While BERNOULLI dropped the problem, EULER through exploration of more and more general cases approached the equations of motion of an arbitrary rigid body. He finally achieved them in a paper written in 1750. This paper, *Discovery of a new principle of mechanics*[25], lays down the principle of *linear momentum*, or the "Newtonian" equation $\boldsymbol{F} = m\ddot{\boldsymbol{x}}$, as the axiom which "includes all the laws of mechanics," if applied to every several element of mass m in every body. The end product of the paper is the "Eulerian equations of motion" for rigid bodies (Fig. 90). How can this be? EULER states that the forces "include both such external forces as act upon the body from without, and also the internal forces binding the parts of the body to each other, so as to prevent them from changing their relative positions. But it is to be noted that the internal forces mutually destroy each other, so that the continuation of the motion does not require any external forces, except insofar as those forces do not mutually destroy each other." EULER thereupon obtains the equations of motion by taking moments of the equations of linear momentum for elements of mass, leaving altogether out of account any mutual forces that may be present. His method, then, is now akin to that of DANIEL BERNOULLI, but while he sees that mutual forces have no effect, he does not make any special hypothesis about their functional dependences or their directions. In effect, he *asserts* that (5) follows from the principle of linear momentum when the mutual forces are such as to maintain rigidity[26]. The modern

[25] E 177, "Découverte d'un nouveau principe de mécanique," *Hist. Acad. Sci. Berlin* **[6]** (1750), 185—217 (1752) = *Opera omnia* II **5**, 81—108. Presentation date: 3 September 1750. See § 22 and §§ 40—58.

[26] One modern author has reverted to a similar view. While in the first edition (1934) of his book HOUSTON gave a derivation of the equations of motion of rigid bodies on the basis of particles exerting central forces, in the second edition, cited in Footnote 5, he abandons the particles in favor of smoothly distributed internal forces per unit volume. He states that the vanishing of their resultant moment "may be taken as part of the definition of a rigid body. It would follow from the assumption that the forces between the parts of the body are central

LV. Dans cet état le corps soit sollicité par des forces quelconques, & pour trouver le changement, qui en sera causé dans le mouvement du corps, on n'aura qu'à avoir égard aux momens des ces forces par rapport aux trois axes OA, OB, OC: soit donc le moment qui résulte de ces forces

pour l'axe OA dans le sens BC $= Pa$.
Le moment pour l'axe OB dans le sens CA $= Qa$.
Le moment pour l'axe OC dans le sens CO $= Ra$.

Maintenant égalant ces moments à ceux qui ont été trouvés cy-dessus (§. 49.) nous obtiendrons les trois équations suivantes.

$$\text{I.}\quad \frac{Pa}{2M} = \frac{ffd\lambda}{dt} - \frac{nnd\mu}{dt} - \frac{mmdv}{dt} + \lambda v nn - \lambda\mu mm - (\mu\mu - vv)ll + \mu v(hh - gg)$$

$$\text{II.}\quad \frac{Qa}{2M} = \frac{ggd\mu}{dt} - \frac{lldv}{dt} - \frac{nnd\lambda}{dt} + \lambda\mu ll - \mu v nn - (vv - \lambda\lambda)mm + \lambda v(ff - hh)$$

$$\text{III.}\quad \frac{Ra}{2M} = \frac{hhdv}{dt} - \frac{mmd\lambda}{dt} - \frac{lld\mu}{dt} + \mu v mm - \lambda v ll - (\lambda\lambda - \mu\mu)nn + \lambda\mu(gg - ff)$$

d'où on pourra déterminer les changemens infiniment petits $d\lambda, d\mu$ & dv, qui seront produits dans l'elément du tems dt.

LVI. Mais, puisque la résolution de ces équations nous conduiroit à des formules trop longues, posons, comme nous avons fait auparavant, que le corps tourne à l'instant présent autour de l'axe OA dans le sens BC avec une vitesse angulaire $= v$: & qu'après le tems $= dt$, l'axe de rotation change, en sorte qu'il fasse alors avec l'axe OA un angle $= d\zeta$, avec l'axe OB un angle $= 90° + d\eta$ & avec l'axe OC un angle $= 90° + a\theta$; & que la vitesse angulaire devienne alors $= v + dv$: & nous avons vû qu'il est $d\zeta^2 = d\eta^2 + d\theta^2$. Cela supposé, nous aurons les trois équations suivantes:

$$\text{I.}\quad \frac{Pa}{2M} = \frac{ffdv}{dt} + \frac{nnvd\eta}{dt} + \frac{mmvd\theta}{dt}.$$

theorist of mechanics, who has constantly in mind the general case of a deformable body, when (5) does not generally follow from the principle of linear momentum, sees the ridgepole that EULER walked here; as usual, he did not slip off.

EULER's book[27] on the dynamics of rigid bodies is so diffuse as to conceal the line of thought in a multitude of special cases, but it seems to rest upon the same approach to the subject.

7. Euler's Fundamental Laws of Mechanics (1775)

The general problem was considered next by LAGRANGE[28]. His "new solution" of 1773 begins, "I consider the proposed body as an assembly of corpuscles or mass-points joined together in such a way as always to conserve their mutual distances .... I shall have by the principles of mechanics, since the system is supposed free to move about a fixed point but is not subject to any further outside force, — I shall have, I say, on the spot, these ... equations," and LAGRANGE writes down the integrals of moment of momentum and kinetic energy for a system of free mass-points. He says they follow from a "known principle" concerning "every system of bodies acting on one another in any manner". The

---

forces, but this assumption lacks generality and seems unnecessarily restrictive. Frequently [the vanishing of the torque] is justified by the statement that the internal forces within a rigid body are in equilibrium among themselves. It may be considered as a law similar to Newton's third law of motion. It describes the observed fact that isolated bodies do not set themselves into rotation."

[27] E 289, *Theoria motus corporum solidorum seu rigidorum ex primis nostrae cognitionis principiis stabilita et ad omnes motus, qui in huiusmodi corpora cadere possunt, accomodata*, Rostock and Greifswald, Roese, 1765. 2nd ed., *ibid*, 1790 = *Opera omnia* II **3** and **4**. See Chapter 15.

[28] "Nouvelle solution du problème du mouvement de rotation d'un corps de figure quelconque qui n'est animé par aucune force accélératrice," *Nouv. Mém. Acad. Berlin* 1773, 85—120 = *Œuvres* **3**, 579—616.

Fig. 90. EULER's derivation of the equations of motion for a rigid body with respect to its center of mass, from the *Mémoires de l'Académie des Sciences*, Berlin, for 1750

alleged proof is only a restatement. This typical looseness followed by a cloud of calculation moved Euler to reply[29]: "... But when I tried with greatest avidity to follow in detail his extremely profound thoughts, truly I could not get myself to go through all his calculations. Even the first lemma so deterred me that on account of my blindness I could not hope in any way to check through all the analytic devices he used." In this new paper, written in 1775, the old Euler laid down as *fundamental, general, and independent laws of mechanics*, for all kinds of motions of all kinds of bodies, the principles of linear momentum and moment of momentum for each element of mass. In justification, he wrote only, "... by the principles of mechanics it is necessary that ... ." The two principles, integrated forms of which are

$$\boldsymbol{F}=\dot{\boldsymbol{M}} \quad \text{and} \quad \boldsymbol{L}=\dot{\boldsymbol{H}}, \tag{7}$$

may justly be called Euler's *laws of mechanics* (Fig. 91).

What caused Euler in his last work to revert to the ideas of James Bernoulli and abandon the attempt to regard the linear momentum principle, supplemented by restrictions upon the mutual forces, as sufficient for the whole science of mechanics? The answer is to be sought, as always, in the *practice*, not the philosophy, of the subject. The tenet that Newton's Laws or the "Newtonian" equations in any of their forms suffice, can be held only by those who limit their attention to mass-points, rigid bodies, and certain other special systems. Any modern theorist knows this and sees at once the evasions used in typical treatments, such as that of Joos above described, to get the form (6) suitable for deformable bodies. In the eighteenth century, fluid mechanics was limited, as far as concerned developed theory, to frictionless fluids, where, again, the principle of linear momentum sufficed. But in *elasticity* Newton's laws never have been and never can be sufficient: The simplest problems of elasticity rest essentially upon the balance of moments, and if we choose to think of the body as made up of molecules exerting mutual forces, those forces *cannot* be left out of account. In the theory of elasticity,

[29] In the introduction to E 479, "Nova methodus motum corporum rigidorum determinandi," *Novi Comm. Acad. Sci. Petrop.* **20** (1775), 208—238 (1776). Presented 16 October 1775. Reprinted in the 2nd ed., 1790, of the *Treatise on rigid bodies.* The fundamental laws are given in §§ 27—29.

Fig. 91. Euler's general and final statement of the principles of linear momentum and moment of momentum, from the *Novi Commentarii Academiae Scientiarum Petropolitanae* for 1775

ea nullum nascitur momentum pro hoc axe; pro axe autem I B nascetur momentum $= z\, dM\, (\frac{ddx}{dt^2})$ et pro axe I C momentum $= y\, dM\, (\frac{ddx}{dt^2})$. Simili modo ex vi acceleratrice secundum directionem I B, quae est $dM\, (\frac{ddy}{dt^2})$ nascitur momentum pro axe I A $= z\, dM\, (\frac{ddy}{dt^2})$, at pro axe I C momentum $= x\, dM\, (\frac{ddy}{dt^2})$. Denique ex vi acceleratrice secundum I C, quae est $dM\, (\frac{ddz}{dt^2})$ nascitur momentum pro axe I A $= y\, dM\, (\frac{ddz}{dt^2})$ et pro axe I B momentum $= x\, dM\, (\frac{ddz}{dt^2})$. Hinc igitur pro quolibet axe habemus bina momenta elementaria, quae in partes contrarias vergunt; vnde pro axe I A summa omnium momentorum elementarium erit

$$+\int z\, dM\, (\frac{ddy}{dt^2}) - \int y\, dM\, (\frac{ddz}{dt^2}) = i\, S.$$

Eodem modo pro axe I B obtinebimus hanc aequationem:

$$\int x\, dM\, (\frac{ddz}{dt^2}) - \int z\, dM\, (\frac{ddx}{dt^2}) = i\, T.$$

Tertia vero aequatio erit pro axe I C

$$\int y\, dM\, (\frac{ddx}{dt^2}) - \int z\, dM\, (\frac{ddy}{dt^2}) = i\, U.$$

§. 29. Hac igitur ratione sex nacti sumus aequationes, quas hic coniunctim conspectui exponamus

I. $\int dM\, (\frac{ddx}{dt^2}) = i\, P$ IV. $\int z\, dM\, (\frac{ddy}{dt^2}) - \int y\, dM\, (\frac{ddz}{dt^2}) = i\, S$

II. $\int dM\, (\frac{ddy}{dt^2}) = i\, Q$ V. $\int x\, dM\, (\frac{ddz}{dt^2}) - \int z\, dM\, (\frac{ddx}{dt^2}) = i\, T$

III. $\int dM\, (\frac{ddz}{dt^2}) = i\, R$ VI. $\int y\, dM\, (\frac{ddx}{dt^2}) - \int z\, dM\, (\frac{ddy}{dt^2}) = i\, U.$

In

forces described by Joos as "not proportional to the volume of the element, but to the area of its surface" are of the essence, and Joos would have done better to adduce them instead of the law of BIOT and SAVART as casting doubt upon the generality of binary central forces. A molecular approach to elasticity is fraught with difficulty from start to finish, as anyone knows who has looked at classical and modern theories of the solid state. *Everyone* in the eighteenth century who studied problems of elasticity invoked the principle of moments. Thus, finally, if we are to find the origin of the independent principle of moment of momentum, we must search the history of theories of elastic materials.

Elsewhere I have done so[30]. Throughout his long life EULER studied flexible or elastic lines in various cases, slowly moving towards greater generality. Continually he strove for a *single mechanics* that would include all kinds of systems. He succeeded in showing that for a perfectly flexible line, either the balance of forces or the balance of moments sufficed to obtain the equations of equilibrium; for such a system, the two principles are equivalent. For an elastic line, balance of moments had been used since the beginning, and no effort or hypothesis ever enabled him to dispense with torques in favor of forces. Furthermore, from the equation of equilibrium for an elastica obtained by balance of moments, the D'ALEMBERT-EULER principle of reversed accelerations does not lead to equations of motion. What then *are* the equations of motion for an elastica? In 1771, after the publication of his book on rigid bodies but before his statement of the general laws of motion (7), EULER finally saw that all the results obtained on the elastica up to that time flowed from special properties of the constitutive equation rather than from the principles of mechanics. In effect reverting to the program of JAMES BERNOULLI[31], he sought the equations of equilibrium and motion for *an arbitrary plane continuum*, independently of its material constitution. For this, both the balance of forces and the balance of moments are necessary; neither by itself suffices. Application of the D'ALEMBERT-EULER principle to the equations of

[30] *L. Euleri opera omnia* (II) **11_2**, 1960. Also "Outline of the history of flexible or elastic bodies to 1788," *J. Acoust. Soc. Amer.* **32**, 1647—1656 (1960).

[31] "Véritable hypothèse de la résistance des solides, avec la démonstration de la courbure des corps qui font ressort .... Lettre du 12 mars 1705," *Mem. Acad. Sci. Paris*, 1705, 4to ed., Paris, 176—186 (1706) = 12mo ed., Amsterdam, 230—244 (1707) = *Opera* **2**, 976—989. This paper is analysed on pp. 105—109 of my book, cited in the preceding footnote.

equilibrium of forces then yields equations of motion, as desired, namely[32] Equations (23), p. 234, above. Corresponding to (6) but more general in being kinetic, more special in that the continuum is assumed to be a plane curve, these equations furnish the first example of *general laws of mechanics, independent of the nature of mutual forces and materials.*

Thus it may well have been the need to apply (5) as a principle independent of NEWTON's laws, in a case to which NEWTON's laws have never been shown to be relevant or useful, that caused EULER to abandon all reference to mutual forces in setting up his general laws of mechanics (7). Hence I assert

Conjecture 4. *The general principle of moment of momentum, as independent of the principle of linear momentum and as applicable to every part of every body, was first proposed by* EULER *in 1775. He was led to it through studies of elastic lines, culminating in 1771 in his general theory of the plane linear continuum.*

Moreover, growth of understanding of the nature and importance of moment of momentum is inextricable from the growth of the concept of stress in a continuum[33]. Here, however, I shall not follow the matter through the researches of CAUCHY and down to the present day, when the very definition of moment of momentum is being subjected to scrutiny because, as shown by work of J. L. ERICKSEN, it needs to be generalized.

8. Transition to the Newtonians (1833 onward)

Our search of the sources has not yet turned up Statement A, although we have seen steps toward it in work of DANIEL BERNOULLI (1744), EULER (1752), and LAGRANGE (1788). The earliest occurrence I have found is a textbook by POISSON[34]:

Conjecture 5. *Statement and proof that a system of pairwise equilibrated, central forces exerts no resultant torque is due to* POISSON (1833).

---

[32] E 410, "Genuina principia doctrinae de statu aequilibrii et motu corporum tam perfecte flexibilium quam elasticorum," *Novi Comm. Acad. Sci. Petrop.* **15** (1770), 381—413 (1771)= *Opera omnia* II, $\mathbf{11}_1$, 37—61. Presentation date: 14 January 1771.

[33] For this growth, see the preceding essay in this volume.

[34] *Traité de Mécanique*, 2nd ed., 2 vol., Paris, Bachelier, 1833. See §§ 552—554.

Of course it is difficult to be sure that the earliest occurrence we happen to find of so ele-

POISSON's proof is that found in modern texts; he mentions "the general law that action always equals reaction," which "always holds in nature," and he says "one always notes" that $(\boldsymbol{r}_k - \boldsymbol{r}_j) \times \boldsymbol{F}_{jk} = \boldsymbol{0}$. (*Cf.* Figs. 92, 93.)

With POISSON we are still some distance, however, from the Newtonians of today. The above remarks occur on page 447 of Volume 2, so it does not seem that POISSON can have considered them fundamental to mechanics as a whole. For example, the theory of rigid bodies is given much earlier[35]. Other than the above-quoted statement of what is often called "NEWTON's third law", *viz*, $\boldsymbol{F}_{jk} = -\boldsymbol{F}_{kj}$ and $(\boldsymbol{r}_k - \boldsymbol{r}_j) \times \boldsymbol{F}_{jk} = \boldsymbol{0}$, there seems to be in the whole book no other reference to NEWTON's laws, either as stated by NEWTON or as found in physics texts today, nor have I seen NEWTON named.

After all this, we can ask, Who first had the idea that the "Newtonian" equations sufficed as a basis for mechanics? Certainly none of the creators of mechanics,

---

mentary an argument marks the first observation of it. In this case, at least, we can see that POISSON himself did not have the matter fully worked out in 1811, when the first edition of his *Traité de Mécanique* was published. In § 456 of that edition he wrote in connection with the resultant torque, "Thus in forming it one will disregard the mutual attraction of the bodies of the system, provided one supposes, as indeed holds in nature, that action equals reaction between two material points of equal mass," and he concludes that this is "easy to verify by substitution." In the passage cited from the second edition, he carries out the verification and thus sees, apparently, that the condition of *central* forces is essential, while that of equal masses is unnecessary. There is no counterpart in the first edition for § 552 of the second, where the "general law of action and reaction" is stated explicitly.

[35] See § 382 of the first edition, § 412 of the second. It is not even clear in these passages that mutual forces among the parts of a rigid body are to be left out of account; POISSON simply never mentions them. That we are in fact to omit them, may be inferred from his adopting D'ALEMBERT's principle as the basic axiom of dynamics (§ 332 of the first edition, § 350 of the second): The effective forces due to constraints form a statically equilibrated system.

Fig. 92 (p. 265). P. 447 of the second edition of POISSON's *Traité de Mécanique*, 1833, showing what seems to be the earliest occurrence of the idea that the principle of action and reaction refers to central forces

Fig. 93 (p. 266). P. 453 of the second edition of POISSON's *Traité de Mécanique*, 1833, showing what seems to be the first proof that the principle of moment of momentum follows from the principle of linear momentum, for pairwise equilibrated central forces

Fig. 94 (p. 267). SIMÉON-DENIS POISSON (1781—1840), after a lithograph by FRANÇOIS DELPECH, following a drawing by NICOLAS MAURIN, before 1825

hautes mers sont les plus considérables dans les *syzygies*, et les moindres dans les *quadratures*.

§ II. *Principes de la conservation du mouvement du centre de gravité, et de la conservation des aires.*

552. Puisque le mouvement du centre de gravité d'un système entièrement libre est le même que si les masses de tous les mobiles y étaient réunies, et que leurs forces motrices y fussent transportées parallèlement à leurs directions, il s'ensuit que les forces données dont les composantes parallèles à chaque coordonnée seront égales et contraires, n'entreront pas dans les équations différentielles de ce mouvement. Or, ce cas est celui des forces motrices provenant des actions mutuelles des points du système, en vertu de la loi générale de l'*action égale à la réaction*, qui s'observe toujours dans la nature, ainsi qu'on va l'expliquer.

Si un point matériel situé en M agit sur un autre point situé en M′, et lui imprime ou tend à lui imprimer, dans un instant, une quantité de mouvement infiniment petite, que je représenterai par μ, on observe toujours :

1°. Que cette action est dirigée suivant la droite menée du point M′ vers le point M, ou suivant son prolongement au-delà de M′;

2°. Qu'en même temps le point situé en M′ réagit sur le point situé en M, suivant la droite qui va de M vers M′, ou suivant son prolongement au-delà de M.

vers m pour le point m'. En appelant ρ la distance de ces deux points, les cosinus des angles que fait la droite mm', avec des parallèles aux axes des x, y, z, menées par le point m, seront

$$\frac{x'-x}{\rho}, \quad \frac{y'-y}{\rho}, \quad \frac{z'-z}{\rho};$$

relativement à la force F, on aura donc

$$mX = \frac{(x'-x)F}{\rho}, \quad mY = \frac{(y'-y)F}{\rho}, \quad mZ = \frac{(z'-z)F}{\rho},$$

pour les composantes de la force motrice du point m; et l'on trouvera de même

$$m'X' = \frac{(x-x')F}{\rho}, \quad m'Y' = \frac{(y-y')F}{\rho}, \quad m'Z' = \frac{(z-z')F}{\rho},$$

pour les composantes de la force motrice de m', provenant de cette force F. Or, d'après ces valeurs, nous aurons

$$m(xY - yX) + m'(x'Y' - y'X') = 0,$$
$$m(zX - xZ) + m'(z'X' - x'Z') = 0,$$
$$m(yZ - zY) + m'(y'Z' - z'Y') = 0;$$

par conséquent, les termes provenant de l'action mutuelle des points du système, se détruisent deux à deux dans les seconds membres des équations (9) du n° 539.

Si donc il n'y a pas d'autres forces motrices qui agissent sur les points m, m', m'', etc., le mouvement de rotation du système autour de l'origine des coordonnées, sera uniquement dû aux vitesses

POISSON.

except, perhaps, NEWTON himself[36]: Such an idea is contradictory[37] to the contents of almost any page of the works of the BERNOULLIS, EULER, D'ALEMBERT, or LAGRANGE, but it fits in well with views held a century later. In 1867 KELVIN & TAIT[38] asserted the supremacy of NEWTON's laws in terms soon to be echoed by MACH, although after a few preliminary passes they choose instead as their axiom LAGRANGE's principle of virtual work. KIRCHHOFF[39] wrote down as comprising the "most general case to be considered in the mechanics of material points" the "Newtonian" equations subject to a particular class of forces, namely, forces that are assigned functions of the positions, velocities, and time, augmented by forces arising from holonomic constraints. Between 1833 and 1867 a fundamental change had occurred in mechanics, not in the subject itself but in the *teaching* of it. To describe this change would require another essay, perhaps a boring one, but nevertheless one that ought to be written, for, unlike most historical studies, it would show us how a subject dies. In all fields except science, modern thinkers tend to doubt their grandparents' myth of "progress". Perhaps science, too, deserves to turn its doubts inward.

I am obliged to Professor VAN DER WAERDEN for valuable discussion.

Appendix: Recent Views on the Principles of Mechanics

Since neither historians nor physicists are likely to have had much contact with modern research on mechanics, I append here a sketch of a newer approach to-

---

[36] I see no reason to think that NEWTON had any such idea; certainly it is not borne out by examination of the *Principia*. See my summaries of the *Principia* in § 1 of the second essay in this volume and in §§ 1—2 of the third. The Newtonian myth seems to have originated with a group of British disciples, most of whom had insufficient mathematics to follow any proof in the *Principia*. It may account for the fact, at first sight astonishing, that no major result in theoretical mechanics was found in England between 1740 and 1815. Since the beginning there has been a tendency to idolize NEWTON as a literary saint rather than to follow him as a leader of science.

[37] Recall the definitions of terms in Footnote 5. Of course, nearly the entire successful effort of mathematical philosophy in the eighteenth century was directed towards the more general subject of "Newtonian mechanics" as distinct, let us say, from Peripatetic mechanics or Cartesian mechanics and as extending some of the mechanical theories of GALILEO and HUYGENS.

[38] §§ 242—264 and § 293 of their *Treatise on Natural Philosophy*, Oxford, Clarendon Press, 1867.

[39] Eq. (17) of Lecture II of *Vorlesungen über mathematische Physik*: *Mechanik*, Leipzig, Teubner, 1876.

ward the ends which seem to be sought in most cases when NEWTON's laws or "Newtonian" equations are invoked, mentioning its historical origins.

A. Forces and torques are undefined objects, along with mass, position, and time. This view seems to be consonant with that of NEWTON, EULER, and current physics teaching, while opposed to that of D'ALEMBERT, LAGRANGE, and MACH. However, in contrast to earlier work, it is recognized now that *mathematical axioms* to be satisfied by forces and torques must be laid down[40], just as real numbers are undefined objects, not regarded as sufficiently described until we have stated the axioms they satisfy.

B. EULER's laws (7), relating the *load* $(\boldsymbol{F}, \boldsymbol{L})$ to the *reaction* $(\dot{\boldsymbol{M}}, \dot{\boldsymbol{H}})$ in inertial frames, express the fundamental axioms of mechanics.

C. All forces and torques arising from the mutual actions of bodies are *indifferent* in the sense that they are invariant under change of observer[41].

D. From the axioms A and EULER's equation $(7)_1$ alone may be proved an action-reaction principle, as follows: The sum of the mutual forces exerted by any two disjoint parts of a body free of singular loads is zero.

E. If the masses are supposed discrete, and if no couples are applied, the strong restriction called "NEWTON's third law" by some authors but rejected by others[42] can be proved as a consequence of the result stated in D and of EULER's equation $(7)_2$: The mutual forces between mass-points are central. Thus, if we are content to confine mechanics to systems of mass-points subject to binary forces obeying the principle of linear momentum, this restrictive "NEWTON's third law" is *equivalent* to the principle of moment of momentum[43]. I remark that those persons who consider the equation $(\boldsymbol{r}_i - \boldsymbol{r}_j) \times \boldsymbol{F}_{ij} = \boldsymbol{0}$ to express the content of NEWTON's third law or to be otherwise a law of mechanics therefore have no ground for regarding it as either more or less fundamental than the principle of moment of momentum, so long as they confine attention to systems of mass-points insusceptible to couples.

---

[40] W. NOLL, "The foundations of classical mechanics in the light of recent advances in continuum mechanics," *The Axiomatic Method, with Special Reference to Geometry and Physics*, Amsterdam, North Holland, 1959, 266—281. *Cf.* § 5 of the seventh essay in this volume, below.

[41] For discussion of this *principle of material indifference* and for an outline of its history, which goes back to work of POISSON in 1829, see § 19 A of *The Non-linear Field Theories of Mechanics.*

[42] *Cf.* Footnotes 5 and 26.

[43] Proofs of this statement and that under D are given by NOLL in the work cited in footnote 40 and are outlined in § 196 A of *The Classical Field Theories.*

F. In NEWTON's own statement of his third law, there is no explanation of what kinds of "bodies" he had in mind or what he meant by their "actions" on each other. He neither stated that all forces in the universe are binary nor laid down any condition whatever on any other kind of "actions". Not only in continuum mechanics do actions of a different sort occur. ERICKSEN has drawn my attention to work of CAUCHY[44] in which, apparently, is considered a system of mass-points having a potential energy depending upon the positions of all the masses, not necessarily the sum of pair potentials. ERICKSEN shows that by aid of CAUCHY's theorem on isotropic functions it follows from the requirement of material indifference, stated under C above, that the resultant torque of such a system of forces is zero. Therefore, for binary mutual forces the principle of material indifference is *equivalent* to the restriction to central forces, and the principle of moment of momentum follows. For more general mutual forces, not envisioned in (1) or in the statements about NEWTON's third law or action and reaction in any physics book I have ever seen, *the principle of moment of momentum continues to follow from the principle of material indifference*. Similar theorems hold in the classical theory of finite elastic strain[45] and in the theory of linearly viscous fluids.

G. The foregoing remarks show that all the formulations of mechanics discussed here are equivalent, as far as each goes. The principle of moment of momentum and the principle of material indifference are infinitely more inclusive than any statement about mutual forces between or among mass-points.

H. In modern researches on mechanics all the axioms listed under A, B, and C are needed, and for pure mechanics they suffice. This statement holds even when the definition of moment of momentum has to be generalized, as is the case in GRAD's theory of polyatomic gases[46] and ERICKSEN's theory[47] of liquid crystals.

---

[44] "Mémoire sur les systèmes isotropes de points matériels," *Mém. Acad. Sci. Paris* **22**, 615—654 (1850) = *Œuvres* (1) **2**, 351—386.

[45] W. NOLL, "On the continuity of the solid and fluid states," *J. Rational Mech. Anal.* **4**, 3—81 (1955). See § 16, Coroll. to Th. I.

[46] H. GRAD, "Statistical mechanics, thermodynamics, and fluid dynamics of systems with an arbitrary number of integrals," *Comm. Pure Appl. Math.* N.Y.U. **5**, 455—494 (1952). See § 4.

[47] J. L. ERICKSEN, "Anisotropic fluids," *Arch. Rational Mech. Anal.* **4**, 231—237 (1959), and later papers.

It seems to me that the approach to mechanics outlined by the following steps is a practical one today:

1. The axioms satisfied by forces and torques are motivated, and to some extent verified by experiments, in statics. These axioms are taken over unchanged in kinetics.

2. The principles

$$\boldsymbol{F}=0, \qquad \boldsymbol{L}=0 \tag{8}$$

as necessary and sufficient for equilibrium are motivated, and to some extent verified by experiment, in statics. Following the path opened by JAMES BERNOULLI, we assume that motion has the effect of giving rise to apparent forces that are equal, per unit mass, to the negatives of the accelerations in those frames where (8) are the statical equations. Application of this D'ALEMBERT-EULER principle leads to (7) as the general equations of dynamics, including both statics and kinetics.

3. The true and general idea imperfectly expressed by NEWTON's third law and the literature concerned with it, is stated by *the principle of material frame-indifference*: The properties of materials ("actions" of bodies on each other) are the same for all observers.

Bibliographical Note

This lecture was delivered at Columbia University and at the University of California, Los Angeles, in 1963 and at the University of Washington, Seattle, in 1964. It was published in the *Mélanges Alexandre Koyré,* in the collection "Histoire de la Pensée," Paris, Hermann, Volume 1, pp. 149—158, 1964. To the editors and publishers of that volume, I am indebted for their kind permission to reprint it.

A partial translation into German, with a different appendix, and taking account of couple stresses and spin angular momentum, was delivered in 1963 at the Universities of Freiburg i. Br., Berlin-Charlottenburg, and Aachen and was published as "Die Entwicklung des Drallsatzes," *Zeitschrift für Angewandte Mathematik und Mechanik*, Volume 44, pp. 195—205 (1964).

VI. Early Kinetic Theories of Gases

1. The Eighteenth Century

That matter is composed of molecules in violent agitation, and that heat is a manifestation of that motion, are ideas of great age. The first definite measure of the heat of intestine motion was proposed by HERMANN in 1716[1]. He stated that

[1] § 659 of J. HERMANN, *Phoronomia*, Amsterdam, 1716.

I do not consider a kinetic *theory* the bare statement that heat is a mode of motion of the parts of a body, or that pressure is the result of such motion, both of which statements may be traced back to the ancients. The molecular speculations of the baroque period, which attribute pressure to the forces exerted by particles on each other or to the effects of molecular vibration, are if anything further from the truth. As for HOOKE, we can hardly expect that a man who could not prove correctly the isochrony of vibrations of a mass attached to a simple spring would be in a position to initiate the kinetic theory of matter. HOOKE's remarks on the latter are as vague as those of DEMOKRITOS, without a hint toward calculation of momentum transfer or kinetic energy. There are, however, those who cannot distinguish between a vote for the only true church and the proof of a mathematical theorem, and these include not only professional historians of science but also professional physicists. *Cf.* the completely uncritical accounts by P. G. TAIT, "Hooke's anticipation of the kinetic theory, and of synchronism," *Proc. R. Soc. Edinb.* 1885 = [in expanded form] *Sci. Papers* **2**, 122—123, and E. N. DA C. ANDRADE, "Robert Hooke," *Proc. Royal Soc. London* A **201**, 439—473 (1950), especially pp. 460—461. It is not surprising that neither TAIT nor ANDRADE notices the fallacy, although it had been recognized as such at first glance by HUYGENS and LEIBNIZ, in HOOKE's "proof" of isochrony; ANDRADE even labels it "a capital discovery".

Other early authors sometimes mentioned in connection with a molecular explanation for the TOWNELEY-POWER-BOYLE law are NEWTON, JAMES BERNOULLI, and JOHN BERNOULLI. Their approaches are reflected in §§ 339—348 of HERMANN's book.

Fig. 95. The kinetic theory of JAMES HERMANN, from his *Phoronomia*, 1716

tionalis quantitati $EM.\ EB.\ Gg : EG.\ FG = 2EM.\ EB.\ GO : GN^2$, erit momentum sollicitationis acceleratricis, ut $2EM.\ EB.\ GO. - dGO : GN^2 = 2EM.\ BE.\ GN.\ dGN : GN^2 = 2EM.\ EB.\ dGN : GN$ proportionale momento celeritatis acquisitæ, seu VdV, vocando hanc celeritatem V, & dV ejus elementum, & $-dGO$ significat elementum decrescens lineæ GO, alterum vero dGN elementum ordinatæ GN. hinc $EM.\ \log.\ (GN^2 : BQ^2)$ est, ut quadratum celeritatis puncti progredientis G.

CAPUT XXIV.

De Motu intestino fluidorum.

658. Hoc nomine non intelligitur hoc loco internus moleculanum motus fluidi cujuscunque in suo statu naturali consistentis; sed is particularum motus, qui in fluidis à caussis externis & accidentalibus excitari solet, quo calor præsertim est referendus, qui dubio procul ex concitatiore particularum motu in corpore calido à caussis externis producitur. Utut vero ejusmodi motus intestinus admodum perturbatus sit, nihilo tamen minus regula physice satis accurata pro ejus mensura media tradi potest.

PROPOSITIO LXXXV. Theorema.

659. *Calor, cæteris paribus, est in compositâ ratione ex densitate corporis calidi, & duplicatâ ratione agitationis particularum ejusdem.*

Agitatio particularum est celeritas media inter celeritates particulares, quibus calidi corporis particulæ agitantur. Vocetur hæc celeritas media V, & corporis densitas D. Jam, quia calor consistit in concitatiore particularum motu, calor erit, ut impressiones particularum corporis calidi in quopiam objecto corpore calorem excipiente, sed hæ impressiones sunt in composita ratione ex duplicata celeritatum & simplâ densitatum, seu, ut $D.\ V^2$. Ergo etiam calor est ut $D.\ V^2$. Quod erat demonstrandum.

Dictum in propositione *cæteris paribus*, id est, in corporibus similis texturæ.

a fluid's molecules, which may be infinitely various, "may be thought to be cast here and there in various and irregular agitation." His fundamental "theorem" is, "Heat in bodies of like composition is in the compound ratio of the density of the hot body and the square of the agitation of its particles." His proof is no more than a reassertion of the statement, except that it makes clear what he means (Fig. 95). First, the "velocity" is the *mean speed* of the molecules. Second, the "heat" is what we should now call the pressure, which is indeed a manifestation of heat. HERMANN's assertion, then, amounts to

$$p = K\varrho\bar{v}^2, \tag{1}$$

where p is the pressure, v is the molecular speed, the bar denotes an average, ϱ is the density, and K is a constant depending only on the particular type of body.

The first to formulate a real kinetic theory was EULER (1729)[2]. In a paper written before his twenty-first year and originating from JOHN BERNOULLI's revision of DESCARTES' vortex cosmology, EULER conceived the air as made up of whirling spherical molecules closely packed together. Upon a spherical core of aether fits a shell of the "true substance of air", covered in turn by a shell of water. Air pressure is a manifestation of the centrifugal force which accompanies the whirling. The proportion of water to the true substance of air is a measure of humidity. The apparent density of the air is the mean density of these composite molecules. Assuming that at any given temperature the linear speed v of the vortex motion is the same for all the air and water particles, EULER derived an *equation of state* connecting p, ϱ, v, and the humidity, moderated by several molecular constants (Fig. 96). For dry air which is not too dense EULER obtained as an approximation

$$p \approx \tfrac{1}{3}\varrho v^2. \tag{2}$$

Even NEWTON slipped here. Not only is his treatment of a gas composed of "particles fleeing each other" thoroughly unconvincing, but also he claims to prove that if the TOWNELEY-POWER-BOYLE law holds, the intermolecular forces must be inversely proportional to the intermolecular distance. See the *Principia*, Lib. II, Prop. XXIII, Theor. XVII (XVIII in the second and third editions).

[2] L. EULER, "Tentamen explicationis phaenomenorum aeris," *Comm. Acad. Sci. Petrop.* 2 (1727), 347—368 (1729). Second edition (Bologna), 303—322 (1741).

Fig. 96. EULER's derivation of an equation of state from his kinetic theory, from the *Commentarii Academiae Scientiarum Petropolitanae*, Volume 2, for 1727

vnde $g^3 = ia^3 : (m - mp + np)$. Erit ergo pondus vi centrifugae aequiualens $= 2\pi nkaa\ [(\frac{m-i+pi-mp+pn}{m-pm+pn})^{\frac{2}{3}} - (\frac{m-pm+pn-i}{m-pm+pn})^{\frac{2}{3}}] = \frac{2\pi nkaa}{\sqrt[3]{(m-pm+pn)^2}}(\sqrt[3]{(mi+pi-pm+pn)^2} - \sqrt[3]{(m-pm+pn-i)^2})$.

XVI. Cum vi centrifuga efficiatur, vt bullulae aereae ſeſe continuo extendere conentur, erit ea aequalis vi elaſticae aeris; ex inuenta igitur aequatione, quanta ſit aeris elaſticitas, inueniri poterit. Verum cum hoc loco primum legem duntaxat, qua aeris vis elaſtica pro diuerſis denſitatis, humoris et celeritatis gradibus immutetur, perſequi conueniat, factor $2\pi n$ vtpote conſtans negligi poterit, eritque vis aeris elaſtica vt $\frac{kaa}{\sqrt[3]{(m-pm+pn)^2}}(\sqrt[3]{(m-i+pi-pm+pn)^2} - \sqrt[3]{(m-pm+pn-i)^2})$ Cum autem ſit $ia^3 = (m - pm + pn)g^3$, loco a in computum duco g, vt ea tanquam conſtans abiici poſſit; Erit ergo $aa = gg\sqrt[3]{(\frac{m-pm+pn}{i})^2}$, quo ſubſtituto erit vis elaſtica aeris vt $k(\sqrt[3]{(\frac{m-i+pi-pm+pn}{i})^2} - \sqrt[3]{(\frac{m-i-pm+pn}{i})^2})$. Coeteris igitur paribus eſt aeris vis elaſtica vt altitudo k, velocitatem materiae ſubtilis in bullulis gyrantis graui deſcendenti imprimens.

XVII. Verum cum vires aeris elaſticae inter ſe comparantur, id fit aeris vim expanſiuam in eandem baſin agentem explorando. Quamobrem, vt menſuram aeris vis elaſticae, vt conſuetum eſt, exhibeam, neceſſe eſt,

vt

Hence he concluded that at constant temperature $p \propto \varrho$, which is the TOWNELEY-POWER-BOYLE law; however, (2) is but an approximation to EULER's general formula taking account of an ultimate density beyond which the air cannot be compressed. EULER used BOYLE's data (1662) to conclude one relation between the various hypothetical molecular constants and to discuss deviations from the simple proportionality inferred by TOWNELEY and POWER. He commented on such experiments as would be needed to determine the remaining constants in his formula, and he made some plausible guesses about them.

If the approximate formula (2) suffices, v is easily calculated for each value of p and ϱ. At mean atmospheric conditions EULER obtained $v = 477$ m./sec. This value is very great, but, as EULER remarked, much larger speeds, such as that of light, exist in nature. Finally, EULER noticed that (2) gave

$$v = \sqrt{3}\sqrt{\frac{p}{\varrho}} = \sqrt{3}\, s_N, \tag{3}$$

where s_N is NEWTON's value for the speed of sound; that is, the molecules whirl at about the same speed as sound travels.

We all know that EULER's molecular model is not in accord with the current view of the gaseous state. However, his results serve as a pattern of what to expect from a rudimentary kinetic theory. Pressure and temperature are gross manifestations of molecular actions; the molecular model represents a gas as heavy, tending to expand indefinitely unless restrained, and susceptible of very great if not indefinite compression when subject to compressing forces. The outcome of the theory is to be an *equation of state*, enabling us to refine and correct the empirical gas laws as well as to identify the specific properties of different gases in terms of the nature and dimensions of their molecules. Particularly remarkable for its day is EULER's theory of humidity. According to modern ideas, the effect of humidity can be explained only in terms of gas mixtures.

Moreover, EULER's procedure, especially in its judicious mixture of facts of experience with a few simple calculations, while far from typical of his work in rational mechanics, affords an early specimen of the type of thinking now identified with "physics", and nearly all researches on the kinetic theory of gases for a century and a half were to follow this pattern.

EULER's theory, while indeed kinetic, is not statistical. The modern view, crudely blocked out by HERMANN, requires a theory which is both kinetic and statistical. It was first put into mathematical form by DANIEL BERNOULLI in an isolated, brilliant passage in his book on hydrodynamics[3], written by 1733. In the cylinder beneath the piston (Fig. 97) are "very minute corpuscles, which are driven hither and thither with a very rapid motion," and "practically infinite in number".

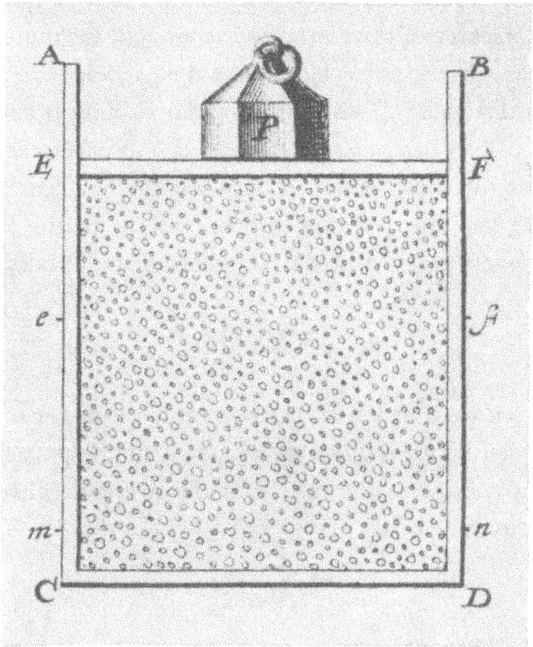

Fig. 97. DANIEL BERNOULLI's figure illustrating his kinetic theory of a gas, in the *Hydrodynamica*, 1738

In proposing this model, BERNOULLI wrote, "Let us therefore substitute for the air a fluid that conforms to the principal properties of elastic fluids, and thus we shall explain some of its properties already discovered and throw light on others not yet sufficiently thought through." The weight on the piston represents the pressure of the superincumbent atmosphere; it is balanced by the impacts of the particles on the piston. Now let the piston move downward. A greater weight must be supplied, first because the number of particles is greater in relation to

---

[3] Section 10, §§1—8 and 26 of *Hydrodynamica, sive de Viribus et Motibus Fluidorum*, Agentorati, 1738. Most of the material forming the contents was worked out in 1728—1733, and BERNOULLI upon his departure from Petrograd in 1733 left behind him the manuscript of the whole volume. Portions of this manuscript, which differs considerably from the text printed five years later, are included in the Russian translation of the whole book, published by Isdat. Akad.

the volume, and second because each particle strikes the piston more often. To calculate the effect from the first cause, we may consider the particles as if they were at rest (Figs. 98a, b). Let the height of the piston decrease in the ratio ζ; then the volume decreases, and the number of particles per unit volume increases, in the same measure; since the area of the piston is unchanged, the number of particles in contact with it increases in the ratio $(1/\zeta)^{\frac{2}{3}}$. Take the particles as spheres of diameter d, initially at "mean distance" D, so that the mean distance between their surfaces is $D-d$. When the piston has fallen, the mean distance between the centers has become $D\zeta^{\frac{1}{3}}$, so that the mean distance between the surfaces is now $D\zeta^{\frac{1}{3}}-d$. BERNOULLI assumes[4] that the pressure is proportional to the surface density of the particles and inversely proportional to the mean distance between them. Hence he infers that

$$\frac{p_0}{p} = \zeta^{\frac{2}{3}} \frac{D\zeta^{\frac{1}{3}}-d}{D-d}. \tag{4}$$

For the ratio of D to d we may substitute one which is easier to understand. Let the piston be depressed by an infinite weight, so that the particles are in mutual contact, occupying a volume in the ratio α to the original. Then $D/d = \alpha^{-\frac{1}{3}}$, and hence (4) may be written

$$\frac{p}{p_0} = \frac{1-\alpha^{\frac{1}{3}}}{\zeta - \alpha^{\frac{1}{3}}\zeta^{\frac{2}{3}}}. \tag{5}$$

Experiments indicate that air can be so greatly compressed that we can put $\alpha = 0$ and hence infer the TOWNELEY-POWER-BOYLE law: $p \propto \varrho^{-1}$. However, the analysis

---

Nauk SSSR, 1959. This volume contains also an essay on the works of DANIEL BERNOULLI by V. I. SMIRNOV.

DANIEL BERNOULLI is one of the most mysterious figures in the history of science, mysterious for no other reason than that nobody troubles to study through his works, while historians continue to attribute to him fame or blame for things he did not do or say. Much of the passage on the kinetic theory is printed in English translation in W. F. MAGIE's *A Source Book of Physics*, McGraw-Hill, 1938, pp. 247—251; most of it, in an inaccurate English version by J. P. BERRYMAN, is included in *Kinetic Theory* **1** (ed. S. G. BRUSH), Pergamon Press, 1965. A German translation, with a note by P. DU BOIS-REYMOND, was published in the *Annalen der Physik* **107** = (4) **17**, 490—494 (1859) and by A. SPEISER, pp. 139—143 of *Klassische Stücke der Mathematik*, Zürich and Leipzig, 1925.

[4] I do not think his reasoning is sound.

Fig. 98a (p. 279). DANIEL BERNOULLI's derivation of an equation of state from the kinetic theory, in the *Hydrodynamica*, 1738

Fig. 98b (p. 280). The same, continued

dus P operculum detinens in situ E F non differt à pressione Atmosphæræ superincumbentis, quam proinde per P in sequentibus designabimus.

Notetur autem hanc pressionem minime æqualem esse ponderi absoluto cylindri verticalis aërei operculo E F in atmosphæra superincumbentis, quod hactenus inconsiderate affirmarunt auctores: sed est pressio ista æqualis quartæ proportionali ad superficiem terræ, magnitudinem operculi E F & ponderi totius atmosphæræ in superficiem terræ.

§. 4. Quæratur jam pondus π, quod aërem E C D F in spatium e C D f condensare valeat, positis velocitatibus particularum in utroque aëre, naturali scilicet & condensato, iisdem: sit autem E C $=$ 1 & e C $= s$: Cum vero operculum E F transponitur in $e f$, majorem à fluido patitur nisum duplici modo: *primo* quod numerus particularum ratione spatii, cui includuntur, major nunc est, & *secundo* quod quævis particula sæpius impulsum repetit: ut recte calculum ponamus incrementi, quod à *prima* pendet causa, particulas considerabimus ceu quiescentes, atque numerum earum, quæ operculo in situ E F sunt contiguæ, faciemus $= n$, & erit numerus similis pro situ operculi in $ef = n : \left(\frac{eC}{EC}\right)^{\frac{2}{3}}$, seu $= n : s^{\frac{2}{3}}$;

Notetur autem fluidum à nobis considerari non magis condensatum in parte inferiori, quam in superiori, quale est, cum pondus P veluti infinitè majus est pondere proprio fluidi: Perspicuum hinc est, hoc nomine vim fluidi esse, ut sunt numeri n & $n : s^{\frac{2}{3}}$, id est, ut $s^{\frac{2}{3}}$ ad 1. Quod vero attinet ad alterum incrementum à *secunda* proveniens *causa*, invenitur id respiciendo motum particularum; atque sic apparet impulsus eo sæpius fieri, quo propius ad se invicem sitæ sunt particulæ: Erunt scilicet impulsuum numeri reciproce ut distantiæ mediæ inter superficies particularum: Istæque distantiæ mediæ ita determinabuntur.

Particulas ponemus esse sphæricas, distantiamque mediam inter centra globulorum pro situ operculi E F vocabimus D; diametrumque globuli designabimus per d: ita erit distantia media inter superficies globulorum $=$ D $-$ d: patet vero in situ operculi ef fore distantiam mediam inter centra globulorum $= \mathrm{D}\sqrt[3]{s}$, atque proinde distantiam mediam inter superficies globulorum $= \mathrm{D}\sqrt[3]{s} - d$. Igitur respectu secundæ causæ erit vis aëris na-

 tu-

turalis E C D F ad vim aëris compressi e C D f ut $\frac{1}{D-d}$ ad $\frac{1}{D\sqrt[3]{s}-d}$, seu ut $D\sqrt[3]{s}-d$ ad $D-d$: Conjunctis vero ambabus causis erunt prædictæ vires, ut $s^{\frac{2}{3}}\times(D\sqrt[3]{s}-d)$ ad $D-d$.

Rationi D ad d aliam substituere possumus magis intelligibilem: nempe si putemus operculum EF pondere infinito depressum descendere usque in situm mn, in quo particulæ omnes se tangunt, atque lineam mC vocemus m, erit D ad d ut 1 ad $\sqrt[3]{m}$, quâ ratione substituta, erunt tandem vires aëris naturalis E C D F & compressi e C D f ut $s^{\frac{2}{3}}\times(\sqrt[3]{s}-\sqrt[3]{m})$ ad $1-\sqrt[3]{m}$, seu ut $s-\sqrt[3]{mss}$ ad $1-\sqrt[3]{m}$. Est igitur $\pi=\frac{1-\sqrt[3]{m}}{s-\sqrt[3]{mss}}\times P$.

§. 5. Ex omnibus phænomenis judicare possumus aërem naturalem admodum condensari posse, & fere in spatiolum infinite parvum comprimi; facta igitur $m=0$, fit $\pi=\frac{P}{s}$, ita ut pondera comprimentia sint fere in ratione inversa spatiorum, quæ aër diversimode compressus occupat; quod multiplex experientia confirmavit. Et potest certe hæc regula tuto accipi in aëre rariore quam est naturalis; an vero etiam possit in aëre admodum densiori, non satis exploratum habeo: nec dum enim fuerunt experimenta ea accuratione, quæ hic requiritur, instituta: unico opus est ad definiendum valorem litteræ m, sed eo accuratissime instituendo & quidem cum aëre vehementer compresso; gradus autem caloris in aëre, dum comprimitur, sollicitè invariatus conservetur.

§. 6. Elasticitas interim aëris nonsolum à condensatione augetur, sed & ab aucto calore, & quia constat calorem intendi ubique crescente motu particularum intestino, sequitur, elasticitatem aëris spatium non mutantis auctam, intensiorem arguere motum in particulis aëris, quod cum hypothesi nostra recte convenit: perspicuum enim est, eo majus requiri pondus P ad continendum aërem in situ ECDF, quo majori velocitate particulæ aëreæ agitantur: Imo non difficile est videre pondus P secuturum rationem duplicatam istius velocitatis, ideo quod ab aucta velocitate tum numerus impetuum tum intensitas eorundem æqualiter crescat, utrumq; vero seorsim proportionale sit ponderi P.

Igitur

suggests that for extremely dense air we could determine α by experimental deviation from that law; but very high pressures and great accuracy, with special care to keep the temperature constant, would be required. Thus BERNOULLI calculates a *correction* to the gas laws by taking account of the finite size of the molecules [5].

BERNOULLI states further that adding heat increases the internal motion of the particles. Increasing the velocity requires a greater weight. "It is not difficult to see that the weight should be proportional to the square of this velocity because, when the velocity increases, not only the number of impacts but also the intensity of each of them increases equally." Perhaps influenced by the earlier work of his colleagues HERMANN and EULER, he concludes in effect that

$$p_0 = K v^2. \tag{6}$$

Thus (5) becomes an equation of state of the form

$$p = K \varrho v^2 f\left(\frac{\varrho}{\varrho_0}, \frac{\varrho_m}{\varrho}\right), \quad f(x, 0) = 1, \tag{7}$$

where K is a dimensionless constant, f is a dimensionless function, ϱ_m is the mean density of the molecules, and ϱ_0 is a reference density. In BERNOULLI's work, as in HERMANN's, v is the speed of translation of the molecules. If $\alpha = 0$, or, equivalently, $f(\zeta, \eta) \equiv 1$, BERNOULLI's result reduces to EULER's first approximation (2), except that for EULER v was the speed of rotation. BERNOULLI interpreted his analysis as showing that pressure increments produced by equal changes of temperature are proportional to the densities, in accord with the law of AMONTONS. BERNOULLI proposed to regard v^2 itself as a scale of temperature; to measure it, he projected and constructed an air thermometer.

The motion of gross bodies on the earth is always seriously impeded by friction; heavenly motions are not, but in heavenly motions collisions are rare, while in BERNOULLI's model pressure is a manifestation of innumerable collisions with the vessel walls. Nevertheless BERNOULLI had good experimental grounds for supposing all these collisions perfectly elastic: ROBERVAL had imprisoned air in a bottle for fifteen years and upon releasing it had detected no loss of spring.

HERMANN and BERNOULLI conceived the molecules and their translatory speeds as infinitely various; EULER considered the molecules identical, their whirling speeds all the same, and only their axes of rotation random; the modern view of

[5] As suggested by the preceding footnote, his result is not the one now generally accepted.

a pure gas regards the molecules as identical. All three theories are in qualitative accord with basic gas phenomena, and all three yield results which are more or less the same in a certain range. Indeed, it is obvious that entirely different molecular models may predict many gross phenomena in common. This inherent insecurity of molecular theories I will not labor, but I must remark that in many a positive assertion regarding molecular models it is nowadays forgotten.

Thus by 1738 three members of the Basel school of mathematicians [6], all of them working in Petrograd, had laid out in mathematical form the elements of the modern molecular concept of the gaseous state and had derived equations of state. In the next hundred years, other kinetic ideas were put forward. From 1750 to 1760 LOMONOSOV [7] made a number of statements about molecular structure. He mentioned translation, vibration, and rotation of the molecules and speculated about the effect of collisions. At much the same time, BOSCOVICH [8] chose to regard the molecules as mathematical points and conjectured elaborate laws of central intermolecular force which are attractive at great distances but repulsive at moderate ones. These views, like the many molecular speculations which preceded EULER's, were too murky and too entangled to form the basis of mathematical theories. With the recent rise of the profession of non-mathematical historians of non-mathematical physics, LOMONOSOV and BOSCOVICH have become heroes for those who consider a well-painted word picture just as good as a mathematical calculation or a conclusive experiment.

---

[6] Late in life, fifty years after he had written his first paper on the kinetic theory, EULER came back to it in his "Conjectura circa naturam aeris, pro explicandis phaenomenis in atmosphera observatis," *Acta Acad. Sci. Petrop.* 1779: I, 167—187. The idea is much the same, but the analysis is simpler. In this late work EULER, perhaps influenced by DANIEL BERNOULLI and others, remarks that as v^2 in (2) increases, so does the ratio of p to ϱ. Since in experience this ratio increases with the temperature, greater heat (*i.e.* temperature) must correspond to a greater whirling speed. Since accurate measures of heat did not then exist, EULER proposed that henceforth $\frac{1}{3}v^2$ be taken as the definition of heat. This second paper is analysed in greater detail in Part IIIB of my introduction to L. EULERI *Opera Omnia* II **13**, 1956.

[7] See pp. 8—55, 106—139, 145—163 of Volume 2 and pp. 378—409 of the Полн. Собр. Соч. of LOMONOSOV.

[8] R. J. BOSCOVICH, *Theoria Philosophiae Naturalis...*, Vienna, 1758. I have consulted only the revised edition of 1763, republished, with an English translation, by J. M. CHILD, Open Court, 1922. Beyond that fact that KELVIN cited BOSCOVICH as a predecessor in atomics, quite evidently only so as to make a generous gesture toward the past and without owing anything from his reading (if any) of what BOSCOVICH wrote, I am unable to find any influence, let alone concrete achievement, in the treatise of the learned Slovenian Jesuit.

2. Herapath (1816—1821)

The kinetic theory remained dormant, indeed forgotten, until 1816, when HERAPATH[9] proposed a theory which is essentially BERNOULLI's. Unfortunately he chose to define temperature as being proportional to the momentum rather than the kinetic energy of the molecules. HERAPATH was the first to show, more or less, that the kinetic theory can provide simple explanations for changes of state, diffusion, and the propagation of sound. While his ideas were not entirely correct, the hostility they aroused seems to have arisen rather from the difference between the scene in which he found himself and that of a century earlier. By the 1800's, in England at least, original speculation of any kind and mathematical theory in particular were distrusted. Mechanics had been established and exhausted by NEWTON. Heat was explained in gross by the caloric theory, and on the molecular scale by the vibratory motion of atoms within an aetherial substance, not a fit subject for mathematics.

HERAPATH was self-educated, having learnt mathematics and natural philosophy from the writings of NEWTON and the great French mathematicians of his own day. He submitted his main paper to the Royal Society in 1820 after having made the necessary tactful inquiries of influential members. DAVIES GILBERT, the Fellow to whom he submitted it, replied within a fortnight: " [while] I was much pleased with the great ingenuity displayed throughout the whole, ... I entertained strong doubts on the propriety of laying before the Royal Society, anything so abstruse and metaphysical. I, therefore, desired two of the best mathematicians in London to look at the premises; and their opinion confirmed my doubts. They say, that such a work should be laid before the public in a separate form."

HERAPATH offered to explain the mathematics (in fact merely elementary, even primitive) and so as to confirm his predictions went on to perform some experiments on mixing of liquids at various temperatures. GILBERT and his august colleagues stood fast; "some members of the Council, who are usually looked up to on such occasions, ... considered the investigations too theoretical for the

---

[9] J. HERAPATH, "On the physical properties of gases," *Annals of Philos.* (1) **8**, 56—60 (1816).
"A mathematical investigation into the causes, laws, and principal phaenomena of heat, gases, gravitation, etc.," *Annals of Philos.* (2) **1**, 273—293, 340—351, 401—416 (1821).
"On Mr. Tredgold's 'refutation' of Mr. Herapath's Theory," *Annals of Philos.* (2) **1**, 303—307 (1821).
Mathematical Physics ..., London, 1847.

Transactions, without taking on themselves to judge of the mathematics." GILBERT proposed that the revised paper, including the experiments, be put in the hands of the new president and council, to be elected a month later.

HERAPATH expected to find a backer in the new president, Sir HUMPHRY DAVY, who was an avowed atomist. To support himself, HERAPATH was just then setting up "a school for mathematics, and for preparing young men for the navy." He wished "to strain every nerve to establish my scientific reputation, as well for the sake of maintaining myself in the good opinion of my friends, as for the benefit which it may be of to my family, in my business." He was neither the first nor the last to learn how fragile is the hope of help from rajahs of science.

DAVY wrote to HERAPATH that he had

"read those parts of it which are intelligible, without profound mathematical study, with attention; and highly ingenious as I find your views, I must say I am not impressed with a conviction of their truth .... Before I return the paper, I shall take the liberty of submitting it to the most acute philosopher (I know) in this country....

"There is so much ingenuity and so minute an acquaintance with the progress of discovery displayed in your paper, that I cannot help wishing that its views and objects had been limited to matters of pure experimental inquiry. For instance, the doctrine of heat, and the investigation of its laws, supposing it to be motion. Such a preliminary paper, if satisfactory, would prepare the philosophical world for greater and more abstruse researches. You refer, in your last letter, to some experiments which you have concluded, on heat, and which confirm your general views. Any investigation of this kind, I shall have great pleasure in communicating to the Royal Society. Should anything call you to London, I hope you will favour me with a call; I shall be glad to see you at my house any Wednesday evening, after nine o'clock. You will always find some of our most distinguished men of science with me on that day. I am, Sir, with much consideration,

Your obedient humble servant,
H. DAVY"

HERAPATH paid the visit at the very next opportunity, four days later. He was introduced to THOMAS YOUNG, "who, on my asking him, said, he had not seen or heard of the paper, yet, in a few minutes after, discussing the mode of making some experiments, he observed, 'but you, in your paper,' (referring to this very one) 'say ....'"

DAVY wished to reject the theory and publish the experiments. HERAPATH refused but got into an argument about the experiments, resting on his unfortunate

Fig. 99. JOHN HERAPATH (1790—1869), by courtesy of SPENCER D. HERAPATH, Esq., after a portrait painted about 1815, destroyed in 1951

definition of temperature. Here HERAPATH was in the wrong, but DAVY, with a talent common among great and important officers, chose to attack him instead on a point where he was right and invincible. DAVY wrote, "... having considered a good deal the subject of the supposed real zero, I have never been satisfied with any conclusions respecting it. I cannot see any necessary connexion between the capacity of bodies for heat, and the absolute quantity they contain; and temperature does not measure a quantity, but merely a property of heat."

HERAPATH's paper, withdrawn from the Royal Society, was published in a minor journal in 1821, but since it had been damned already by the great of British science, it attracted little notice in Britain and none whatever abroad. The little British notice was hostile, and HERAPATH in the course of answering some objections published an extract from the letter in which DAVY had referred to comprehending the paper "without profound mathematical study". DAVY, self-condemned but self-righteous, wrote to HERAPATH not to expect anything kind of him in the future. The Royal Society refused to witness HERAPATH's experiments, and non-official journals began to reject his papers about them. HERAPATH considered himself persecuted by "the insidious propagation of reports". Five years later, in 1826, he laid out the whole matter in letters to *The Times*, publishing most of the correspondence, connected by his account of the whole affair. No answer. On the 10th of January, 1826, HERAPATH challenged DAVY:

"Problem of Defiance to Sir H. Davy and the Royal Society.
From two postulates only, of less than fifty words, the one on the nature of heat, the other on

---

Fig. 100. First and last pages of a letter from HERAPATH to an unknown correspondent on February 23, 1864.

HERAPATH states that his theory was "perfected in 1814," before he had any idea that others had touched the subject. He concludes, "... There are other parts of my labours in relation to gases and their chemical combinations, &c which I think, if I can ever find leisure to publish what I have had by me for many years, will change the ideas entertained respecting elastic particles in gases and some other points of Physics ... Mine is strictly a mechanical system of Physics, based on the property of absolute hardness in the component atoms or particles." The two pages not shown explain HERAPATH's views on the impact of bodies. The original letter belongs to C. TRUESDELL. The concluding remarks about electricity suggest that the letter may be addressed to MAXWELL, who took up the kinetic theory between the appearance of the February, 1859, issue of the *Philosophical Magazine*, and 30 May, 1859, on which date he mentioned in a letter to STOKES "Clausius' (or rather Herapath's) theory". In his paper of 1860, MAXWELL had mentioned DANIEL BERNOULLI and HERAPATH as having derived equations of state from the hypothesis of uniform rectilinear motions, and then CLAUSIUS in 1862 had given a list of early authors on the kinetic theory, such as that which HERAPATH in this letter acknowledges having received in a letter of February 8, 1864, to which he is replying.

Catford Bridge,

LEWISHAM, KENT, S.E.

Feb 23 1864

Dear Sir

Indifferent health and many pressing [illegible] must I hope plead for my delay in replying to your obliging note of the 8th inst.

I thank you most cordially for the trouble you have taken & the information [illegible] respecting the [illegible] theory of gases [illegible] a particle of which I had never heard of before I published my book, except a slight allusion [illegible] to [illegible] idea of gravi- [illegible] many years ago [illegible] after my theory of gravitation was perfected in 1814.

I think I published something about it in Dr Thomson's Annals of Philosophy in 1816, but of it I have no account of it by me. In 1821 however some papers of mine were published in the Annals on the collision of perfectly hard bodies & the theory of heat and gases.

The theory of the division of absolutely hard bodies rests on the [illegible]

of the [illegible] particles.

The theory of elastic particles is touched on in Vol. I pp. 245, & 87 Vol I Ma. Ph. but there are other parts of my labours in relation to gases and their chemical combination, &c which I think, if I can ever find leisure to publish what I have had by me for many years, will change [illegible] elastic particles in gases & some other parts of Physics.

You will perceive that mine is strictly a mechanical system of Physics, based on the property of absolute hardness in the component atoms or particles. The cause of the opposition I think [illegible] what you have I perceive [illegible] though [illegible] I [illegible]

[illegible]

J. Herapath

that of a gas, to determine from theory alone in a function of the altitude, the law of the diminution of temperature and elasticity of any air surrounding a globe of uniform temperature, whose action varies according to any law whatever."

While this is but the beginning, I will not detail the rest. If you are expecting a glorious Hollywood finale in which HERAPATH is given a golden Base Prize and then elected Fellow of the Royal Society, installed at a special ceremony in which his chastened and converted old detractors rise to applaud him, while he is toasted from the galleries packed with beautiful women, that finale never occurred. Official silence greeted HERAPATH's challenge problem. A second and a third met the same response. DAVY resigned from command, but the war went on. HERAPATH complained that the Fellows of the Royal Society whispered he could not solve his own problems, and he ought to enter the field in pure mathematics as being fairer to all. (Since he had done original work in the theory of gases, in it he plainly had a head start on the Fellows, most of whom had done no original work of any kind, so it was only decent, in a sporting trial of skill, to choose from the common school background, in which HERAPATH had not shown any special distinction.) HERAPATH was drawn in. He proposed a mathematical problem and backed his stand with a substantial bet. No reply. In 1828 HERAPATH wrote a final letter to *The Times*, mentioning that the time for solving all his problems had expired without a single written reply or comment. His letter concludes, "I hope never to appear before the public again on such business."

Perhaps DAVY's attitude is reflected in his brother's biography of him, published not long afterward. In discussing the onerous burdens of a President of the Royal Society, JOHN DAVY wrote in regard to disappointed authors, "The man of real ability or of true dignity would be above the Royal Society, and would not condescend to resent any act of injustice toward him .... He has the world for his tribunal; and it is only necessary for him to publish the results of his inquiries, and he is sure to have justice done to him ... ." This cozy attitude seems not to have been confirmed in HERAPATH's case. Today, DAVY remains a general hero of science, while even specialists in statistical mechanics often have never heard HERAPATH's name. Time has proved that the opinion of a President of the Royal Society, in a case when he was not only wrong but also foolish, outweighs pages of sound, logical calculation by an original but unknown outsider. Neither DAVY nor anyone else then in the Royal Society, it seems, knew enough mathematical physics to appreciate the earlier work of HERMANN, BERNOULLI, and EULER, which was never mentioned in the course of the controversy.

3. Joule and Krönig (1851, 1856)

HERAPATH'S theory joined its predecessors in oblivion (Fig. 100), but not for long. In 1851 JOULE[10] published a note in which, without writing down an equation or using a symbol, he calculated the speed of the molecules.

"Let us suppose an envelope of the size and shape of a cubic foot to be filled with hydrogen gas, which, at 60° temperature and 30 inches barometrical pressure, will weigh 36.927 grs. Further, let us suppose the above quantity to be divided into three equal and indefinitely small elastic particles, each weighing 12.309 grs.; and, further, that each of these particles vibrates between opposite sides of the cube, and maintains a uniform velocity except at the instant of impact; it is required to find the velocity at which each particle must move so as to produce the atmospherical pressure of 14,831,712 grs. on each of the sides of the cube." To solve this problem JOULE calculated the momentum imparted to the wall at each impact. Then, "since it is manifest that the pressure will be proportional to the square of the velocity of the particles," the speed itself may be obtained directly. JOULE'S result is 6255 ft./sec., like EULER'S more than a century earlier an astonishingly large speed, which, had his paper been noticed, might well have changed the apathy with which British scientists greeted his work into active hostility.

On the whole, JOULE simply repeated in cruder form the calculations of BERNOULLI and EULER. He had begun in 1844 with ideas much like EULER'S but had seen HERAPATH'S paper in 1848 and had adopted HERAPATH'S hypothesis of molecular translation rather than spin. Like EULER, he concluded from (2) that temperature is proportional to v^2, but since v is now the speed of translatory motion, it must follow that temperature is proportional to *vis viva*. Immediately

---

[10] J. P. JOULE, "Some remarks on heat and on the constitution of elastic fluids" (1848), *Mem. Manchester Lit. Phil. Soc.* **9**, 107—114 (1851) = (with added notes) *Phil. Mag.* (4) **14**, 211—216 (1857) = *Papers* **1**, 290—297.

In his article on JOULE, cited at the end of this essay, MENDOZA states that JOULE'S unpublished notebooks show that in 1844—1845 his ideas were close to EULER'S of a century before (though of course not reaching the same level of mathematics and concreteness), while it was his reading of HERAPATH'S papers that caused him to adopt HERAPATH'S hypothesis in October, 1848.

Like HERAPATH, JOULE was self-educated and unspoiled by the rotten Cambridge tradition of his day. Unlike HERAPATH, JOULE found within the stream of new men who were enthusiastically transplanting to Britain the French way of looking at mathematical physics a strong and unswerving defender, adherent, and collaborator: WILLIAM THOMSON.

JOULE concluded that "the zero of temperature is 491° below the freezing point of water," such being the temperature at which the pressure falls to zero if we extrapolate by the law of AMONTONS, using an experimental value for the constant of proportionality in hydrogen. JOULE went beyond his predecessors in proposing a definite theory of specific heats, but it is wrong.

I have taken pains to reproduce the arguments of BERNOULLI and JOULE in their original forms. Their work was not noticed when it was published. BERNOULLI's calculation appears in the middle of a treatise on cisterns, conduits, and fountains; moreover his is but one of several kinetic theories of the period, and that it should turn out to be the only one in accord with later views of the gaseous state, was hardly to be expected. JOULE was a private individual without university training or associations, unsupported by any of those powerful groups of eminent authorities which control science by bullying scientists, not only in his day, alas, but also in ours. There are persons who claim that work which is without influence does not belong in a history of science. Perhaps they transfer the operationalist philosophy of modern physics to the historical domain, regarding papers that do not happen to be read by physicists, like events that do not happen to be observed by physicists, as not existing at all. For such persons the history of the kinetic theory begins in 1856 with a paper by KRÖNIG[11].

KRÖNIG's ideas, while much the same as those of BERNOULLI and JOULE, are a little easier to follow and a little less correct.

Though KRÖNIG stated that there is a statistical element in the problem, he frankly expressed a view of the theory of probability which has been shared, though with greater reserve, by nearly all his successors in the kinetic theory until recently and which is implied in many textbooks for physicists. After claiming that gases consist of perfectly elastic spheres moving in random fashion in empty space, he wrote, "By the laws of probability theory we can replace this complete chaos by complete regularity." His mechanical model, slightly generalized, is based on the following ideas:

1. A chemically homogeneous gas in equilibrium is represented as a system of identical mass-points or rigid bodies, called "molecules", moving at random, so that the number moving in any one direction is on the average the same as that moving in any other.

---

[11] A. KRÖNIG, "Grundzüge einer Theorie der Gase," *Annalen der Physik* **99** = (4) **9**, 315—322 (1856).

2. The volume of the molecules is negligible with respect to the volume of the void.

3. Intermolecular forces are negligible except at the instant of impact; thus the molecules move at uniform velocity until they impinge upon other molecules or upon the walls.

4. Since a confined and insulated gas shows no tendency to lose its pressure, the impacts of molecules upon each other and upon the walls must be perfectly elastic. In particular, if the molecules are rigid spheres, molecular collisions may be neglected altogether, for when two such spheres collide, the subsequent motion is the same as if they had passed through each other without mutual influence.

There is no need for me to repeat here the arguments of KRÖNIG. As may be seen in any good freshman book[12] on physics or chemistry, these ideas lend themselves easily to calculation of the pressure resulting from transfer of momentum from impacts of the molecules upon the walls. The right answer is easy to get:

$$p = \tfrac{1}{3}\varrho\overline{v^2}, \tag{8}$$

but KRÖNIG did not get it. He first made a mistake in the calculation of momentum lost, and then he threw all factors of proportionality away, obtaining $p = \varrho\overline{v^2}$, and he made other errors as well.

While the work of HERMANN, EULER, and BERNOULLI had been forgotten, and HERAPATH was ignored as a crank, KRÖNIG's rudimentary paper, not even correct in detail, was published, read, and respected, and to this day is generally regarded as the starting point of the modern kinetic theory. Indeed, KRÖNIG's paper made no advance. To account for the favor bestowed upon it, we can remark that in the years just before it was printed the mechanical theory of heat had been established experimentally, and that KRÖNIG, however small his name may sound today, in his own time was a person important enough for other scientists to pay attention to him.

---

[12] *E.g.*, §§113—117 of MILLIKAN, ROLLER, & WATSON, *Mechanics, Molecular Physics, Heat, and Sound*, Boston, 1937. Since this book has recently been the object of an oblique jibe by a famous nuclear physicist in a course of lectures intended to replace it, and since there is little chance I shall ever again have occasion to cite it, I record here an expression of my life-long gratitude for my good luck in having been a freshman at the California Institute of Technology in 1938—1939 and thus privileged to learn the elements of mechanics from it. Its sound and clear physics, clean if simple mathematics, generous scope and detail, good writing, apposite choice of problems, and high historical scholarship give it a unique place among textbooks.

4. Waterston (1843—1846)

If the work of HERMANN, EULER, BERNOULLI, HERAPATH, and JOULE was unnoticed, what shall we say of WATERSTON'S? Perhaps none of you has heard his name before. On December 11, 1845, six years before JOULE'S publication and eleven years before KRÖNIG'S, the Royal Society received from Bombay a memoir by WATERSTON entitled "On the physics of media that are composed of free and perfectly elastic molecules in a state of motion". The next year, the Royal Society published a half-page abstract[13] of the paper, with the author's name misspelled; in it we read only that WATERSTON used a molecular model like BERNOULLI'S and that he discussed the gas laws, the effect of different molecular masses, condensation and dilatation and the value of the *vis viva*, the resistance of media to a moving surface, the equilibrium of the atmosphere, and the velocity of sound. In 1851 in connection with a meeting of the British Association WATERSTON published a more descriptive one-page abstract[14] (Fig. 102). Among other things it states that "the *vis viva* of those [random] motions in a given portion of gas constitutes the quantity of heat contained in it," and "equilibrium of pressure and heat between two gases takes place when the number of atoms in unity of volume is equal, and the *vis viva* of each atom equal. Temperature, therefore, in all gases, is proportional to the mass of one atom multiplied by the mean square of the velocity of the molecular motions, being measured from an absolute zero[15] 491° below the zero of Fahrenheit's thermometer." Thus WATERSTON was the first to publish the modern kinetic-theory definition of temperature and the simplest case of the law of equipartition.

The original memoir of WATERSTON is a masterly exposition of all the kinetic-theory notions we have explained thus far and of related ones. It is easy and persuasive reading. The results are summarized in sweeping italicized statements at the end of each argument. By a slip in calculation WATERSTON obtained $\gamma = \frac{4}{3}$ rather than $\frac{5}{3}$, rendering invalid certain other numbers, but the reasoning is

---

[13] J. J. WATERSTON, "On the physics of media that are composed of free and perfectly elastic molecules in a state of motion" (abstract), *Proc. R. Soc. London* **5**, 604 (1846) = *Phil. Trans. R. Soc. London* A **183**, 78 (1893) = *Papers* 317—318. This abstract can scarcely have been composed by WATERSTON himself.

[14] WATERSTON, "On a general theory of gases," *Rep. Brit. Assn. Adv. Sci.* 1851, 6 = *Phil. Trans. R. Soc. London* A **183**, 79 (1893) = *Papers* 318—319.

[15] According to other work of WATERSTON, this must be a misprint for 461°.

Fig. 101. JOHN JAMES WATERSTON, 1811—1883, after a photograph at the age of 46, taken by his nephew, GEORGE WATERSTON, reproduced by courtesy of ROBERT WATERSTON, Esq.

right. WATERSTON was familiar with all the relevant experimental data known at the time, and his paper contains many numerical results and comparisons with experiment. Included is a calculation of the mechanical equivalent of heat. The treatment of the resistance encountered by a moving surface is faulty for the same

solids; seeing that any amount of rigidity, however small, will account for the phænomena, if we adopt certain suppositions as to molecular forces.

V. Our knowledge of molecular forces is not as yet sufficiently advanced to enable us to use experiments on the velocity of sound as a means of determining accurately the coefficients of elasticity of solids.

If we adopt for them the hypothesis already stated with respect to liquids, a theoretical investigation given in an Appendix shows that the velocity of sound in a cylindrical rod of an uncrystallized substance, whose surface is absolutely free, will be less than that in an unlimited expanse in a ratio which is sensibly $1 : \sqrt{2}$ for a very slender filament, and approaches $\sqrt{2} : \sqrt{3}$ as the diameter of the rod increases; but the absolute freedom of the surface cannot be realized in practice; the means used in fixing the rod tend to restrain the lateral oscillations and accelerate the velocity of sound. The ratios ascertained by experiment range from $\sqrt{2} : \sqrt{3}$ to near equality; but they are not sufficiently numerous to form data for any definite conclusions.

The oscillations treated of in the special problems of the body of the paper being of a kind called *nearly longitudinal*, a second Appendix is added, containing the general equations of another kind, called *nearly transverse*, in uncrystallized bodies.

On a General Theory of Gases. *By* J. J. Waterston, *Bombay.*

The author deduces the properties of gases, with respect to heat and elasticity, from a peculiar form of the theory which regards heat as consisting in small but rapid motions of the particles of matter. He conceives that the atoms of a gas, being perfectly elastic, are in continual motion in all directions, being restrained within a limited space by their collisions with each other, and with the particles of surrounding bodies. The *vis viva* of those motions in a given portion of gas constitutes the quantity of heat contained in it.

He shows that the result of this state of motion must be to give the gas an elasticity proportional to the mean square of the velocity of the molecular motions, and to the total mass of the atoms contained in unity of bulk; that is to say, to the density of the medium. This elasticity, in a given gas, is the measure of temperature. Equilibrium of pressure and heat between two gases takes place when the number of atoms in unity of volume is equal, and the *vis viva* of each atom equal. Temperature, therefore, in all gases, is proportional to the mass of one atom multiplied by the mean square of the velocity of the molecular motions, being measured from an *absolute zero* 491° below the zero of Fahrenheit's thermometer.

If a gas be compressed, the mechanical power expended in the compression is transferred to the molecules of the gas increasing their *vis viva*; and conversely, when the gas expands, the mechanical power given out during the expansion is obtained at the expense of the *vis-viva* of the atoms. This principle explains the variations of temperature produced by the expansion and condensation of gases—the laws of their specific heat under different circumstances, and of the velocity of sound in them. The fall of temperature found on ascending in the atmosphere, if not disturbed by radiation and other causes, would correspond with the *vis viva* necessary to raise the atoms through the given height.

The author shows that the velocity with which gases diffuse themselves is proportional to that possessed by their atoms according to his hypothesis.

Light, Heat, Electricity, Magnetism.

On the Conduction of Electricity through Water. *By* F. C. Bakewell.

Mr. Bakewell stated the results of some experiments on the conduction of electricity by water, made with a view to prove that an electric current may be transmitted for a considerable distance through unprotected wires immersed in water. The experiments were conducted on Saturday last in one of the Hampstead ponds. A thin copper wire (No. 20), three hundred and twenty feet long, was stretched across the pond, and two copper plates ten inches square, to which wires were soldered,

reason as was NEWTON'S, *viz*, neglect of the action of reflected molecules on those about to impinge, and naturally the solution to the problem of equilibrium of the atmosphere, an elaborate matter still imperfectly understood, is not satisfactory. Finally in an appended note WATERSTON constructed a theory of chemical combination of gases differing from that now accepted only in that he did not recognize molecules as composed of indivisible atoms. At the end of the note is a great chart of the molecular properties of specific substances. To say that WATERSTON in 1845 had worked out all of the kinetic ideas and results that were to be published by others in the next eleven years thus fails to do justice to his great paper.

In submitting it to the Royal Society, he wrote: "Whether gases do consist of such minute elastic projectiles or not, it seems worth while to enquire into the physical attributes of media so constituted, and to see what analogy they bear to the elegant and symmetrical laws of aeriform bodies. Some years ago I made an attempt to do so, proceeding synthetically from this fundamental hypothesis .... The results have appeared so encouraging, although derived from very humble applications of mathematics, that I have been led to hope a popular account of the train of reasoning may not prove unacceptable to the Royal Society."

In 1845 physicists believed either that heat is a fluid which can pass through continuous matter, or that heat is a vibratory motion of molecules. They did not wish to think about other possibilities. Moreover, as there developed, first in Britain and then on the Continent, a straitening of the channel of physics until its walls nearly met at the center line of tables of data from experiments, anything speculative, unless written by a recognized authority, had already become distasteful, being relegated to mere philosophy. Then as now, docile and laborious calculation of details within the accepted views was preferred above search of principle. The first referee objected to the underlying assumption that the gas pressure results from molecular impacts, which he asserted was "by no means a satisfactory basis for a mathematical theory." While he conceded that the paper "admits much skill and many remarkable accordances with the general facts as well as numerical values furnished by observation," he did not think it worthy to be published in the *Philosophical Transactions*. The second referee wrote that "the paper is nothing but nonsense, unfit even for reading before the Society." The paper was rejected. The system of secret refereeing, still in almost exclusive

---

Fig. 102. WATERSTON'S abstract of 1851, giving his statement of equipartition

use today, has protected the anonymity of the eminent referees, while burying the genius of WATERSTON in permanent oblivion[15a].

According to the Society's rules, manuscripts sent to it became its property forever. WATERSTON submitted a written request that his manuscript be returned, but it was not. In 1891 Lord RAYLEIGH found it in the archives; he caused it to be published in 1893 in the journal to which it had been submitted forty-eight years earlier[16]. He furnished it with an introduction, in which he defended the conscientiousness of the second referee, "one of the best qualified authorities of the day", adding the counsel that "highly speculative investigations, especially by an unknown author, are best brought before the world through some other channel than a scientific society ...," and "a young author who believes himself capable of great things would usually do well to secure the favourable recognition of the scientific world by work whose scope is limited, and whose value is easily judged, before embarking upon higher flights." We are to presume that these words, while if published today they would be read as sarcasm, were written in Victorian earnest. RAYLEIGH praised the "marvellous courage" of the work and the great advance it attained, according WATERSTON priority in the law of equipartition on the basis of the abstract from 1851. By this time WATERSTON had been dead for ten years[17]. That he made no claim of priority later when the kinetic theory became a recognized field, is not hard to explain. First, he learned that a part of his views had been put forth earlier by BERNOULLI and by HERAPATH. Second, he felt himself bound by the rule of the Royal Society in impounding as its own his manuscript, since he had by implication acknowledged this rule in sending anything to the Society. Third, his religious faith[18] forbade contention

[15a] The names of the referees were revealed in 1965: The Reverend Mr. BADEN POWELL, Savilian Professor of Geometry at Oxford, and Sir J. LUBBOCK, who was considered to be an expert in the field of gas theory. I am unable to find any contribution to science by either of these men.

Priority fifty years after the fact is legal priority honored only in the breach. While the work of WATERSTON — including his *unquestionable* priority not only in discovery but also in publication of the principle of equipartition — has been discussed again and again by writers on the history of physics, his name remains totally unknown even among specialists on the kinetic theory and statistical mechanics. Physicists, apparently, are as little likely to take note of what historians write as they are to do historical reading themselves.

[16] WATERSTON, "The physics of media that are composed of free and perfectly elastic molecules in a state of motion" (with an introduction and notes by Lord RAYLEIGH), *Phil. Trans. R. Soc. London* A **183**, 1—79 (1893) = *Papers* 207—331.

[17] JOHN JAMES WATERSTON, 1811—1883.

[18] He was a Sandemanian.

for personal eminence, and his sole object was the truth. His critical attitude toward the acknowledged leaders of science, such as Lord KELVIN (who seems in fact to have been innocent in the affair of WATERSTON's great paper), caused him in later life to be regarded as a crank. One of his relatives reported that "any mention of the Royal Society led to very strong and contemptuous language ... he showed a restlessness and dislike at the mention of scientific men, except FARADAY, and he used very strong language in respect to some who bulk largely in public estimation."

RAYLEIGH's suggestion of "some other channel" calls for a further remark. Some thirty years before WATERSTON, HERAPATH, after his similar experience with the Royal Society, had gained oblivion through publishing in the channel of a minor journal. In 1843 WATERSTON had employed the channel of a minor press for a book on the physics of the mind; at the end is a note which is virtually an abstract of the long paper afterward rejected by the Royal Society. We are not surprised that publication in this channel attracted no notice whatever; if the leaders of science are hostile to a new idea when it is called to their official attention, there is little likelihood they will choose to go out of their ways to search for it in obscure places. In this note of 1843 WATERSTON introduced the concept of *mean free path*[19] (Fig. 103), rediscovery of which by CLAUSIUS in 1857 started a new period in the kinetic theory. To CLAUSIUS is due also the first explicit assignment of a probability in physics, if on the vaguest of grounds. In 1860, only fifteen years after the disgrace of WATERSTON, appeared the first of the four great memoirs by MAXWELL which raised the whole kinetic theory to a new level of abstraction, rigor, and applicability and created the science of statistical mechanics. To describe how this work of MAXWELL, supplemented by the mass of detailed if not always clear or fortunate calculations of BOLTZMANN, cast all previous work on the kinetic theory into the shade, would carry us beyond the time of a single lecture.

For his first paper on the kinetic theory, MAXWELL, though nineteen earlier papers on other subjects had given him a high reputation, chose an informal

[19] WATERSTON, Props. I and II of "Note on the physical constitution of gaseous fluids and a theory of heat," at the end of *Thoughts on the Mental Functions*, Edinburgh, 1843 = *Papers* 183—206. Of course, nobody could be blamed for not noticing something published so obscurely, but the moguls of science cannot have it *both* ways, advising the youngsters with original thoughts both to keep out of major journals *and* to gain their initial recognition by publishing in pages nobody reads.

will continually impinge against each other by reason of their *vis insita* alone, and will suffer no loss of collective momentum. The surface of impact is supposed to be spherical, and may either be the actual surface of the solid molecule, or a surface of powerful molecular repulsion surrounding it.

I. The distance traversed by a molecule, after impinging on one and before encountering another, is inversely as the density of the medium.

If all the molecules were arranged equidistant, the number at any distance from a central point will be as the square of that distance, but the intercepting effects of each of that number is as the inverse of the square of the distance; hence the intercepting effects are uniform, and accumulate as the distance increases. Let $d =$ diameter of molecular surface of impact; then *as* the distance $(x\,d)$, beyond which a molecule cannot go without impinging on another, *is to* the mean distance of the molecules $(n\,d)$, *so is* the number of circles (n^2) (whose common diameter is equal to that of the molecular surface of impact) that can be inscribed in an equilateral triangle, the side of which is equal to the mean distance of the molecules *to* unity. Hence, $x\,d \doteqdot n^3 d$, and $x \doteqdot n^3 \doteqdot$ to the inverse of the number of molecules contained in a given volume. [Upon the accuracy of this reasoning depends almost all that follows.]

II. The density of the medium being constant, the impinging distance will vary reciprocally as the square of the diameter of the molecular surface of impact.

For $n\,d$ being constant, $n \doteqdot \frac{1}{d}$, and $x\,d \doteqdot n^3\,d \doteqdot \frac{1}{d^2}$.

III. The impinging distance being constant, the density will vary directly as the diameter of the molecular surface of impact.

For $x\,d$ being constant, $x = \frac{1}{d} = n^3$, hence $d = \frac{1}{n^3} \doteqdot$ density.

IV. Let there be two elastic parallel immovable planes, and between them, in a common perpendicular, let two molecules that have unequal velocities continually impinge, then shall the sum of the squares of their velocities be always equal, and the velocities will con-

journal[20] in which to make "the hypothesis that ... minute parts are in rapid motion" the "subject of rational curiosity". Six years later, in 1866, and twenty-one years after it had scornfully rejected WATERSTON's hypothesis of the molecular constitution of gases, the Royal Society printed in its *Transactions* a great memoir of MAXWELL starting from just this same hypothesis. With its specific calculations of probabilities and its detailed reckonings of the effects of collisions on an infinite assembly of molecules subject to that very strange and special inverse fifth-power law of repulsion, this paper is vastly more mathematical and vastly more speculative than the rejected essay of WATERSTON. Partly, it was a difference of persons: an established Fellow, not just some nobody out in Bombay. More than this, it was the difference of twenty-one years: The Royal Society was ready to catch up with the Germans.

5. In Praise of Disorganization of Science

The older historians loved to point a moral from their tales, sometimes going so far as to alter history in order to conform it with their own preconceptions. With the recent rise of history of science as a profession, not only alteration of the facts but also all kinds of moralizing based on history have fallen into disrepute. Not being a historian of profession, I feel myself privileged to remain a reader of history in the simple old way. I am not ashamed to agree with HERODOTOS and MACHIAVELLI and MONTAIGNE and LEIBNIZ in believing that the greatest use of history is for the lessons it teaches us for our conduct today. The recent work in continuum mechanics has drawn inspiration from fresh, direct study of the classics of mechanics: NEWTON, the BERNOULLIS, EULER, CAUCHY, STOKES, and MAXWELL in particular were consulted for their methods of inquiry, standards of inference, and philosophies of natural science, so as to corrrect the

[20] The *Philosophical Magazine*, which in the first half of the century had published, along with many fine papers, a good deal of plain rot and outright error. I do not know the circumstances of this journal and cannot guess why HERAPATH and WATERSTON did not submit their papers to it.

In the second sentence of his paper, MAXWELL names BERNOULLI, HERAPATH, and JOULE as the first to explain the gas laws on the basis of rectilinear motions and impacts.

Fig. 103. A page from WATERSTON's *Thoughts on the Mental Functions*, 1843, explaining his concept of mean free path

bad morals grown common and even exalted through the so-called applied mathematics and applied mechanics of the first half of this century. This afternoon's lecture on the early kinetic theory furnishes a lesson of a different sort, a lesson about the organization of science.

We have seen the Royal Society twice in thirty years with one righteous voice stifle the truth in favor of the wrong, twice bury a great man in contempt while exalting tame bustling boobies whose whole lives add nothing to the science passed on into our day. To see how ludicrous was the position of the Royal Society, it is necessary only to state it. The Society, or at least its officers, regarded the Society itself as to some extent committed in support of any paper it published. Consequently, it was dangerous to accept a "speculative" paper, even admitting the chance that some such papers turn out to be right. Responsibility for the decisions was delegated to referees. Today, it is hard to find something less important about the kinetic theory of gases than what two anonymous referees who had never touched the subject thought about it in 1816 or 1845. If that thing hard to find can be found, however, it is whether or not the Royal Society as a body was to commit itself to the kinetic theory in 1816 or 1845. Nevertheless, the Society's position in those two years was so strong as to stop dead two careers in mathematical physics which at their short beginnings gave promise second to few in the century, and of which even the exhumed truncations stand high among the monuments of mechanics, while for most of the honored Fellows of that body from 1800 to 1850 it would be like pulling hens' teeth to find a single paper or experiment worth a moment's notice now. Such is not the picture the layman has of a great academy of science. He is taught that scientific truth is demonstrable, irrespective of persons, and that those who sit in the great academies are the great scientists of the time. Since the stories I have just told do not bear out either of these beliefs, they may be taken as criticism of the Royal Society, but they are not so intended. I could tell of similar happenings in the French Institut, in the Berlin Academy, and in the Academy at Petrograd — in a word, in *every* famous academy down to 1850. (I should not dare to come any closer to today.)

That my moral is of a different kind may be seen from the fact that no such story can be told of any of our own numerous academies of science in America. These range from local clubs ever eager in recruiting new dues-payers to a semi-official body of influential scientific administrators whose advice is sometimes asked but rarely heeded by the State Department and the President. No-one could justly claim that the American academies uphold standards of qualification for membership anything like those of their great European counterparts, yet no-one can justly impute to our more modest bodies such fatal mistakes as have been made again and again by the august official Olympuses of Europe. I have good reason to doubt that our academicians are wiser or more kindly[21]. The difference is to be found, not in the persons, but in the importance attached to them. *Science* is independent of passions and titles; *scientists* are not. That the Royal Society twice unanimously considered the kinetic theory of gases to be nonsense from top to bottom, has no more bearing on the truth or falsity of that theory than the true value of π would have been affected, had a certain mid-western legislature passed the bill it indeed seriously considered so as to establish a standard of $\frac{4}{3}$ in law. What is, unfortunately, not only ludicrous but also poisonous about European academies is the importance which they attach to themselves and which others allow them to enjoy. Up to the present time, scientists in the United States have been happier. Few of them care enough about our academies to learn who is and who is not a member, and the academies, as yet, do not set themselves up as arbiters over the truths of science and the lives of scientists.

The difference, starting at the top, goes down all through our work. The European assistant who disagrees openly with his professor risks losing all chance of going on with his research, not to mention failure ever to get a decent job. In the United States, a paper is no more esteemed if it appears within covers sealed by an academy or professional society, no less so if it has been rejected by such a body before being published in a private journal, and for the young giant, trampling upon his professors is a more honorable path to fame, promotion, and such modest prosperity as the scientific trade allows than is the fawning filial piety the European professor expects and receives from his disciples as long as he lives. Our academic life presents to the foreigner a lamentable scene of chaos. No-one knows who is on top. If in University 1 Professor A is a demi-god, we have only to consult Professor B in University 2 to learn that in *his* department A would not qualify even as an assistant. True, A belongs to six national committees, has a million dollar grant from the Central Spy Bureau, and has published eight successful textbooks, but B, who points to A's textbooks as models of nonsense, has written 216 research papers with twenty-three co-authors and also is consultant for four

[21] They are, however, less rancorous and more given to forgive and forget. They would, I am sure, one day have elected HERAPATH, not, of course, for his off-beat odd-ball kinetic theory, but for the second half of his life, which was devoted to good sound railway science.

major corporations, assistant editor of five journals and second vice-president of a professional union. The courses *A* sets down as minimum requirements for a degree under his aegis are not even given in *B*'s university. Then there is always *C*, who upholds "good teaching" by refusing to allow his students any contact with anything less than fifty years old. Even our administrators, frantic though they are to bring to heel their own faculties, add their part to the exterior disorganization in refusing to allow full "credits" to the transferring student who has drunk his mediocrity in another grove of the Muses. Finally, in their precaution lest one of their hirelings usurp any semblance of rule or become so notable that the university could lose fame by losing him, they blow now hot, now cold on each local hero in turn, so as to keep their pantheons fluid.

It is bad taste to praise one's own land and dispraise the foreign without strong reason. I am doing so because today, the day of massive unanimity, when social democracy triumphantly levels us all in informally uniform, cordially timorous, voluntary servitude, there are pressures from all sides to unify science. The societies and academies begin to demand a uniform format for papers as well as a "style" of polysyllabic baby-talk after Dutch patterns. Officials, accustomed to dealing only with other officials, perplexed in a world where the applications of science hold the fortunes and lives of all in balance, turn to the heads of these bodies and impute to them an authority which in fact is theirs only by default if not arrogation. The government, which two decades ago began to give and accustomed us to receiving, begins now to grasp. The public, ever fearful in ignorance, recognizes the size of the budget as the measure of excellence in science as in welfare, and turns the spigots of the common purse toward those readiest to spend, be it to dig to China, to carry syphilis to the moon, or to choke the human mind in a dense dust of digits. Clearly, great social activities such as these, like the building of the pyramids, require a central organization and a clear chain of command. While they themselves touch science at most on its margins, they are so gigantic and so poisonous that even marginal contact may be fatal.

The early history of the kinetic theory of gases illustrates the principle that in science, the majority is always wrong. This principle is difficult for a social democracy to swallow. We cannot move a step without being reminded that the minority of persons who prefer silence to soft music is in the wrong; they must hear the soothing syrup because "most people like it." The great society, the affluent society, sets for itself the humane doctrine of nourishing the weak and the maimed, of subsidizing the incompetent and the foolish; it looks with suspicion and fear on the strong, the thoughtful, the wise. An unsurrendered mind proclaims an antisocial man. "Protest", indeed, is met with high tolerance if not full capitulation, so long as it is the stupid mob cry of the unlucky, the self-unemployed, and the jacobins. What modern society cannot stomach is a man who thinks. Even if social democracy is now so enlightened as to stop short of deciding a good value for π by plebiscite, it takes refuge in polls of "experts". We all hear of what "doctors say" and what "scientists say." A scientist who lets his laymen friends know that he rejects some tenet the newspapers have published as "accepted by science today" will arouse their suspicion that he himself is a crank.

All this is of no importance so long as it is only the public that plays the booby. The danger comes when scientists allow *themselves* to be organized, when they begin to *respect and obey* pronouncements on science by academies, universities, societies, and, finally, governments. May that day never come: May our academies remain clubs, may our universities remain disjoint snarls of local red tape and campus-club poolroom politics, may our professional societies remain subscription lists for advertisers! So long as we preserve the disparate, entwined, mutually suspicious inefficiency which rules the circumstances of scientific work in this country today, a HERAPATH or a WATERSTON, should he arise, may eat and work and discover.

Bibliographical Note

The first half of this lecture coincides largely with the first half of the first lecture in the course on statistical mechanics I have been giving every two or three years since 1950. It is based, of course, on my own reading of the sources.

Especial notice is deserved by the studies of S. G. BRUSH on the history of the kinetic theory and statistical mechanics[22]. When these, the first substantial publi-

---

[22] S. G. BRUSH, "The development of the kinetic theory of gases, I. Herapath," *Annals of Science* **13**, 188—198 (1957).
ibid. "II. Waterston," *ibid.* **13**, 273—282 (1957).
ibid. "III. Clausius," *ibid.* **14**, 185—196 (1958).
ibid. "IV. Maxwell," *ibid.* **14**, 243—255 (1958).
ibid. "V. The Equation of State," *American Journal of Physics* **29**, 593—605 (1961).
"John James Waterston and the kinetic theory of gases," *American Scientist* **49**, 202—214 (1961).

cations on the subject of this lecture, began to appear, of course I made use of them, and my text has been enriched thereby. In particular, some personal details of the affairs of HERAPATH, WATERSTON, and JOULE were first published in the papers by BRUSH and MENDOZA.

This lecture, in substantially the form printed, was delivered as a Centennial Lecture at the University of Kentucky (1966) and at the Center for the Philosophy of Science, University of Minnesota (1967).

"The Royal Society's first rejection of the kinetic theory of gases (1821), John Herapath versus Humphry Davy," *Notes and Records of the Royal Society of London* **18**, 161—180 (1963).
"Foundations of statistical mechanics 1845—1915," *Arch. Hist. Exact Sci.* **4**, 145—183 (1967).
Cf. also E. MENDOZA:
"A sketch for the history of the kinetic theory of gases," *Physics Today* **14**, 36—39 (1961).
"The surprising history of the kinetic theory of gases," *Mem. Manchester Lit. Phil. Soc.* **105**, 1—14 (1962/3).

VII. Reactions of the History of Mechanics upon Modern Research

Abstract:

It is shown by examples that knowledge of the history of mechanics can lead to new discoveries in the mechanics of today.

The examples adduced concern: (*a*) The strongest form of column for axial thrust. (*b*) The form to be given a beam in order that when loaded it shall take on a desired shape. (*c*) Invariance requirements for the constitutive equations of non-linear materials.

A parallel is then drawn between the seventeenth century's futile attempts to define force and the nineteenth century's futile attempts to define entropy.

As a second and related parallel, the special devices used by seventeenth-century mathematicians to solve particular problems of motion by quasi-static means, before the discovery of general laws of motion, are compared with twentieth-century attempts to describe irreversible energetic phenomena by some small perturbation of thermostatic methods and concepts. It is suggested that an attempt be made to create and organize the logical structure of a true thermodynamics of irreversible processes along the lines successfully employed two hundred years ago by EULER and others in converting the unorganized special methods and principles of seventeenth-century mechanics into the general theory we know today.

1. Proem

My topic was suggested by the Program Committee. It is a difficult one.

In the first place, no one could justly contend that even a faint idea of the true historical development of mechanics is *necessary* in order to do good research today. There are eminent colleagues who demonstrably combine skill in research with disdain for history and those who trace it; with benign indifference; with fantastic distortion or reversal of historical facts; or with naive historical credulity akin to belief in Santa Claus, the principle of least action, or the greatest good for the greatest number. Many, however, are drawn, it would seem compulsively, to make historical statements that, while perhaps not inconsistent with disdain and certainly not with ignorance and naivety, undeniably refute any alleged claim to indifference. I could name a book on elasticity where the author seems to try to restrain himself from entering into the history of the subject but cannot resist informing the historically curious reader that the Lamé constants are named "after GABRIEL LAMÉ (1795—1870), a French mathematician." The same authority, however, does not disclose who it is that YOUNG's modulus is named after. Another well known author writes in his preface, "The coupling of a particular name with a theorem or method simply indicates an association in my own mind. The proper historical setting I must leave to those who have the time and taste for such research." The text that follows is everywhere dense with theorems and methods named after persons, often with references. Then there is an eminent colleague who by design names quantities and theorems after his teachers and friends; to any objection that these things really belong to much prior authors, he replies, smiling graciously, that when he himself learned about stress by reading a paper of ADOLF BRAUN, why should he mention EULER or CAUCHY? These examples may be interpreted variously in terms of caprice, carelessness, disorder, irregularity, ignorance, or even lying with respect to history, but certainly they suggest a sense of guilt rather than indifference.

Of *historical indifference* there is, truly, little. The writers of elementary textbooks openly value history as a paedagogical aid; as is usual in mass paedagogy, it does not seem to matter whether the statements made are true or false, for one writer copies from another and then embroiders. This practice, which goes back to MACH himself, seems to have some propaganda purpose, usually to instil in us a belief that all science is basically experimental, often to beat that favorite dead horse of popular science, the Middle Ages, and sometimes to cast ridicule onto religion or philosophy. As a result, in our student days upon each of us was ingrained a certain picture of the history of mechanics. Who has not read and heard how GALILEO shattered the tradition of ARISTOTLE by hearkening to the simultaneous thuds of balls he had dropped from the Leaning Tower of Pisa? In some popular works these and other experiments of GALILEO are described in detail. Our teachers urged us to question mathematical proofs and experimental demonstrations but never suggested that it even mattered at all whether their historical statements were true or false. How many of us, in our days of falling grey hairs, have gone back to the origins of freshman mechanics and have learned that the whole Leaning Tower story is a literary legend? [1][1]. If we have come

---

[1] The experiment was already more than 950 years old when GALILEO is popularly alleged to have invented it. JOHN THE GRAMMARIAN, a very learned commentator on ARISTOTLE in the late 5th or early 6th century A. D., asserts that "Aristotle wrongly assumes that the ratio of the times required for motion through various media is equal to the ratio of the densities of the media .... But this is completely erroneous, and our view may be corroborated by actual observation more effectively than by any sort of verbal argument. For if you let fall from the same height two weights, one of which is many times as heavy as the other, you will see that the ratio of the times required for the motion does not depend on the ratio of the weights, but that the difference in time is a very small one." See pp. 219—220 of COHEN & DRABKIN's *A Source Book in Greek Science*, Harvard University Press, 1958. There is, of course, neither any indication that JOHN THE GRAMMARIAN was the first to perform this experiment, nor that he attached great importance to the result. In the passage from which the quotation is taken, he shows more interest in estimating the effects of the resistance of media than in the properties of fall in a vacuum.

In the western world, the experiment had already been repeated in Holland before GALILEO assumed the chair at Pisa. In a work published in Dutch in 1586, in Latin in 1608, in French in 1634, and in English on p. 511 of his *Principal Works*, Vol. 1 (Amsterdam, 1955), STEVIN wrote: "Let us take (as the very learned Mr. JAN CORNETS DE GROOT, most industrious investigator of the secrets of Nature, and I have done), two spheres of lead, the one ten times larger and heavier than the other, and drop them together from a height of 30 feet onto a board or something on which they give a perceptible sound. Then it will be found that ... they fall together onto the board so simultaneously that their two sounds seem to be one and the same rap. The same is found also to happen in practice with two equally large bodies whose gravities are in the ratio of one to ten ... ." However, in his *de Motu*, written by 1590, GALILEO wrote that he "often made a test" of free fall "from a high tower", by which he found that wood starts to fall faster than lead, but ultimately lead outstrips wood! In the same work he wrote, "It is of no matter if now and then experience contradicts the well grounded theory." In this early passage, GALILEO comes off poorly in comparison with JOHN THE GRAMMARIAN, both as an observer of nature and as a theorist. That GALILEO was essentially a Neo-Platonic idealist rather than an empiricist, and that he was the author *neither* of the mathematical theory of uniformly accelerated motions *nor* of the idea that gravity induces uniform acceleration in all bodies *nor* of the classic experiments on free fall *nor* of repeatable experiments on motions along inclined planes, has not displaced him from his heroic leadership of science for the masses. It

to know this much, what have we thought afterward of the famous hard-headedness of scientists? Along with criticism and doubt toward assertions about dead nature, have we not been taught a gullibility toward human experience as naive as any monkish scribe's toward miracles? Do we not then blush at our own naivety when we learn of the scrupulous chain of evidence and reasoning by which the humanist LORENZO VALLA in the fifteenth century proved the falsity of the Decretals of Constantine? The fable of the Leaning Tower would have fallen at the first probe of such a mind. And when we read of GALILEO's experiments on bodies sliding down an inclined plane, do we kneel before the shrine of the saint, or do we, as good empiricists should, run to repeat the experiments? This is precisely what MERSENNE did (c. 1634); the results of his own trials led him to conclude [2], "I question whether Lord GALILEO ever did the experiments of falls along the plane, since he nowhere says so, and the proportion he gives often contradicts experiment."

In pure mathematics it is no surprise when an old problem, long unsolved, is attacked again, or when an old theory, long known, is given new life by new application or new foundation. In explanation it is usually said that mathematics is a very old science, and that its truths are eternal. The same holds for mechanics, too. While biologists and psychologists may have little reason to consult any source more than ten years old, as several have told me, in mechanics it is otherwise. The science of mechanics, as a mathematical discipline, is almost as old as

---

is harder to summarize in a cozy capsule the philosophical and social motives of a grammarian converted to Christianity who wrote 3000 pages of commentary on ARISTOTLE in the Dark Ages, a group of clerics from the High Gothic, and two Dutch Protestants, one of whom is not known for anything else except his Greek and Latin poetry and his being the father of the lawyer GROTIUS.

Dr. JOHN P. MORAN has called my attention to the paper of T. B. SETTLE, "An experiment in the history of science," *Science* **133**, 19—23 (1961). SETTLE reconstructed the apparatus of GALILEO as nearly as can be determined and repeated the experiments under conditions as similar as possible to those specified by GALILEO's text. SETTLE confirms GALILEO's assertions that the distances vary as the squares of the times of fall on a given plane, and that for various planes the times vary jointly with distance and inversely as the square root of the vertical drop. He notes that these results are not equivalent to the formula $s = \frac{1}{2} g t^2$ for free fall, and that no determination of g is involved. SETTLE adds the interesting remark, "By the time both the theory and the experiment had evolved to the level implicit in the *Discorsi*, GALILEO would have had sufficient confidence in the worth of each independently, irrespective of their mutual confirmation." To this I add that if experiment had really played the part in the *original acceptance* of "classical" mechanics that many philosopher-scientists seem to believe, we should not have had to wait until 1961 for a critical analysis and *repetition* of an allegedly fundamental experiment first reported in 1638.

mathematics itself. We must not think that Greek mechanics was limited to the statement of the few simple principles we customarily attribute to it. As in any great science, the principles arose from experience with cases, and ancient statics became a developed subject at the same time as its principles came to be understood. ARCHIMEDES determined the positions of equilibrium, and their stability, for any floating right segment of a paraboloid of revolution of arbitrary specific gravity. I doubt if a modern professor of mechanics, applying integral calculus and relying upon 2000 years' experience in hydrostatics, will solve this problem without at least a week of hard work. EULER, for example, using only an infinitesimal criterion of stability instead of ARCHIMEDES' finite geometrical argument, gave up the problem and did not include ARCHIMEDES' results in his own treatise. I am not recommending ARCHIMEDES as a fruitful author for the modern research engineer, but I am adducing his work to show that old mechanics is not necessarily easy mechanics. Neither is old mechanics concerned only with problems by now solved or out of date. Not all of us are compelled to search for truth nowhere but in the nucleus; the principles of mechanics, like the real number system, even in the "atomic age" continue to help us understand nature by casting new light upon old problems and by showing new problems cleared through old solutions.

2. The Elastica of Optimum Section

An old problem closer to modern interest is that of finding the buckling load of a Bernoulli-Euler elastica subject to end thrust. In the historical introduction to his treatise on elasticity, LOVE writes that LAGRANGE used this theory "to determine the strongest form of column"; PEARSON in his history of elasticity tells us also that LAGRANGE's result "may fairly be said to have shaken the then current architectural fallacies," but he does not disclose what those fallacies were. Referring to the entasis of columns made in Greek times, LAGRANGE in 1770 proposed to find the form that should be given to a column of assigned height and volume in order that it bear the greatest possible thrust without buckling. He claimed to show that the best form is a circular cylinder; in other words, entasis cannot increase strength. Doubting this result, I looked up LAGRANGE's paper in his Collected Works, but I read there that over ninety years ago SERRET, the editor, had found in the analysis many mistakes he could not set right, and also it seemed to me that LAGRANGE's mathematical formulation was insufficient.

I then proposed the problem to H. F. WEINBERGER and J. KELLER; each of them found the solution corresponding to LAGRANGE's hypothesis that the cross-sections are circles centered on the axis; the solution is not the uniform column but rather one swollen at the middle; suitable entasis (Fig. 104) does, in fact, increase the strength by $\frac{1}{3}$. It turned out that the maximizing profile, somewhat like a cycloid, had been found by CLAUSEN in 1851:

$$\frac{r}{l} = \frac{2}{\sqrt{3}\,\alpha} \sin\theta, \quad \frac{z}{l} = \frac{1}{\pi}\left(\theta - \frac{1}{2}\sin 2\theta\right), \tag{1}$$

where l is the length of the column, r is the radius of the cross-section, z is the distance along the axis, and α is the ratio of l to the constant radius of the cylindrical column of equal volume. Meanwhile, however, KELLER pursued the matter

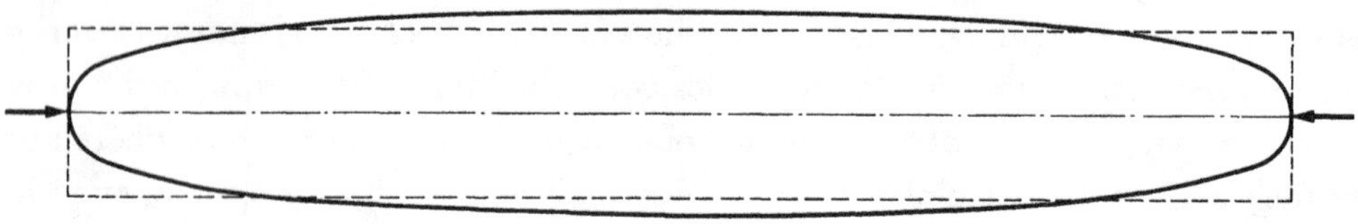

Fig. 104. Profile of the strongest column having the bulk of the cylinder shown dashed

further. Dropping the assumption of circular cross-sections and considering columns of arbitrary form, he showed that twisting cannot increase strength, but that a column whose cross-sections are equilateral triangles with mutually parallel sides and collinear centroids, following a profile with entasis as determined by CLAUSEN, has a buckling load greater by 61.2 per cent than that of the corresponding circular cylinder [3]. He conjectured that such a column is the strongest of all having given height and volume, and he adduced some partial results, but no proof has yet been given[2].

3. The Elastica of Desired Shape when Loaded

While this individual problem is of interest in itself, another old problem in the theory of the elastica, little known until recently, is of greater implication. Having to referee a paper on beam theory for an engineering journal in 1952,

[2] KELLER's conjecture is true. Professor PÓLYA has shown me a simple argument deriving it as a corollary to a theorem proved in his paper, "More isoperimetric inequalities conjectured and proved," *Comm. Math. Helv.* **29**, 112—119 (1955).

I followed a historical hunch that EULER must have known all that, and I began to read his treatise of 1744. Not only was my hunch right, but as I read further I saw a few lines on a problem of a kind not then treated in the modern literature, namely, what form must a body have in order that it shall assume a given shape under given load? In the linearized theory this problem is equivalent to the usual one of determining the deformation caused by given loads: One has only to apply to the given body displacements equal to the negatives of those produced in it by the given forces. In the finite theory, such a solution by superposition holds no longer. That the two problems are distinct, was seen by JAMES BERNOULLI, who, in the very first paper on the mathematical theory of elasticity, published in 1694, with his usual insight into principles and ideas fruitful for future study included among the problems "left to the industry of our readers" that of finding "what should be the curvature of the band in order that from an attached load or from its own weight or from both together it assume the form of a straight line (this would be useful in designing the arms of balances and axles...)." Not one to propose to others problems he could not himself solve, JAMES BERNOULLI had in fact set up the solution to this one, but his fragment concerning it was not published until the same time as EULER's treatise appeared, almost fifty years later.

Both BERNOULLI and EULER considered the case of a beam such as to be bent out straight by an assigned terminal load. EULER found the shape in terms of what have since come to be called FRESNEL's integrals, and he proved that it is a spiral (Fig. 105). In looking at this work, I noticed how much easier it is to solve this inverse problem than the usual direct one, and I set up a general method [4]. Instead of a second-order differential equation, one gets a double quadrature, and the general solution is exhibited. As an example I took the case when uniform pressure is to be applied. While the free shape is again spiraliform, a notion of the initial curvatures a beam must have in order to bear various pressures is gotten more easily by looking at Fig. 106, where the dashed lines indicate the erroneous results following from the linearized theory. Another special case yields the solution of the general piston-ring problem, where a shape is sought such that the ring when put into a cylinder of given form shall fit snugly [5] (Fig. 107). The applications, however, are less important than the idea, which lends itself to the general theory of finite elastic strain, formulated some hundred years ago and in this last fifteen years the object of striking and lucrative rediscovery and exploitation. In this theory we now have numerous solutions both exact and approximate, as well as some powerful general theorems. In this theory, *a fortiori*, solution of the

$$x = \int ds \sin. \left(\frac{s}{b} + \int \frac{ds}{r} \right), \ \& \ y = \int ds \cos. \left(\frac{s}{b} + \int \frac{ds}{r} \right);$$

unde curva quæsita per quadraturas construi poterit.

Fig. 17. 51. Hinc determinari potest figura amB, quam lamina in situ naturali habere debet, ut a potentia P in directione AP sollicitante in lineam rectam AMB explicetur. Sumpta enim longitudine AM $= s$; erit momentum potentiæ sollicitantis pro puncto M $= Ps$; radius osculi autem in M, per hypothesin, erit infinitus seu $\frac{1}{R} = 0$. Sumpto jam in statu naturali arcu am $= s$, positoque radio osculi in m $= r$; quia hæc curva convexitate sua rectam AB spectat, in calculo præcedente poni debet r negativum. Hinc erit $Ps = \frac{Ekk}{r}$, seu $rs = aa$; quæ est æquatio naturam curvæ amB complectens.

52. Cum igitur sit $\frac{1}{r} = \frac{s}{aa}$; erit $\int \frac{ds}{r} = \frac{ss}{2aa}$; seu erit amplitudo arcus am ut quadratum ipsius arcus. Hinc coordinatæ orthogonales x & y pro hac curva amB ita definientur, ut sit $x = \int ds \sin. \frac{ss}{2aa}$, & $y = \int ds \cos. \frac{ss}{2aa}$: Scilicet in circulo, cujus radius $= 1$, abscindi debet arcus $= \frac{ss}{2aa}$, cujus sinus & cosinus ad coordinatas determinandas assumi debent. Ex eo autem quod radius osculi continuo decrescit, quo major capiatur arcus am $= s$, manifestum est curvam in infinitum non protendi, etiamsi arcus s capiatur infinitus. Curva ergo erit ex spiralium genere, ita ut infinitis peractis spiris in certo quodam puncto tanquam centro convolvatur, quod punctum ex hac constructione invenire difficillimum videtur. Non exiguum ergo analysis incrementum capere existimanda erit, si quis methodum inveniret, cujus ope, saltem vero proxime, valor horum integralium $\int ds \sin. \frac{ss}{2aa}$ & $\int ds \cos. \frac{ss}{2aa}$ assignari posset, casu quo s ponitur infinitum; quod Problema non indignum videtur, in quo Geometræ vires suas excerceant.

53. Sit

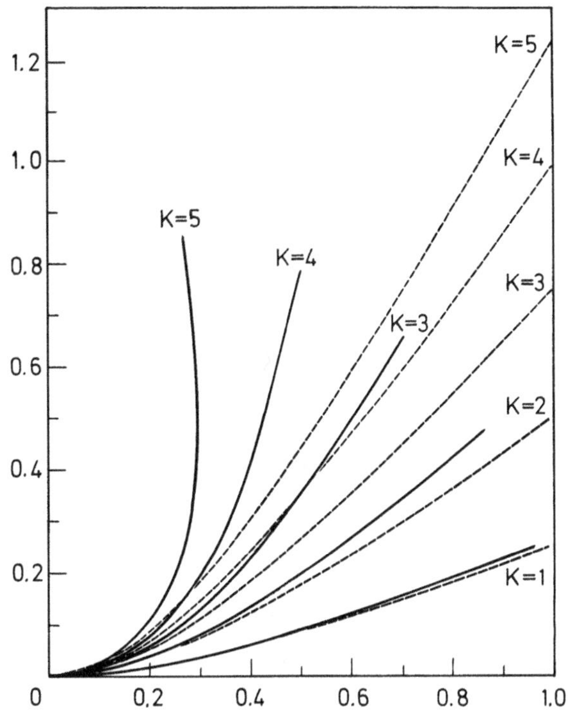

Fig. 106. Free shape of a band such as to be bent out straight by assigned uniform pressure (The dashed lines show the results according to the linearized theory, while the heavy lines are the true shapes. K is the dimensionless load factor:

$$K = \frac{\frac{1}{2} p L^3}{EI}.)$$

Fig. 105. Euler's solution for the free shape of an elastica such as to be bent out straight by an assigned terminal load

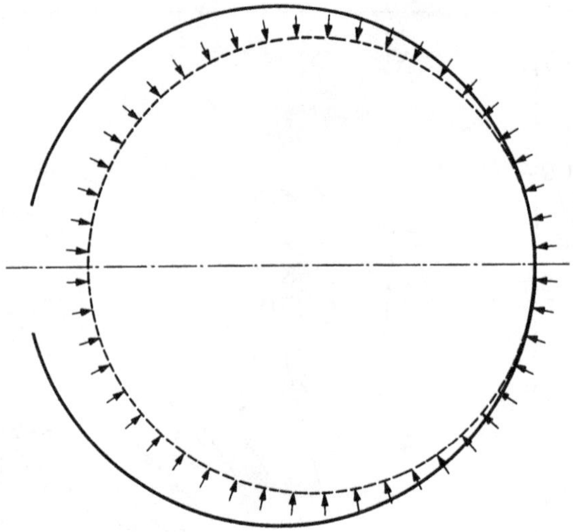

Fig. 107. The piston ring problem (The dashed line is the desired form under load, while the heavy line is the free shape to be determined.)

free-shape problem is not given by solution of the standard boundary-value problems. It should be easier to attack in general terms[3]. So far practically nothing is known about the effect of body forces, and the direct effect of surface forces has been calculated only by means of series expansions. The reason is that in the usual boundary-value problems the points at which the body and surface forces are applied are themselves unknown *a priori*. In the free-shape problem, this difficulty disappears, for the forces act at known points. If, for isotropic materials, we take the stress-strain relations as $\boldsymbol{t}=\boldsymbol{t}(\boldsymbol{c})$, where $\boldsymbol{t}$ is the Cauchy stress tensor and $\boldsymbol{c}$ is Cauchy's deformation tensor ($c_{km}=g_{\alpha\beta}X^{\alpha}_{,k}X^{\beta}_{,m}$, where $\boldsymbol{X}$ is the initial position and $\boldsymbol{x}$ is the corresponding position in the loaded configuration), then the differential equations and boundary conditions for the stresses are of the form

$$\begin{aligned} \frac{d\boldsymbol{t}}{d\boldsymbol{c}}\cdot\nabla\boldsymbol{c}+\varrho\boldsymbol{b}&=\boldsymbol{0} \quad \text{in the body} \\ \boldsymbol{t}\cdot\boldsymbol{n}&=\boldsymbol{s} \quad \text{on the boundary,} \end{aligned} \tag{2}$$

where $\boldsymbol{b}$, $\boldsymbol{n}$, and $\boldsymbol{s}$ are all given functions of position, $\boldsymbol{x}$, in the loaded configuration. This is a second-order differential system to be satisfied by the initial positions $\boldsymbol{X}$

[3] There is no use looking for exact solutions valid for arbitrary isotropic materials subject to surface loads alone, since the corresponding deformation tensors are determined once and for all by Ericksen's Theorems, which can be applied either forward, as usual, or backward, as in the present case.

[*Note added for the reprinting.* The results of Ericksen for incompressible bodies were not summarized as a theorem by him and do not exhaust all cases.]

as functions of $\boldsymbol{x}$. While special methods have had to be discovered, or remain to be discovered, so as to treat the direct boundary-value problems of finite elasticity, for the system (2) all the usual theorems and methods for non-linear differential systems apply at once. In the ten years since I first made this remark, no one has followed it up, but I expect that he who does will not be disappointed by what he finds[4]. Here we see not so much a single problem as a whole program of research that is suggested by reading a work 200 years old.

4. Methods of Approach to General Questions

Both the cases I have adduced point to particular problems or groups of problems, but giving leads of this kind is not the main service of historical study to the student of the mechanics of today. Recall ABEL's statement that he had reached the front rank quickly "by studying the masters, not their pupils." Among the masters he selected was NEWTON, a difficult author, and then antiquated by a century. A century before our own day lived CAUCHY and STOKES. It was with them that my own reading of the masters began, later, I am sorry to say, than my student days. Employed to work in gas dynamics, I was trying to take the first step by finding out what the NAVIER-STOKES equations meant. LAMB [6] writes that they express "the simplest hypothesis we can frame," that "there is considerable *a priori* probability that it will represent the facts accurately in the case of infinitely small motions," and that "this hypothesis has been put to a very severe test in the experiments of POISEUILLE and others." In view of the British empirical tradition, I presumed his reference to framing hypotheses was a joke, nor could I see how *a priori* probability was used to find constitutive equations for materials. Since, as LAMB remarked a few lines earlier, the "rates of distortion" are not dimensionless, I wondered what rate should be taken as a standard for comparison in order to decide when rates are "small". No textbook I could find said a word about this. When I appealed to some eminent senior colleagues, they told me disdainfully that everyone knew those equations were established by experiment. I became curious about experiments that could establish an isotropic linear functional relation between two tensors, so I looked up LAMB's references to the derivations by NAVIER, POISSON, and STOKES, and to the experiments of POISEUILLE

[4] *Note added for the reprinting.* This idea was not fruitful. See § 44, Footnotes 1 and 2, of *The Non-Linear Field Theories of Mechanics.*

and others. I found in the work of NAVIER and POISSON my first experience with what I since have learned to be a common phenomenon in the history of science:

$$\text{Falsehood} \Rightarrow \text{Truth.}$$

Both NAVIER and POISSON started from hypotheses of atoms unlike any mentioned in the freshman physics course at Cal Tech in my day; each of them, despite a bitter lifelong quarrel about differences between their respective imaginary atoms, a prototype, perhaps, of a modern conference on dislocations, came out with the since-accepted answer. The memoir of STOKES [7] was a burst of sunlight. STOKES tells us exactly how he found his equations: "In reflecting on the principles according to which the motion of a fluid ought to be calculated..., I was led to construct the theory...." On the basis of one hypothetical statement after another, frankly laid before the reader, STOKES concluded, "I shall assume the following principle: —

"That the difference between the pressure on a plane in a given direction passing through any point p of a fluid in motion and the pressure which would exist in all directions about p if the fluid in its neighborhood were in a state of relative equilibrium depends only on the relative motion of the fluid immediately about p; and that the relative motion due to any motion of rotation may be eliminated without affecting the differences of the pressures above mentioned."

Pages of purely kinematical analysis follow so as to provide the tools necessary to make concrete use of this principle, and further hypotheses are needed before the NAVIER-STOKES equations can be derived.

Here it is not the detail but the method that tells. Another passage, referring to elasticity, reads, "Are we then to suppose that when a solid is constrained it has no *tendency* to relieve itself from the state of constraint, in consequence of its molecules *tending towards* new relative positions.... It appears to me to be much more natural to suppose *a priori* that there should be some such tendency." Both for method and for style, this paper would be rejected by the secret referees of any society's journal today. The language of STOKES is plain *English*; today's crab-dance of German noun piles tottering among Dutch passives and impersonals with quotation marks, dashes, and parentheses to string together street slang, Latin fustian, and ad-men's gabble was still to be invented as the "scientific style". To modern official modesty, the use of the personal pronoun "I" may suggest the pontificating of an elderly authority; in fact, it was the honesty of an unknown beginner, twenty-six years old, who was *thinking*, and who admitted it.

Fig. 108. GEORGE GABRIEL STOKES (1819—1903), from a photograph taken in 1857

George G. Stokes

In the argument leading to the basic equations there are several references to experience, but none to any specific experiment. As OSGOOD [8] wrote,

"Mechanics is a natural science and like any natural science requires for its comprehension the observation and knowledge of a vast fund of individual cases ...

"But Mechanics is not an empirical subject in the sense in which physics and chemistry, when dealing with the border region of the human knowledge of the day, are empirical... The laws of Mechanics, like the laws of Geometry, so far as first approximations go — the laws that explain the motion of the golf ball or the gyroscope or the skidding automobile, and which make possible the calculation of lunar tables and the prediction of eclipses — these laws are known, and will be as new and important two thousand years hence, as in the recent past of science when first they emerged into the light of day ...

"The world in which the boy and girl have lived is the true laboratory of elementary mechanics. The tennis ball, the golf ball, the shell on the river; the automobile — good old Model T, in its day, and the home-made autos and motor boats which youngsters construct and will continue to construct — the amateur printing press; the games in which the mechanics of the body is a part; all these things go to provide the student with rich laboratory experience ..."

DANIEL BERNOULLI said it more briefly: "... there is no philosophy which is not founded upon knowledge of the phenomena, but to get any profit from this knowledge it is absolutely necessary to be a mathematician." The motto of HUYGENS is still better: "From experience and from reason."

But no one learns from mottos. A motto is a summary of what one has learned, or tried to learn. The true school of scientific method is study of the masters.

Contrary to the belief of many scientists, historians of science are not much interested in establishing priorities. When, however, a priority is set backward or forward by a century, it is shifted into a different scientific scene, and we may be certain of a difference in method. It is customary to think of YOUNG as a non-mathematical, largely empirical scientist, and it is natural to regard "YOUNG's modulus" as therefore the discovery of an experimenter. In truth, the so-called "YOUNG's modulus" was invented almost a hundred years earlier by EULER as a boy of nineteen, studying mathematics under JOHN BERNOULLI's direction. It was invented for a definite reason: EULER tried to derive JAMES BERNOULLI's law of bending of a beam from HOOKE's law of extension for the longitudinal fibres. This is a definite mathematical problem, the solution to which we can all read in any elementary book on the strength of materials. To solve this problem, as EULER did, it is *mathematically necessary* to introduce a measure of elasticity that represents a material property, independent of the size of the specimen. When we look into the details, we find that YOUNG, despite all his talk about "useful knowledge", bungled by defining the modulus as the ratio of force to strain, as had nearly everyone else for a century before him, thus obtaining a quantity that varies from specimen to specimen of the same material. YOUNG knew EULER's work, but he did not always understand it. The scene in which YOUNG lived was not conducive to anything of great originality or even to comprehension of the solid mathematical results that had come earlier from a more rational, imaginative milieu. It was YOUNG's favorite author, the engineer ROBISON, who called EULER's buckling theory "a dry mathematical disquisition, proceeding on assumptions which (to speak favourably) are extremely gratuitous..." and "evidences of this wanton kind of proceeding..."

Reading of STOKES' original paper on viscosity and of CAUCHY's researches on solids and fluids helped to recreate a scene in which it became possible, once again, to *think* clearly about the nature of materials as we all encounter them in everyday life, the true laboratory of mechanics. I say "recreate", for we had all been brought up in an era when "applied" mathematics was synonymous with the art of leaving out appropriate terms in long-known equations considered too difficult to solve. In this period rational mechanics dwindled almost to nothing from neglect by the pure mathematicians, who identified it with the problem, too dirty to touch, of integrating a system of ordinary differential equations named after HAMILTON, and by the physicists, who rejected as "merely mathematics", *i.e.* bad physics, any argument not too obscure to be followed in logical steps. Elsewhere [9] I have described the growth of the earlier theories of non-linear viscosity from 1945 onward. More recently, they have been brought to a measure of completeness in one direction by the "simple fluid" of NOLL [10], while another direction has been opened by the "anisotropic fluids" of ERICKSEN [11]. The considerations of these savants are quite in the style of STOKES, though of course, with the advantage of a century's harvest of experience, they rest on a broader physical platform and bring to bear mathematics both stronger and briefer.

The relation of these theories to experiment needs to be developed. I have remarked above on the positive claims made in earlier years in favor of the NAVIER-STOKES equations. Their logical weakness was seen by my old teacher, BATEMAN [12], who seems to have been the only expert on mechanics in the period between the wars to have both knowledge of the literature and some critical faculty; he remarked that the supposed tests would confirm equally any other theory leading to the same relations among dimensionless parameters in the special flows considered. In other words, he referred to the possibility that

$$\text{Falsehood} \Rightarrow \text{Truth},$$

usually not recognized by empiricists. When experiment confirms a formula derived by a theory, it does not confirm the hypotheses of the theory; it only fails to contradict them. Experimental facts are facts indeed, incontrovertible, but in themselves they do not offer test of a theory unless they contradict it. Experiment teaches us what to reject. We can accept, provisionally, only those theories not yet shown to contradict experiment. To be sure of a test with a definite outcome, we must have at least two theories leading to *different* results for the experiment in question; experiment must then contradict at least one of the two theories, thus enabling us to reject at least one set of hypotheses. Not just *an* experiment, but a *discriminating experiment* is needed. Moreover, we cannot know whether an experiment is discriminating or not except by prior knowledge of theory. Theory plays a characterizing, unremovable part in the test of theories. For example, if the viscous stresses in fact depend upon higher spatial derivatives of the velocity field, the HAGEN-POISEUILLE efflux formula will not be affected, since it pertains to a particular flow in which those derivatives are all zero. Even if no effect is allowed to higher derivatives, the fact that the HAGEN-POISEUILLE law emerges from the NAVIER-STOKES equations has no force in logic to make us think it can *not* come out from other plausible theories as well. In fact, it does. In viscometry, only since 1959 have we been in a position to see the problems clearly, for in that year COLEMAN & NOLL [13] obtained the exact solutions for viscometric flows of all simple fluids. Among their results is a general, explicit formula for the efflux Q in terms of the defining equations of the fluid. We can rewrite their efflux formula as follows:

$$Q = \frac{\pi a R^4}{2\mu} \int_0^1 u^3 \zeta\left(\frac{\frac{1}{2} a R \tau}{\mu} u\right) du, \tag{3}$$

where a is the driving force, R is the radius of the tube, μ is the linear viscosity of the fluid, τ is the natural time of the fluid, and $\zeta(\ )$ is a dimensionless material function of its dimensionless argument. According to the NAVIER-STOKES equations, $\zeta(\ )=1$, and then (3) yields the famous law of the fourth power and the numerical coefficient $\frac{1}{8}\pi$. Conversely, if Q varies as R^4, we can conclude that $\zeta(\ )=1$, but this form of the shear viscosity function does not imply the truth of the NAVIER-STOKES equations. In fact, the typical normal-stress effects of non-linear viscosity may be present without affecting the form of $\zeta(\ )$. Thus, as

BATEMAN suspected on purely logical grounds, verification of the HAGEN-POISEUILLE formula by measurements is far from sufficient to establish the physical truth of the NAVIER-STOKES equations.

Among all older authors on mechanics, it is CAUCHY who offers us the greatest number of good ideas not yet fully exploited. Hundreds of pages in his works concern very general dynamical systems or materials and obtain mathematical principles making it possible to begin to consider them, but CAUCHY himself did not go far along the lines he opened, and I doubt anyone else has. In a paper [14] published in 1952 I tried to summarize theories presented up to that time in which elastic stress is regarded as arising from differential increments rather than from finite strain. ERICKSEN showed me that my statement was wrong, inconsistent, in fact, with a formula of CAUCHY I had given on a previous page as concerning a more special problem. Reading the original paper of CAUCHY that I had cited but not properly understood, ERICKSEN found there a fairly clear though special and unemphatic statement of the requirement that constitutive equations should be invariant under rigid motions. I then recalled having seen something relevant by ZAREMBA (1903). These historical sources, and the discussions and discoveries based upon them, gave rise first to the theory of hypo-elasticity [15] and then to the beautiful, inclusive *principle of material indifference* formulated by NOLL [16] [5] and used, often without explicit recognition, by several other authors before and since as the main tool in reducing constitutive equations of nonlinear materials to manageable and explicit forms.

5. The Axioms of Forces

As originating in study of history I have adduced a single problem, a class of problems, a method, and a concept. It remains to show you that by historical study the language itself of mechanics may change and, ultimately, a new law may be created.

[5] I do not mean to give the impression that all the others who contributed to the statement and elaboration of these ideas from 1950 onward were influenced by historical study. My purpose in this lecture is not to write the history of non-linear viscosity or of recent developments in mechanics, but to bring forward cases in which study of the old literature did help in modern researches.

We are all used to mechanics as based upon a fundamental, primitive concept of force, but it was not always so. The word for force is older than all its measures. Since the earliest dynamical inquiries, motions and forces were associated, but what is the force of motion? What is the measure of force? The seventeenth century saw many debates on these equations, debates confusing to the participants because they felt rather than specified the problems they sought to solve, and confusing to the typical modern reader because, with his slanted vocabulary, he is not even aware that there was ever something to argue about. The scientists of the seventeenth century had not been taught in school that when you have a problem of mechanics, the "forces" are "given", and you go ahead and set it up. When they saw a body moving on an inclined plane, or a body swung at the end of a rope, or a top spinning on a point, or fluid split apart by the prow of a ship, they knew that "forces" were manifested, but how should those forces be measured? One was the "force of inertia", defined by KEPLER and mentioned by EULER, D'ALEMBERT and many others; this was equivalent to the modern word "inertia", a property of matter, but in the early literature it was an example of a "force", much as is "force of habit" in common speech today. For a long time there was an effort to *define* force in terms of motion: the weight times the velocity, the weight times the distance moved, the velocity times the distance moved, *etc.* It was not clearly seen that *something new*, in addition to kinematical quantities, had to be brought in if a workable, useful mechanics was to result. An *a priori* concept of force was, perhaps, NEWTON's main contribution, but in his writing it is rather illustrated than presented, explained, or defended explicitly, and at first only his own British disciples were converted to his way of looking at things. More than half a century after the publication of the *Principia*, D'ALEMBERT, expressing the view that what we call force is only a manifestation of motion, spoke of Newtonian forces as "obscure and metaphysical beings, capable of nothing but spreading darkness over a science clear by itself." It was through the dozens of examples worked out by DANIEL BERNOULLI and EULER that the fruitfulness of taking forces as quantities given *a priori* came gradually to be accepted. By common consent, nowadays, scientists and non-scientists rarely say anything concrete to each other; a scientist who uses the word "force" in its Newtonian meaning to musicians or poets will notice at once that understanding ceases, since to them "force" has only its pre-Newtonian senses even today.

While few scientists know that any non-Newtonian view of force ever existed, even fewer are able to tell us what a Newtonian force is. If we ask, we shall be told by a modern expert[6], "Force? Well, everybody knows about force intuitively. We all feel pushes and pulls, and we see experiments in which forces are applied. In the theory, we represent forces as vectors, and we assign them the dimensions of mass times acceleration, so they can go into NEWTON's second law." This kind of talk would certainly not have impressed LEIBNIZ or D'ALEMBERT. It is accepted now for no better reason than that *we are all used to it.* We don't listen to the talk, really. Each of us in his own mind has a further association with forces that no book states. When we ask a mathematician what a real number is, he does not mumble, "Well, we all know about real numbers from our income-tax forms and from solving equations; they belong to an ordered field, you know, and we use them when we study calculus." Rather, he replies that a real number is an *undefined object* or *primitive*, described only by a definite set of *axioms*, axioms which he can state precisely and without hesitation[7]. Now forces, too, are undefined objects; they are vector measures, but saying that they are vector measures no more characterizes forces than saying real numbers form an ordered algebraic field characterizes real numbers. (In fact, a system of forces involves two vector measures: one giving the forces exerted on a given body by all bodies exterior to it, and another giving the forces exerted on the subbodies of a given body by a single exterior body, and these two measures must be compatible with each other.) For about a century we have known that there are many kinds of ordered algebraic fields, but that real numbers satisfy an axiom of continuity; moreover, any field that does satisfy that axiom is indistinguishable from the real numbers. With forces we are now at about the place where we were a hundred years ago with numbers. Only recently, it seems, has anyone felt the need [17] of defining forces precisely by listing *all* the axioms they satisfy, and whether the axioms so far proposed are sufficient to characterize forces as distinct from more general kinds of vectors remains unknown.

---

[6] Or, if we prefer, a modern text book. *Cf., e.g.*, pp. 15—16 of SYNGE & GRIFFITH's *Principles of Mechanics*, 2nd ed., McGraw-Hill, 1949. SYNGE & GRIFFITH compound confusion by deciding finally to measure force by "some standard extension in some standard spring". Fortunately their book does not contain any theory of elasticity. Thus they are saved the embarrassment of telling us what they mean by "force" when HOOKE's law is stated so as to explain their "standard spring".

[7] Of course, there are alternatives: The mathematician may choose to define real numbers in terms of rationals, rationals in terms of integers, and integers in terms of sets. The point is that *somewhere* he begins with undefined objects and precisely stated axioms; he does not try to pull himself up by his bootstraps, inferring his definitions from the results of the theory he is constructing, in the way many experts on mechanics try to pull a definition of force out of some corner of the whole edifice of mechanics such as the theory of linear springs.

6. Seventeenth-Century Parallels to the Obscurity of Thermodynamics and Thermostatics

The long, slow, and still not quite complete growth of the concept of force helps us to understand the troubles in another subject: thermodynamics. Just as mechanics is the science of motions and forces, thermodynamics is the science of forces and entropy. What is entropy? Heads have split for a century trying to define entropy in terms of other things. I tell you, entropy, like force, is an undefined object, and if you try to define it, you will suffer the same fate as the force-definers of the seventeenth and eighteenth centuries: Either you will get something too special, or you will run around in a circle. CARNOT seems to have seen this when, apparently rejecting the notion that heat (chaleur) is a substance, he proposed a theory in which heating or cooling occurs according as a certain new substance, called caloric (calorique), increases or decreases. This caloric may be entropy in some cases. Just as force can be defined kinematically in certain special and common systems, such as a pendulum, entropy can be, and is, defined thermally for certain special systems, such as a perfect gas. Just as force can never be measured directly — for all measurements of force presume some special mechanical constitutive equation such as that of the rigid body or the linear spring or the gravitational field — so entropy cannot be measured except indirectly through use of special thermal constitutive equations. While we can never define entropy except in cases where we do not need it, and while we cannot measure it directly, we can, in time, get used to it, as we have, despite the opposition of some of the greatest of all scientists, gotten used to an *a priori* concept of force. We are not yet sufficiently used to entropy to state, with any measure of agreement, what axioms it obeys.

Here a second parallel to seventeenth-century mechanics may be drawn. You may be thinking that the text-book definitions and descriptions of entropy are good enough. The older books deal exclusively with *thermostatics*; how well they do with it is a matter of taste, but I admit the evidence that a lot of people accept what is said, repeat it without stumbling, and use it to design good steam engines. In the seventeenth century, statics was a well developed subject, and it was applied in a way then acceptable to many persons in many cases where any modern engineer would require laws of motion, then unknown. For example, we may cite BORELLI's book, *On the Motion of Animals* (1685), where the parallelogram of forces seems to be the only quantitative basis for two volumes on the subject named, and where, despite the title, we look in vain for any laws of motion. I do not mean at all to ridicule the book; it is not only truly scientific but also ingenious in many places; I adduce it as an example to show that work both intelligent and extensive can be done on a wobbly foundation, and that the existence of serious literature in a domain, leading to some measure of success, does not necessarily imply that the structure is sound.

Laws of motion do not spring Minerva-like from the head of anyone, whether his name be NEWTON, GALILEO, LEONARDO, or NOSTRADAMUS. The properties of uniformly accelerated motions, derived by scholars of Merton College [18] between 1328 and 1350, became widely known through the writing of GALILEO, 300 years later. Despite some uncertain attempts by BURIDAN and ORESME in the fourteenth century, LEONARDO seems to have been the first to state fairly clearly that all bodies have a common, uniform acceleration in a vacuum, whether falling freely or in oblique motion [18]; these ideas, too, were widely diffused through GALILEO's publication a century later. In the mid-seventeenth century more general laws of motion were sought. The two most famous examples are given by HUYGENS' solutions for fall along a cycloid and for oscillation of a rigid body about a fixed axis. As his basic principle for the former problem, he assumed that the laws of uniformly accelerated motion hold as an approximation for motion along the tangent at each instant; for the latter, he assumed that each corpuscle has such a velocity that if released from all the rest, it will by itself rise to the height from which it has fallen. DESCARTES and LEIBNIZ proposed laws of motion that may be regarded as ancestors of the principles of conservation of total linear momentum and total energy. NEWTON's scheme, better known today, in effect gave rules for solving many more problems. None of these principles or methods is wrong, but none is general enough to include all the rest. By 1700 it could fairly be said that many important special problems of motion had been solved correctly, but dynamics remained half an art, half a science.

That is where we are in thermodynamics today. Thermal processes in nature are no more reversible than mechanical processes are static. To try to describe thermal phenomena by means of thermostatics has led to about the same measure of success and failure as did the attempts to discuss motion in terms of mechanical statics. Since physicists of the nineteenth century laid less weight on logical precision than did their predecessors in the seventeenth — recall that in CLAUSIUS' day mathematics and physics were already distinct professions, both "pure mathematics" and "physical intuition" having been invented by then — nineteenth-century thermodynamics, if simple in essence, is obscure from vague wordiness, while seventeenth-century dynamics is obscure from luxuriance of mathematical detail and proof. The Victorian pioneers created a "second law of thermodynamics", which lays down prohibitions on the outcome of a process but does not tell us a thing about how the process occurs. It is analogous to a bare statement that a rolling ball will never come to rest on the top of a hill. This is true, and once in a while it is useful, but of dynamics we expect to learn not only where the ball won't go but also where it will. In recent years there has sprung up from promotion by a school of specialized enthusiasts a theory named "irreversible thermodynamics", based, be it noted, upon an equation of entropy production[8]. Instead of facing the problem of discovering the true physical axioms governing the production of entropy, this doctrine takes refuge in immediate linearizing. In principle, though not in mathematical ingenuity, the approach reminds us of HUYGENS' theory of bodies falling on a cycloid. In practice, it is not so happy, since its main if not its only tenet, an assertion that the matrices of coefficients are always symmetric, is not self-consistent in the typical cases of diffusion, thermal conduction, thermal diffusion, *etc.* [19]. However vague and insecure its foundations, and however uncertain its concrete achievements, the "irreversible thermodynamics" of today has served to acquaint many people with the phenomena it attempts to describe, and to remind them that the physical laws governing those phenomena are known but in part. Its contribution may be compared to that of DESCARTES to pure mechanics; he solved a few problems, and he interested a lot of people, but most of what he said reduced, at bottom, to words.

I have drawn a parallel between the laws of entropy today and the laws of dynamics in the seventeenth century. At the end of the seventeenth century, a major step was taken by JAMES BERNOULLI (1685, 1703) when, recognizing HUYGENS' results on the free rotation of a body about a fixed axis as true despite the insecurity of the hypothesis from which HUYGENS had obtained them, he groped toward two new principles of dynamics, principles that turned out, eventually, to hold for all motions of all bodies. From those principles, long after his death,

[8] In effect, the authors working in this field take entropy as a primitive variable. In the textbooks they write, it is true, the opening pages are often devoted to a classical description of entropy in terms of heat and temperature, but the concepts and arguments used there are never shown to be applicable to the time-dependent, non-uniform entropy for irreversible phenomena presumed later in the same books.

evolved the modern equations of dynamics for rigid bodies and for more general space-filling materials. I am not suggesting that study of seventeenth-century mechanics will solve the problems of twentieth-century entropy. I am suggesting that seeing where mechanics has been may help us to see where we wish thermodynamics to go. As GASSEND put it, over 300 years ago, "History is surely the light of life, since it gives the mind a handle for learning from the past what may be hoped for the future."

The first step is to separate the equilibrium theory, reorganizing it and presenting it as such, and no more. Here, too, we have a seventeenth-century parallel, for there were several formulations of statics in terms of small motions. Consider, for example, a hanging chain (Fig. 109). A principle sufficient to determine the figure of equilibrium is: The work done when the chain is subjected to quasi-static infinitesimal displacement from equilibrium is zero. This principle, still used in some quarters today, is *sufficient*, I say: We get the right figure of equilibrium.

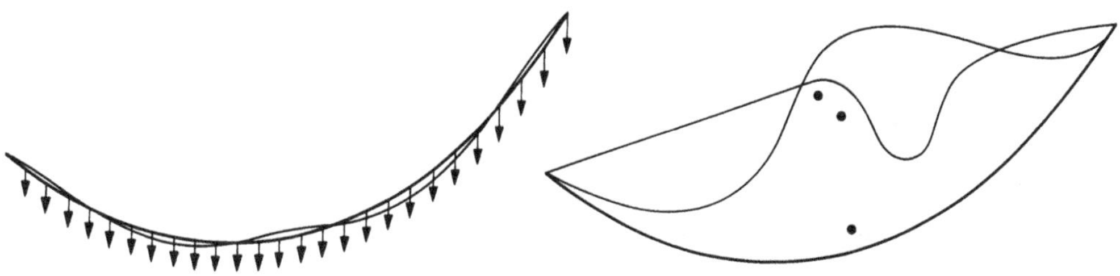

Fig. 109. Equilibrium of the catenary. Left: Quasi-static infinitesimal displacements. Right: True static minimal principle

Nevertheless, this principle does not happen to be *true*, for if we disturb an equilibrated, perfectly flexible chain, no matter how little, according to the usually accepted theory of kinetics we shall have to do work, because we shall provoke a state of perpetual oscillation, a motion perfectly determinate by dynamical laws. In the seventeenth century it was seen to be possible to avoid altogether the use of quasi-static infinitesimal processes in mechanics. For the chain we may say that among all curves of given length between the given endpoints, the figure of equilibrium is that in which the center of gravity hangs lowest. Here are no infinitesimals and no nonsensical "processes"; instead, we have a characterization of the figure of mechanical equilibrium by a true minimal principle that selects it alone among all others equally possible in pure geometry. The laws of motion of a chain are more complicated; different models, those neglecting friction or specifying it in various ways, and those allowing greater or lesser freedom of movement, all are consistent with the same figure of equilibrium; granted that if equilibrium is our concern, we do not need these dynamical theories, at least it is no longer necessary to lie about motion in order to solve the static problem.

The analogous step in thermostatics was taken by GIBBS. Not mentioning the processes, the cycles, and the engines, GIBBS set up true minimal principles for thermostatic equilibrium of fluids, fluid mixtures, and elastic solids. This is not always clear from the language he used, and many presentations today when claiming to follow his methods are in fact closer to the gloomy murk of PLANCK. Until very recently, almost the only author who gave evidence of understanding that the classical theory is truly and exclusively statical was DUHEM. Even today, the student is taught a meaningless ritual involving First and Second Laws, Irreversible and Quasi-Static Processes, Perpetual Motion Machines, and Completely Efficient Heat Engines. In 1959, COLEMAN & NOLL[9] succeeded in putting GIBBS' theory of equilibrium into a logically clear and mathematically precise, definite, and consistent form, and in using it to derive major new results in the theory of finite elastic strain.

Now that we know, at last, what thermostatics really is, we are ready to look for the true thermodynamics. In the theory of entropy production, we have various special but isolated or approximative methods and solutions, which slightly perturb a basically static approach; most of these, perhaps even all of them, lead

---

[9] The work of COLEMAN & NOLL on fluid mixtures, mentioned in Sects. 264—265 of *The Classical Field Theories*, is not yet published. Their theory of equilibrium of elastic solids is given in "On the thermostatics of continuous media," *Arch. Rational Mech. Anal.* **4** (1959/60), 97—128 (1959). This paper is directed specifically toward a problem that originated in historical study. Noticing the thermodynamic method of DUHEM for deriving the inequalities satisfied by the linear elasticities and linear viscosities, in a lecture at Berlin in 1955 I had pointed out that corresponding considerations for the theory of finite elastic strain were not known, and I proposed the problem of finding the restrictions necessarily satisfied by the strain-energy function of an elastic material; *cf.* "Das ungelöste Hauptproblem der endlichen Elastizitätstheorie," *Z. angew. Math. Mech.* **36**, 97—103 (1956), English translation in *Foundations of Elasticity Theory*, N. Y., Gordon & Breach, 1965. COLEMAN & NOLL are the first to have given GIBBS' formulation of thermostatics mathematical precision and generality sufficient to derive inequalities restricting the response of elastic materials in finite strain. Their researches illustrate the wisdom of ABEL's advice to study the masters, not their pupils, providing the right masters be selected (in this case, GIBBS and DUHEM rather than CLAUSIUS, KELVIN, and PLANCK).

to correct results for the special situations they concern, but the true theory, valid for all kinds of irreversibilities, remains to be created.

Parallels between analysis, mechanics, and thermodynamics are shown in the following chart:

Continuous Functions	Mechanics	Thermodynamics
1. Axioms of vector spaces	Axioms of forces	Axioms of entropy
2. Definition of a continuous function	Principles of mechanics	Principles of entropy production
3. Theorems about all continuous functions	Theorems about all mechanical systems	Theorems about all thermomechanical systems
4. Definitions of special functions, and their properties	Definitions of special systems and materials, and their properties	Definitions of special systems and materials, and their properties

If I could now state the axioms and principles of entropy, I should have chosen a different topic today. Toward the end desired, history has one more lesson for us: The historical order has been the reverse of the logical one. Mechanics began with special cases, and special cases are still all that we have in thermodynamics today. We wish to go upward on the right-hand column of the chart. How this is to be done cannot be said positively until it has been done, but from history we know that the middle column was climbed neither by "inspiration" toward the revolutionary nor by complacence with what was already known. The great, inclusive, and definitive principles have been gotten by pondering, neither experiments nor philosophic beliefs, but the incomplete and sometimes even partially wrong previous *mathematical methods for concrete special problems* — by induction, indeed, but induction over theories, induction followed by refinement, slowly piecing together what was known but isolated, and slowly stripping off the accidents from what was special[10].

The true theory will not be a linear one. Had NEWTON, the BERNOULLIS, and EULER shared the modern passion for linearity above even fatherland, we should not have dynamics today. Before the later work of LAGRANGE, no-one seems to

[10] *Note added for the reprinting, 1967.* The way to a general thermodynamics was opened a year after the above statements by a great memoir of COLEMAN & NOLL: "The thermodynamics of elastic materials with heat conduction and viscosity," *Arch. Rational Mech. Anal.* **13**, 167—178 (1963). Of course they take temperature and entropy as primitive concepts and leave processes unrestricted by any relation to equilibrium. From their work a general theory with manifold applications has been created by COLEMAN, A. E. GREEN, GURTIN, MIZEL, and others.

have seen that linear problems are easier. The general theory of small oscillations was found after, not before, the laws of finite motion. Nature has never been shown to favor any particular functional form or any special kind of mathematical quantity, and the creators of mechanics did not try to force any such prejudice upon her.

That numerous authorities are satisfied with thermodynamics just as it now is, need not discourage us. It would be wrong to think that the great unifying and organizing discoveries of the eighteenth century were the outcome of a general, co-operative progress, or that they were acclaimed by science as a whole when they were made. Then as now, most scientists were content with getting themselves a cozy berth by perturbing slightly (whether inward or outward) some corner of the leaky hull of Authoritative Opinion. The cumulative effort of the scientists of the eighteenth century, the academicians and the professors, was nil. Their busy little lives added up to nothing; their voids and vortices and atoms and subtle fluids, that could explain everything but calculate nothing, sound as quaint to us today as the fluxes, humours, menstruums, bleedings, purgatives, and cauterizings of the contemporary physicians, intent upon torturing or killing mankind as an offering to the Science of Medicine. Far from seeing any need for a more organized, more mathematical mechanics, the typical savant of the eighteenth century ignored or even, in countries like England and Prussia, where the empirical bias was strong, sneered at the great researches that are now all we see when look back at their period. But even the names of those influential little men are forgotten; the organized, mathematical mechanics we all learn today is the creation, not of Scientific Progress but of exactly *seven men* of the eighteenth century. The readers who could follow what those men wrote made up but a minority in the scientific scene of that day.

In our day, the existence of a dissatisfied minority is shown by the steady though small flow of literature[11] attempting to reformulate thermodynamics or at least to explain it in a new way. The writers of this literature sense what is wrong, and each thinks he has gotten somewhere, but none of the others will accept what he has done. A recent critic [20] has written of his training in "intense thermodynamic discipline at its ... very stronghold," "Instead of the firm rock of unambiguity and lucid exposition looked forward to, he found thermodynamic

[11] Some of this literature is cited in § 245 of *The Classical Field Theories*, where a further critique is given.

theory a dismal swamp of unsound metaphysics, forced arbitration and expedient compromise." What he then develops as a substitute is only a minor perturbation of the traditional floundering through the trackless bog. It is not enough to see that the cheese is fouled; one has to find and stop the rat holes before the house can be set in order.

The true thermodynamics cannot expect such universal acclaim as greets a new elementary particle or a trip to Mars, nor can its discoverers await reward by a Nobel prize or even election to the eminent mediocrity of an academy. The mathematical theory we now call "classical mechanics" had to overcome the opposition not only of the ruling majority of important nobodies in The Age of Reason but also of some great mathematicians: HUYGENS deplored the calculus, and DANIEL BERNOULLI rejected partial differential equations as abstract pure mathematics. The true thermodynamics will express the drag of the past on the present. This drag obviously depends upon what the past has been. Therefore the natural language of thermodynamics is afforded by the as yet scarcely initiated theory of tensor functionals. Consequently not only those who are satisfied with the quasi-static cult of cycles, engines running backward, and universes in torpid surfeit of entropy, but also the enthusiasts of today's so-called irreversible thermodynamics, are likely to reject the new functional thermodynamics with contempt as soon as, not to say before, it is discovered. Human nature has not changed since the Roman physician GALEN wrote, "The fact is that those who are enslaved to their sects are not merely devoid of all sound knowledge, but they will not even stop to learn! Instead of listening, as they ought ..., they even go so far as to scoff ... ." But it is just as sure that a mathematically and physically clear and right theory of irreversibility will have the same value two hundred years hence as does today the rational mechanics created and organized two hundred years ago — the same, that is, if today's physics permits any rational being to exist two hundred years hence.

Acknowledgment

It is a pleasure to thank Messrs. COLEMAN, ERICKSEN, NOLL, SERRIN, and TOUPIN for their criticism of the first and second drafts of this lecture. The investigation reported here was sustained in part by a grant of the U.S. National Science Foundation.

References

1. Cooper, L., *Aristotle, Galileo, and the Leaning Tower of Pisa*, Ithaca, Cornell University Press, 1935.
2. Mersenne, M., *Harmonie Universelle, Paris*, 1636, p. 112.
3. Keller, J., "The shape of the strongest column," *Arch. Rational Mech. Anal.* **5**, 275—285 (1960).
4. Truesdell, C., "A new chapter in the theory of the elastica," *Proc. 1st Midwest. Conf. Solid Mech.* (1953), 1954, pp. 52—55.
5. Truesdell, C., "Exact theory of self-expanding piston rings," *Ing.-Arch.* **30**, 77—87 (1961).
6. Lamb, H., *Hydrodynamics*, 2nd Ed., Cambridge, 1895, and later editions, see Sect. 326.
7. Stokes, G. G., "On the theories of the internal friction of fluids in motion, and of the equilibrium and motion of elastic solids," *Trans. Cambr. Phil. Soc.* **8** (1844/9), 287—319 (1845) = *Papers* **1**, 75—129; see Sects. 1—5, 19.
8. Osgood, W. F., *Mechanics*, New York, MacMillan, 1937, pp. v—vi.
9. Truesdell, C., "Experience, theory and experiment," *Proc. 6th Hydraulics Conf., Iowa State University*, 1956, pp. 3—18.
10. Noll, W., "A mathematical theory of the mechanical behavior of continuous media," *Arch. Rational Mech. Anal.* **2** (1958/9), 197—226 (1958), Sect. 21. Reprinted in *The Rational Mechanics of Materials*, Gordon & Breach, 1965, pp. 336—367, and in *Continuum Theory of Inhomogeneities in Simple Bodies*, Springer, 1968.
11. Ericksen, J. L., "Anisotropic fluids," *Arch. Rational Mech. Anal.* **4**, 231—237 (1960), and later papers. Reprinted in *The Rational Mechanics of Materials*, Gordon & Breach, 1965, pp. 407—414.
12. *Report of the Committee on Hydrodynamics*, Nat. Res. Council, 1932, Sect. 1.7 of Ch. I of Part II. (This collective volume is indexed by librarians under the name of the author of its first few pages, H. L. Dryden.)
13. Coleman, B. D., & W. Noll, "On certain steady flows of general fluids," *Arch. Rational Mech. Anal.* **3**, 289—303 (1959). Reprinted in *The Rational Mechanics of Materials*, Gordon & Breach, 1965, pp. 368—383.
14. Truesdell, C., "The mechanical foundations of elasticity and fluid dynamics," *J. Rational Mech. Anal.*, **1**, 125—300 (1952), Sects. 55—56. This reference and the next have been combined in a corrected reprint issued by Gordon & Breach in 1966.
15. Truesdell, C., "Corrections and additions to the mechanical foundations of elasticity and fluid dynamics," *J. Rational Mech. Anal.* **2**, 593—616 (1953), Sects. 55 bis and 56, and later papers.
16. "Principle of Isotropy of Space", Sect. 4 of W. Noll, "On the continuity of the solid and fluid states," *J. Rational Mech. Anal* **4**, 3—18 (1955), and later papers. The reference cited has been reprinted in *The Rational Mechanics of Materials*, Gordon & Breach, 1965, pp. 64—81.
17. Noll, W., "The foundations of classical mechanics in the light of recent advances in continuum mechanics," *The Axiomatic Method, with Special Reference to Geometry and Physics* (1957), North Holland Co., Amsterdam, 1959, pp. 266—281.
18. Clagett, M., *The Science of Mechanics in the Middle Ages*, University of Wisconsin Press, Madison, 1959, Chapters 4—5.

19. Coleman, B. D., & C. Truesdell, "On the reciprocal relations of Onsager," *J. Chem. Phys.* **33**, 28—31 (1960).
20. Svantesson, Sven, *Foundations of Thermodynamics with Antitheses and Disputation*, Almqvist & Wiksell, Stockholm, 1958.

General References for further reading

Short essays or surveys

21. Cohen, I. Bernard, "A sense of history in science," *Am. J. Phys.* **6**, 343—359 (1950).
22. Truesdell, C., "The rational mechanics of materials — Past, present, future," *App. Mech. Rev.* **12**, 75—80 (1959). Corrected reprint in *Applied Mechanics Surveys*, Washington, Spartan Books, 1966.
23. Chapter II in this volume.
24. Truesdell, C., "Outline of the history of flexible or elastic bodies to 1788," *J. Acoust. Soc. Amer.* **32**, 1647—1656 (1960).
25. Chapter IV in this volume.
26. Chapter V in this volume.

Treatises

27. Dugas, R., *La Mécanique au 17me siècle*, Neuchâtel, Grifon, 1954.
28. Truesdell, C., "Rational fluid mechanics, 1687—1765," *L. Euleri Opera Omnia* (2) **12**, IX—CXXV (1954).
29. Truesdell, C., "I. The first three sections of Euler's treatise on fluid mechanics (1766); II. The theory of aerial sound, 1687—1788; III. Rational fluid mechanics, 1765—1788," *L. Euleri Omnia* (2) **13**, VII—CXVIII (1956).
30. Truesdell, C., "The Rational Mechanics of Flexible or Elastic Bodies, 1638—1788," *L. Euleri Opera Omnia* (2) $\mathbf{11_2}$, 435 pp. (1960).

Bibliographical Note

This lecture was delivered in 1962 at the Fourth U. S. National Congress of Applied Mechanics, Berkeley, California, and was printed in its *Proceedings*, pp. 35—47. In the same year it was read also at Aberdeen Proving Ground, at the Armament Research Establishment, Fort Halstead, at the Universities of Manchester, Nottingham, Newcastle upon Tyne, Stanford, and Illinois, at the National Bureau of Standards, Washington, and at a Sigma Xi meeting at the Illinois Institute of Technology.

For permission to reprint it here I am grateful to the American Society of Mechanical Engineers.

VIII. Recent Advances in Rational Mechanics (1956)

1. What is Rational Mechanics?

I begin by answering the question most of you asked upon reading the title: What is rational mechanics? It is difficult to define rational mechanics, but no more so than to define chemistry or physics or mathematics. However, a chemist coming before you would not be expected to begin by defining chemistry. The reason most of you wish me to define rational mechanics is that in the United States, at least, it is not a recognized science. Indeed, there are some who disbelieve in its existence[1].

[1] *Cf.* the four volumes of reprints, *Continuum Mechanics* I—IV, Gordon & Breach, New York, 1965—1966, and the two articles in the *Handbuch der Physik*, Volumes III/1 (1960) and III/3 (1965): C. Truesdell & R. A. Toupin, *The Classical Field Theories*, and C. Truesdell & W. Noll, *The Non-linear Field Theories of Mechanics*. *Cf.* also my three introductions to *L. Euleri Opera Omnia* and the book of R. Dugas, *La Mécanique au XVIIme Siècle*, Griffon, Neuchâtel, 1954.

First, *rational mechanics is a part of mathematics*[2]. It is a mathematical science, and in its relation to experience, intuition, abstraction, and everyday life it does not differ in essence from other branches of mathematics. In this audience, I am sure, mathematics itself needs no defence. It is unnecessary to persuade you that mathematics is trying to be physics or trying to be engineering. It should also be unnecessary to point out that mathematics, however abstract and however precise, is a science of *experience*, for experience is not confined to the gross senses: Also the human mind can experience, and we need not be so naive as to see in an oscilloscope an instrument more precise than the brain of man.

That rational mechanics grew out of practical mechanics and co-operated with it, if not always gracefully, to produce applied mechanics and mechanical engineering, is obvious. In writing the first treatise on rational mechanics, NEWTON[3] established its standard of mathematical rigor as precisely that of geometry. Not always has this standard been maintained, but today as in 1687 it remains the ideal. NEWTON's comparison with geometry is enlightening, for geometry, too, grew from physical experience. To those who scoff at geometry for its precise calculations when all measurements are liable to error, the geometer for millenia has replied: Geometry is *mental*, not *instrumental*. The scoffers have always been with us and remain today; not only does the ultimate practical and physical value of geometry need no defense before scientists, but also no-one who has known a geometer needs reminder that practical and physical usefulness seldom has supplied or suppressed a single equation in the progress of geometrical research.

The analogy to geometry is a good one. That rational mechanics speaks not only of space and time but also of mass, force, and energy does not make it any the less precise. Since it deals with a greater number of physical concepts than does geometry, its applications to physical problems may be expected to be more

---

[2] In connection with this paragraph, see my articles, "Experience, theory, and experiment," *Proc. 6th Hydraulics Conference* (1955), *Bull. 36, State Univ. Iowa Studies in Engr.*, 3—18 (1956), and "The modern spirit in applied mathematics," *I.C.S.U. Review of World Science* **6**, 195—205 (1964).

[3] I. NEWTON, *Philosophiae Naturalis Principia Mathematica*, London, 1687; 2nd ed., Cambridge, 1713; 3rd ed., London, 1726. The only available English translation is unreliable, especially in the allegedly modernized reprint. Some of the relevant passages, newly translated, are quoted in Essay III, above. A new edition of the *Principia*, collating the three original ones, and presumably also to be provided with a correct translation, has been undertaken by I. B. COHEN and the late A. KOYRÉ; its appearance is eagerly awaited.

frequent and more far-reaching, but physical applications are not its objective.

But does not rational mechanics deal with quantities of physical experience? Indeed it does; so does geometry, for lengths, surfaces, and volumes are equally related to physical experience. The geometer may visualize a surface in terms of a twisted strip of paper, as in mechanics we may think of a force as a push with the hand, but whatever these motivations, the symbols in the equations of geometry and mechanics are precisely defined mathematical quantities. Origin in broader experience may make mechanics more interesting, but it need not make it any less rigorous.

Might not the same be said of any part of theoretical physics? Indeed, it might; each discipline couched in mathematical symbols is ultimately susceptible of independent mathematical existence. In his famous list of problems set for this century to solve, HILBERT[4] included the forming of a set of axioms for physics, but this problem, like all those he proposed concerning the relation between mathematics and physical experience, has been neglected by mathematicians. Some of the older parts of physics are well on the way toward fully mathematical treatment, but they lie outside my scope.

Finally, does not the emergence of rational mechanics as a purely mathematical discipline imply that we have the freedom to invent new laws of mechanics as the geometers invent new spaces? In fact the geometers are not so wildly speculative as might appear; their inventions, usually, arise from some need of experience, though it may not be experience explicable in terms of eggs and apples. Purely mathematical theories of space have been in existence at least 2500 years, while in mechanics, with some exceptions, we have a tradition of only four or five hundred years, and the rational mechanics of materials, where most modern researches occur, is little over 300 years old. The properties of mechanical laws

[4] D. HILBERT, "Mathematische Probleme," *Nachr. Ges. Wiss. Göttingen* 1900, 253—297. An English translation of an amplified version is printed in *Bull. Am. Math. Soc.* (2) **8**, 437—479 (1902).

Note added for the reprinting. Only two significant attempts to solve the part of HILBERT's sixth problem that concerns mechanics have been published: that of HAMEL, "Über die Grundlagen der Mechanik," *Math. Annalen* **66**, 350—397 (1908), and that of NOLL, mentioned in Essay VII, above, and presented in full in the notes of his lectures at the C.I.M.E. Course on *Recent Developments in the Mechanics of Continua* at Bressanone in 1965, Edizioni Cremonese, Rome, 1966, and issued also as a report of the Mathematics Department of the Carnegie Institute of Technology, 1966.

most immediate to our experience are not nearly so well explored as are the properties of the space most immediate to our experience. Moreover, mechanics is enormously more complicated than geometry. The ordinary system of mechanical laws continues to offer us mysteries and challenges. While indeed we have the liberty to explore other mechanical laws, usually we do not choose to exercise this liberty except in the limited and sober way I will explain later.

That rational mechanics is a mathematical science, does not mean that its standard is one of mathematical difficulty. Indeed, some great researches in rational mechanics are difficult and intricate, but others are simple. An innovation that is simple and easy to understand is especially prized. The problems are set by the subject; hard or easy, they must be solved, and their mathematical difficulty is just as much incidental as is their physical application. After all, in this rational mechanics is not different from other parts of mathematics, for only to the mathematical novice is the complexity or difficulty of a mathematical science to himself a measure of its level.

Is rational mechanics a part of pure mathematics? To most mathematicians today, pure mathematics means topology, abstract algebra, or analysis in abstract spaces. These, most certainly, rational mechanics makes no attempt to imitate. While in spirit it is nearest to geometry, its problems, its aims, and its methods bear little evident similarity to those of other parts of mathematics. A theorem in topology is not evaluated in terms of its bearing on the theory of numbers. It is equally ridiculous, though unfortunately not infrequent, to deprecate theorems in rational mechanics when they do not also contribute to the more popular branches of pure mathematics.

Is rational mechanics a part of applied mathematics? Most certainly not. While in some cases known mathematical techniques can be used to solve new problems in rational mechanics, in other cases new mathematics must be invented. It would be as misleading to claim that each achievement in rational mechanics has brought new light to mathematics as a whole as to claim the opposite, that rational mechanics is a mere reflection from known parts of pure mathematics. It is a mistake to raise the issue of applied versus pure mathematics. Indeed, it is not the aim of rational mechanics to produce methods or results that rate *per se* as "new" in other parts of mathematics; neither is its aim to avoid them. Equally, it is not the aim of rational mechanics to produce applications, be it to physics, to engineering, or to other parts of mathematics; neither is its aim to avoid them.

Rational mechanics is an independent branch of mathematics. When a research in rational mechanics predicts a new physical phenomenon or produces a new analytical method, as occasionally happens, this is so much the better. But such by-products, while most welcome luxuries, are not essential. Rational mechanics, like every other distinct science, has its own aims, its own standards, its own independent problems.

There is a widespread prejudice to the effect that in physics the basic equations governing physical phenomena are established, and then it is the duty of the mathematician, or the applied mathematician, to solve them. When this view is imposed upon mechanics, we are driven to conclude that the basic equations of mechanics are in the textbooks on classical physics, and all that is left is for analysts to solve them, for engineers to apply them. This leaves no place at all for rational mechanics, since rational mechanics rarely concerns itself with theorems of existence and uniqueness or with calculation of numerical answers. Since this is the first time I address this audience, I cannot be sure that all the physicists have stayed away, and thus I must tread carefully. While the simple division of responsibilities between mathematician and physicist stated above may be ideal, it is unreal. Surely we may admit that biologists study biology because they prefer it to physics; thus that physicists nowadays study nuclear physics rather than classical mechanics, need not imply that classical mechanics is a dead field but may have a more subjective explanation which is less unflattering to those persons who do cultivate mechanics. In simple fact, mathematicians studying mechanics today have rejected older views and have formulated the subject afresh; a part of their work has included discovery of new basic equations defining and explaining new classes of physical phenomena. Second, we often hear complaints from the physicists that mathematicians study the wrong problems, neglecting those the physicists have set and instead pursuing flights of their own imagination. Mathematicians readily admit this charge. If the mathematicians have failed so miserably in their half of the ideal bargain, it would be unjust to expect the other end to be supported in full. Third, the historical fact is otherwise: The great achievements in rational mechanics two centuries ago, now imbedded in our instruction as "classical", were wrought by a small group of persons who, calling themselves "geometers", pursued the mathematical solutions of their equations as vigorously as the mechanical principles they embody. That mechanics has its experimental side, is obvious, but the creators of the mechanics we learn today in physics courses resolutely and most openly refused to consider mechanics an experimental science. Had our forebears approached mechanics in the style and spirit nowadays often recommended as the scientific method, what today we call "classical mechanics" would have lain undiscovered.

Why then is rational mechanics so unfamiliar today? The word is not new. From NEWTON's generation until our grandfathers' time it was in fairly common use. The great American physicist GIBBS[5] described his book on statistical mechanics as a contribution to rational mechanics. In several countries of Europe, where academic tradition was fixed earlier, each university has its chair of rational mechanics. The neglect of rational mechanics must be laid to fashion. Especially in this country, about fifty years ago the main course of physics turned to the structure of matter; of mathematics, to abstract collections. Since matter is made up of many small particles, it is natural to expect that understanding of the physics of these particles will lead to understanding of the behavior of ordinary materials. Since the collections of ordinary experience are very special cases of the abstract collections encountered in the mathematics of the last fifty years, it is natural to expect that the new mathematics will explain ordinary phenomena. As is now well known, both these expectations are ideal, not real. The physics and mathematics of the last fifty years have led to most absorbing developments *of their own*, but they have not brought us appreciably nearer to the solutions of the basic problems of mechanics; for example, to the principles governing failure of metals under load or to the solutions of the equations which describe this phenomenon. It would be wrong to condemn the last half century for failure to solve the problems it did not really attempt. But the basic problems of mechanics remain and are being attacked again. It is some of the brilliant successes of the past decade that I wish to describe to you. I say the past decade: The list of references in a modern paper in mechanics is likely to begin with some studies fifty or more years old, then skip to the last six or eight years. With relatively little injustice, I could give this same lecture as a summary of progress in the twentieth century.

I have said that rational mechanics is a branch of mathematics with its own objective, and that this objective is confined neither to the existence theorems typical of the theory of differential equations nor to the calculation of numerical answers for comparison with experiment. Rather, the objective is to *understand* mechanics. Understanding, after all, is the objective of every branch of mathematics. The

[5] J. W. GIBBS, *Elementary Principles in Statistical Mechanics*, 1902, reprinted in Volume 2 of GIBBS' *Collected Papers.*

measure of understanding, in any field, is partly aesthetic, and it is difficult to explain in general terms what a mathematician means when he says he understands or does not understand a given subject.

2. A Classical Example: The Vorticity Theorems of Helmholtz, Kelvin, and Ertel

After these preliminaries, it is best to enter into examples. Let us begin with an old example whose value and quality is uncontestable, because from such you will most readily see the kind of result that is prized. No one example can illustrate every possible excellence in rational mechanics, but I will select a single classic specimen before passing on to the most recent work.

In a fluid motion, each small part of the fluid may be considered as a body in motion. Such a body may or may not be spinning. If it is spinning, it spins about an axis at a definite rate, and thus its spin may be represented by an arrow. The arrow, which points along the axis of spin and has length equal to the angular speed, is the *vorticity* (Fig. 110). In one fluid motion, different particles may have

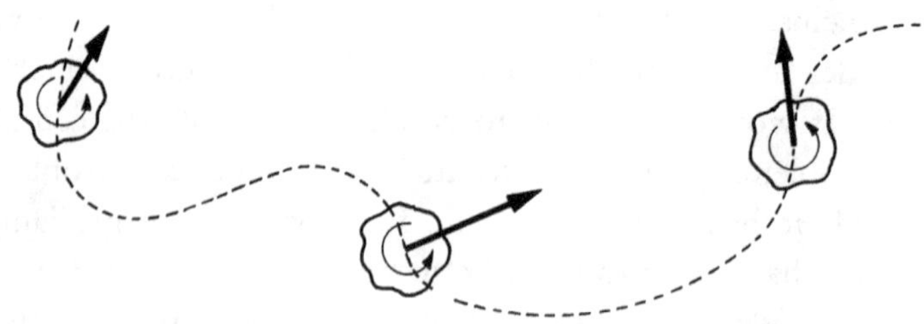

Fig. 110. Spin or "vorticity" of a fluid element

different vorticities, and the vorticity of a given particle may change as the particle moves. On the surface of a fluid, the vorticity may be observed by following a cork upon which a cross is marked. If the arms of the cross do not rotate, the vorticity is zero; if they do rotate, there is vorticity (Fig. 111). Now the theory of fluid motions in which the vorticity is zero is mathematically much easier than that for the general case, when the particles are spinning. While the case of no spin is appropriate to some applications, particularly for waves on water and for aeronautics, on other cases it is not. For example, without vorticity we could have no winds. For a century after the discovery of the fundamental equations for ideal fluids, the nature of spinning motions remained mysterious.

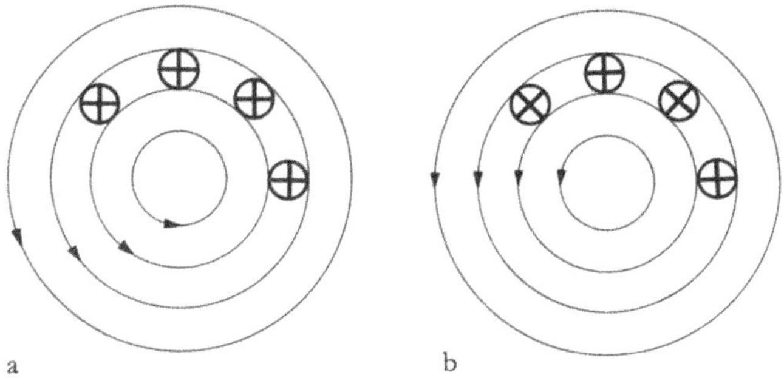

Fig. 111. (a) Vorticity zero (b) Vorticity not zero

In 1858 an entirely new approach was created by HELMHOLTZ [6]. At each point of spinning motion, draw the vorticity vector. Then connect these vectors by curves. Such curves are called *vortex lines* (Fig. 112). If we draw a loop in the fluid, the vortex lines through it will sweep out a tube, called a *vortex tube* (Fig. 113).

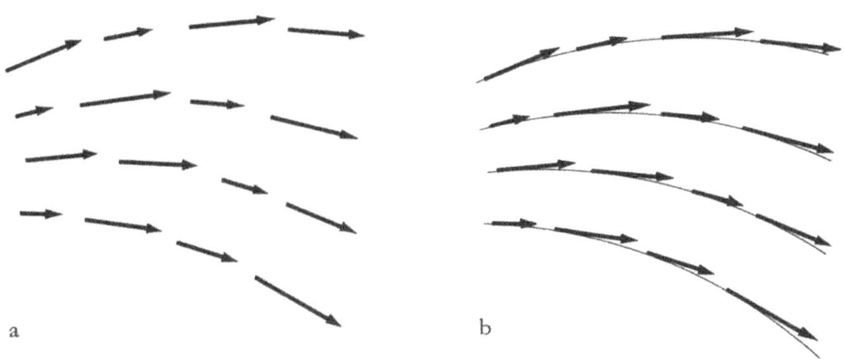

Fig. 112. (a) Vorticity field (b) Vortex lines

Fig. 113. Vortex tube

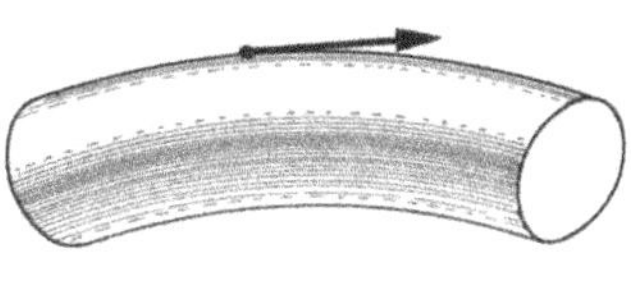

---

[6] H. v. HELMHOLTZ, "Über Integrale der hydrodynamischen Gleichungen, welche den Wirbelbewegungen entsprechen," *J. Reine Angew. Math.* **55**, 25—55 (1858) = *Ges. Abh.* **1**, 101—134. English translation, *Phil. Mag.* (4) **33**, 485—512 (1867).

At a given cross-section, project the vorticity upon the direction normal to the surface, and then add together these projections at each point of the cross-section. The resulting quantity is called the *strength* of the vortex tube at that section (Fig. 114). After introducing these new ideas, HELMHOLTZ proved three great theorems about them:

1. *The strength of each vortex tube is the same at all cross-sections.*

2. *A fluid particle outside a vortex tube never crosses into it, and a fluid particle inside never escapes.*

3. *The strength of the vortex tube remains constant during its motion.*

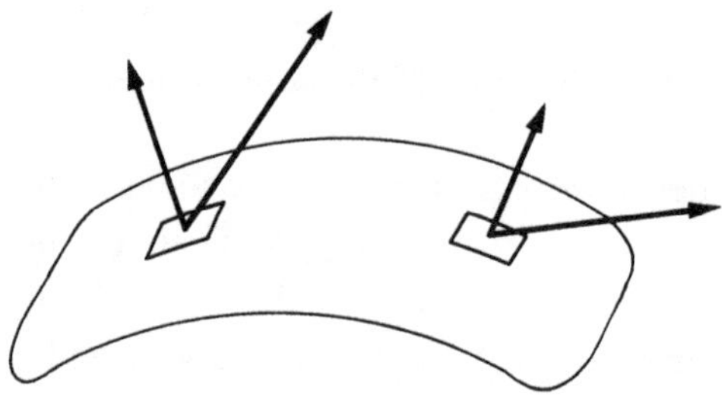

Fig. 114. Strength of a vortex tube

In these theorems, we observe that

a) No new physical laws are proposed. HELMHOLTZ used only the equations of a theory established 100 years earlier.

b) The results are not based on experiment but are mathematical theorems.

c) No boundary-value problem is solved.

d) No numerical predictions for comparison with experiment are obtained.

Nevertheless, these theorems are universally accepted masterpieces of hydrodynamics. In publishing them, HELMHOLTZ remarked that while complete solutions of the problems to which they are related remained possible only in a few of the simplest cases, his theorems made the entire class of these motions "approachable in concept". I think this is what we mean by saying we *understand* a subject mathematically. Understanding is just what HELMHOLTZ's theorems give us. We picture the entire fluid composed of vortex tubes, and as the motion proceeds these tubes may be bent and twisted every which way, yet they continue to separate the fluid into distinct parts. Moreover, each tube preserves its strength.

That is, if the fluid spins faster, the tube must shrink, while if the spin decreases, the tube must swell. Spin once started can never be lost entirely, nor can spin be created in a particle not initially spinning. While HELMHOLTZ's theorems do not solve any particular problem, they have proved treasures in subsequent study of spinning motions. They are taught in every beginning course on the mechanics of fluids. While they do not explain the phenomena of turbulence or atmospheric winds, they furnish tools that are used habitually by every student of those fields. In looking back on the theorems of HELMHOLTZ, we see in them some qualities that are special to them, but also two that are necessary for any great work in rational mechanics:

A. New *concepts*.

B. Strict mathematical *proof*.

HELMHOLTZ's theorems were cast into a new form by Lord KELVIN[7] in 1869. KELVIN introduced a still more important concept. Consider a closed loop of fluid particles, and at each point project the velocity of the fluid onto the tangent to the loop. The sum of all these projections KELVIN called the *circulation* of the loop (Fig. 115). The circulation is a measure of the average rate of turning of the loop. KELVIN proved then that HELMHOLTZ's theorems are equivalent to a single statement:

The circulation of a loop of fluid particles remains the same throughout the motion.

That is, however the fluid loop is turned, pulled, and twisted as the spinning motion proceeds, its circulation is unaltered. KELVIN's theorem plainly exhibits the qualities *A* and *B* above.

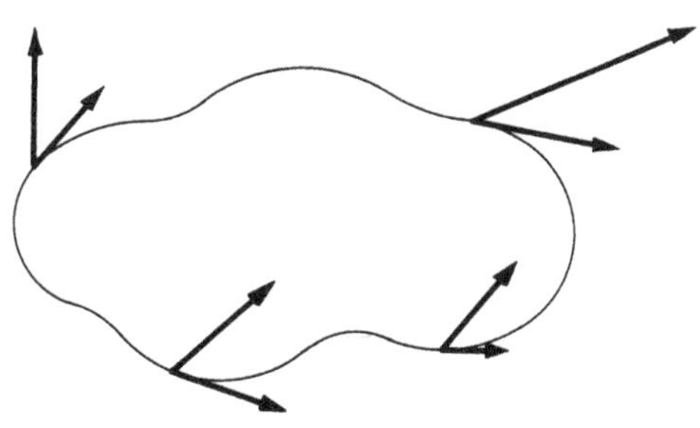

Fig. 115. Circulation of a loop

[7] W. THOMSON, "On vortex motion," *Trans. Roy. Soc. Edinburgh* **25**, 217—260 (1869) = *Collected Papers* **4**, 13—66.

HELMHOLTZ's and KELVIN's theorems refer to a problem 100 years old when they wrote; they are now themselves nearly 100 years old. A beautiful addition to precisely this same subject, spinning motions of ideal fluids, was made in 1950 by the German meteorologist HANS ERTEL[8]. Consider a steady spinning motion whose paths are closed circuits perpendicular to a certain surface (Fig. 116).

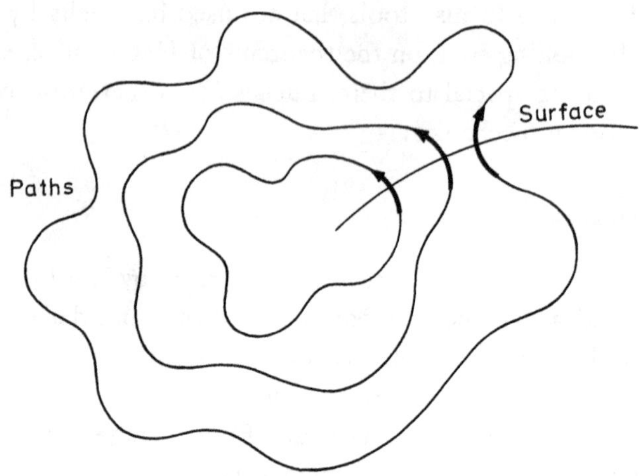

Fig. 116. ERTEL's circulating motions

The character of this motion is assumed not to change in time. That being the case, any one particle on one of the circuits takes as long to go around it as does any other particle on the same circuit. Thus each circuit has a definite time of travel, or *period.* For 200 years it has been known that each path has a definite *energy*; KELVIN's theorem asserts that each path has a definite, constant *circulation.* ERTEL had the insight to perceive that circulation, energy, and period must be related, and for two neighboring circuits he proved that

$$\text{Period} = \frac{\text{Difference of circulations}}{\text{Difference of energies}}.$$

While the theorems of HELMHOLTZ and KELVIN provide a geometrical picture, ERTEL's theorem gives us the first insight into the time a fluid requires to execute a spinning motion.

[8] H. ERTEL, "Ein Theorem über asynchron-periodische Wirbelbewegungen kompressibler Flüssigkeiten," *Misc. Acad. Berol.* **1**, 62—68 (1950).

3. Reissner's Variational Principle in Linear Elasticity

Next to fluid mechanics, the oldest branch of the mechanics of materials is elasticity, the science of the deformation of bodies under load. The basic equations for slight changes of shape have been established for 125 years. For about seventy-five years we have known two alternative ways of stating these fundamental laws of this theory in terms of economy:

1. Among all changes of shape consistent with given displacement on the boundary, that which occurs according to linear elasticity yields the least stored energy.

2. Among all statically possible interior forces consistent with given forces applied on the boundary, those which occur according to linear elasticity yield the least complementary energy.

In the second theorem, "complementary energy" means stored energy less the work done by the surface loads in producing the surface displacements. Both these statements are what are called *variational principles*. In comparing different conceivable states of a deformed body, they assert that the state actually occurring is such as to give to a certain quantity the least possible value. Variational principles are prized for three reasons: (1) they characterize an entire theory in terms of a single simple concept, (2) they are equally valid for all methods of description, (3) they may often be used to prove analytical theorems regarding the subject, (4) they may often be used to calculate numerical solutions in special cases.

For the linear elasticity theory, the two classical variational formulations are different. One refers to displacements, the other to interior forces. If, as often happens, on a part of the boundary the displacement is given, while on another part the loads are given, neither principle can be applied (Fig. 117). For each

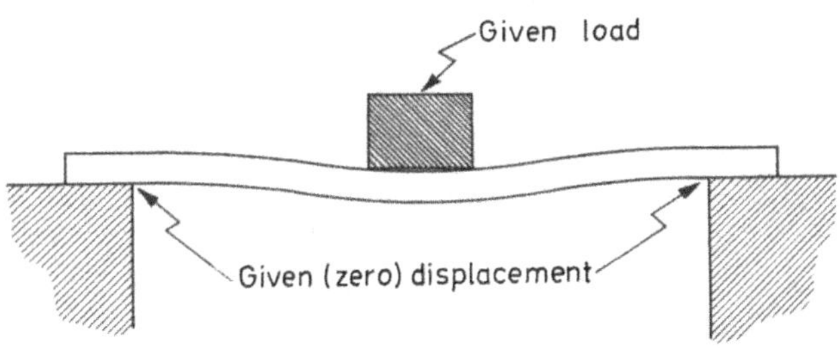

Fig. 117: Boundary of an elastic beam

principle, one half of the equations of the theory are regarded as known, while the other half are consequences of the principle; in the two principles, the two halves of the basic equations play interchanged roles. This is unfortunate, since a variational principle should express the situation in its entirety.

In 1950 ERIC REISSNER[9] of Cambridge, Massachusetts, established a new variational principle in which nothing is presumed. On part of the boundary the loads may be given; on other parts, the displacement; on still other parts, a portion of the load and a portion of the displacement. In the comparison of different possible states, all six measures of internal force and all six measures of deformation are varied. REISSNER defined what might be called a new type of stored energy, depending on all twelve of these quantities. He proved that giving this new energy the smallest value consistent with the given loads and displacements on the boundary is equivalent to satisfying the entire set of equations of linear elasticity theory. Thus a fully general variational expression, including as special cases both the old ones, has been proved.

4. "St. Venant's Principle" according to v. Mises and Sternberg

Just over a century ago ST. VENANT[10] created an ingenious method for finding the interior forces and deformations when a bar is twisted. In this method the total torque twisting the bar may be what you will, but the detailed distribution of load on the ends of the bar is determined by the method and cannot be assigned. Thus the method does not yield a fully general solution of the elastic problem.

---

[9] E. REISSNER, "On a variational theorem in elasticity," *J. Math. Phys.* **29**, 90—95 (1950); "On a variational theorem for finite elastic deformations," *J. Math. Phys.* **32**, 129—135 (1953).
Note added for the reprinting. In fact, the variational principle seems to be due to M. BORN, who gave it for two-dimensional problems, with boundary conditions neglected, in § IV of the Anhang to his Habilitationsschrift, *Untersuchungen über die Stabilität der elastischen Linie in Ebene und Raum, unter verschiedenen Grenzbedingungen*, Göttingen, 1906. In three dimensions, again without boundary conditions, it was given by E. HELLINGER in § 7e of his article, "Die allgemeinen Ansätze der Mechanik der Kontinua," *Enz. Math. Wiss.* $\mathbf{4}_4$, 602—694 (1914). The new contribution of REISSNER lies in the formulation of a boundary integral such as to yield appropriate boundary conditions for all the standard problems of linear elasticity, and in applications to cases.

[10] A. J. C. B. DE ST. VENANT, "Mémoire sur la torsion des prismes ...," *Mém. Divers Savants Acad. Sci. Paris* **14**, 233—560 (1855); also issued separately, "*De la Torsion des Prismes ...*," Paris, 1855.

St. Venant reasoned that this should make no difference. When a bar is twisted in an experiment, it is usually impossible to ascertain the distribution of load. All that is known, usually, is the overall torque (Fig. 118). Near the end itself,

Fig. 118. Equivalent loads twisting a bar

different loadings may produce considerably different effects, but far down the bar no difference is perceptible, provided the total torque be the same. Thirty years later Boussinesq[11] expressed St. Venant's suggestion much as follows: *The difference between the effects produced by two different but equivalent loads applied in a given part of the body becomes very small at great distances from that part.* Two loads are "equivalent" if they would have exactly the same effect on the body if it were rigid. The notion just expressed has been generally accepted and is called *St. Venant's principle.*

In the attitude toward St. Venant's principle we may illustrate a difference between engineering or applied mechanics and rational mechanics. St. Venant's principle can be subjected to partial and indirect test by experiment. It has been so tested and found good. Therefore it is used with confidence by engineers. Every time a bridge or ship is designed, St. Venant's principle is applied repeatedly, if often unconsciously. In rational mechanics, however, elasticity is a mathematical theory, whose basic equations are fixed once and for all. If St. Venant's principle is true, it should be *proved* as a mathematical theorem. Indeed, there have been many unsuccessful attempts to prove it.

If proof of St. Venant's principle were a problem in pure analysis, I should not mention it here. Like other real problems of rational mechanics, it required a preliminary searching of concepts. My statement of it above is rather vague, and several different statements have been proposed. All these were reviewed in 1945 by the late Richard v. Mises[12] of Cambridge, Massachusetts, and he showed all

[11] J. Boussinesq, *Application des Potentiels à l'Étude de l'Équilibre et du Mouvement des Solides Élastiques*, Paris, Gauthier-Villars, 1885.

[12] R. v. Mises, "On Saint-Venant's Principle," *Bull. Am. Math. Soc.* **51**, 555—562 (1945).

to be either trivial or false. MISES proposed to consider the case when loads are applied within a small sphere of radius r, which is then allowed to shrink (Fig. 119). At a fixed point within the body, we should try to show that as r approaches zero the effects of two equivalent loads in the sphere are of more nearly the same order of magnitude than are the effects of two non-equivalent loads. By examples of a special kind, MISES showed that this formulation could not be correct, and he proposed a stronger one.

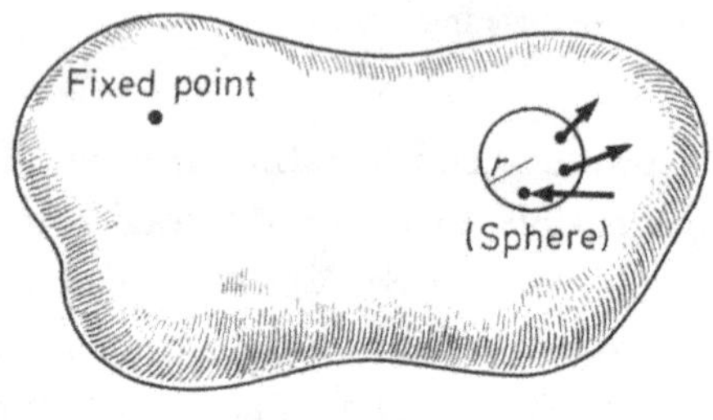

Fig. 119. v. MISES' formulation of ST. VENANT's principle

In 1954 ELI STERNBERG[13] of Chicago put MISES' views into general mathematical form and thus at last produced a definite enunciation, capable of proof or disproof. Notice that what was involved in the first 99 years of this problem's history was search of the *concepts* preliminary to use of the analytical tools usually associated with mathematics. STERNBERG proceeded to construct a strict proof, and he proved ST. VENANT's principle *false*[14]. However, he was able to show that additional restrictions, similar to those suggested by MISES and realized in many cases of application, render it true. For example, if the loads are parallel to a fixed direction, and if the two loadings remain equivalent when rotated, ST. VENANT's principle is true.

---

[13] E. STERNBERG, "On Saint-Venant's Principle," *Quart. Appl. Math.* **11**, 393—402 (1954).

[14] *Note added for the reprinting.* Since this lecture was delivered, I have come to doubt the relevance of v. MISES' formulation of ST. VENANT's principle. It seems to me that the true principle refers to the body as a whole rather than to local behavior of stresses and strains, and in 1959 I stated what seemed to me, and still does seem, a possible precise form of the ideas expressed vaguely by ST. VENANT himself. See my article, "The rational mechanics of materials — past, present, future," *Appl. Mech. Rev.* **12**, 75—80 (1959), reprinted with corrections and additions in *Applied Mechanics Surveys*, Spartan Books, 1966. Since then, an important literature on precise statements related to ST. VENANT's principle has accumulated. The major paper is

5. Large Elastic Deformations

While the foregoing researches fall within the theory of small elastic deformations, that theory itself is insufficient to describe the behavior of such materials as rubber, which may be stretched severely yet springs back to its former shape. Some seventy or eighty years ago, a mathematically proper theory for large elastic deformations was formulated. However, although this theory is an old one, its status is different from that of the only a little older theory of small deformation. University courses in the theory of small deformations are standard; thousands of papers concerning it have been published; for a century there have been text-books and even reliable treatises upon it. The theory of large deformations, on the other hand, has been known by only a few; few have been the papers concerning it, and from 1900 to 1948 general knowledge of it declined to the point that many specialists in mechanics were partially unaware of its existence; and it was not until 1948 that the first book concerning it was published, in Russian.

An exception to the foregoing statements is provided by the work of ANTONIO SIGNORINI[15] in Rome. Of all the persons I shall mention tonight, SIGNORINI is the only one whose position is "Professor of Rational Mechanics". From the 1930's until the present time, SIGNORINI and his pupils have sustained classical knowledge and have made important additions. First I mention a beautiful and simple theory of mean values noticed by SIGNORINI in 1933 and later extended. For any kind of material, whether elastic or not, this theory enables us to estimate the *average* internal forces if we know the loads applied. In elasticity itself, the approach of SIGNORINI has been to conjecture the particular manner in which the stored energy depends upon the deformation. Adopting one or another form as a hypothesis, he explores the properties of such a material and checks them against experimental behavior of real materials. SIGNORINI has obtained some success with certain particular theories. While superficially this view may seem similar to

---

that of R. A. TOUPIN, "Saint-Venant's Principle," *Arch. Rational Mech. Anal.* **18**, 83—96 (1965). TOUPIN's result is of integral form: The energy stored in the part of a cylinder at a certain distance from an end loaded by equilibrated forces is bounded in terms of the smallest proper frequency of free vibration of a section of the cylinder.

[15] The principal results in SIGNORINI's numerous papers are included and related to older and newer work in the treatises by TRUESDELL & TOUPIN, "The Classical Field Theories," *Handbuch der Physik* III/1, 1960, and by TRUESDELL & NOLL, "The Non-linear Field Theories of Mechanics," *Handbuch der Physik* III/3, 1965. See especially §§ 220—222 of the former and § 94 of the latter.

many of the semi-empirical engineering studies on large deformations, in fact SIGNORINI's work is different in character and very precise. In my opinion, later researches show this approach to be inadequate, but there are by-products for which no counterpart is yet known in the more general work to be described presently. These by-products are restrictions on the manner in which the various measures of elasticity of a body may depend upon its temperature. Naturally such restrictions arise in part from the laws of thermodynamics, which are included in the general system of mechanics. In particular, restrictions upon the internal energy are shown to be consequences of the form of the stored elastic energy. Of the numerous results in this domain I state only one, which is of a more general character than the rest: In 1955 PIERO BORDONI[16] of Pisa showed that for any kind of elastic solid, the specific heats and compressibilities are related to each other in precisely the same way as for a perfect gas.

A basically new attack upon large elastic deformations was initiated in 1948 by RONALD RIVLIN[17], then in London. For bodies whose volume cannot change, no matter how they are deformed, he sought exact solutions for simple but important problems. His fundamental departure was to leave the form of the stored energy function *entirely unrestricted.* The equations of the theory are so complicated that an approach of such generality would seem hopeless. In fact, only the idea had been wanting; the calculations turned out to be easy. RIVLIN found the fully general solutions for twisting a circular rod (Fig. 120), for bending a rectangular block (Fig. 121), and for several other cases. These solutions were compared with experimental measurements on rubber, and in this way the form of the stored-energy function for rubber was determined. The measurements showed that none of the forms which had been guessed at various times by various persons were physically correct. In particular, forms suggested by the common argument that the effects of certain "small" quantities may be neglected

---

[16] P. G. BORDONI, "Sopra le trasformazioni termoelastiche finite di certi solidi omogenei ed isotropi," *Rend. mat. e delle sue Applic.* (Roma) (5) **12**, 237-266 (1953). See § 5.
Note added for the reprinting. The thermodynamic basis of this work does not now seem clear to me.

[17] RIVLIN's work is presented in all detail and related to other work on elasticity in the second treatise cited in Footnote 15. Perhaps the most impressive of his numerous papers on elasticity are "Large elastic deformations of isotropic materials," parts IV and V, *Phil. Trans. R. Soc. London* A **241**, 379—397 (1948), and *Proc. R. Soc. London* A **195**, 463—473 (1949). Both these are reprinted in *Problems of Non-Linear Elasticity*, Gordon & Breach, 1965.

were found inadequate. By substituting the experimental form in other particular solutions, RIVLIN was able to predict the behavior of rubber in cases when it was stretched to twice and even three times its initial length, with an accuracy of a few percent. This is a remarkable achievement. Here I must add that "small"

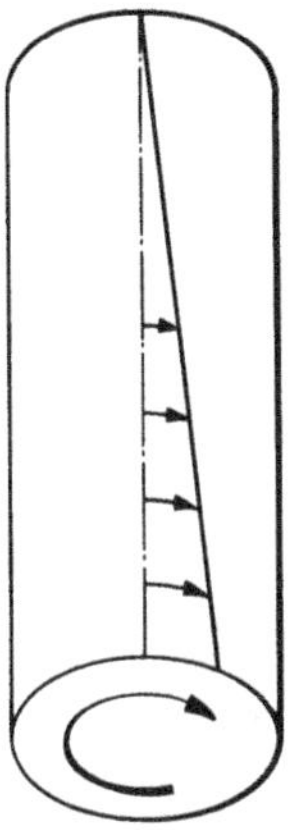

Fig. 120. A twisted circular rod

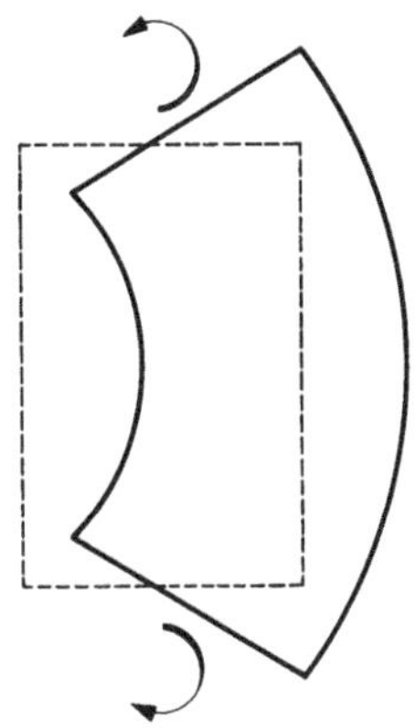

Fig. 121. A bent rectangular block

in the old theory of small deformations means "almost invisibly small": A change in length of one percent is usually too great for adequate description by the theory in books on elasticity. Second, it had been known for a century that rubber does not follow the predictions of the theory of small elastic deformations, and many researches of a physical or chemical nature had been devoted to attempts to get a proper theory for rubber from hypotheses regarding its molecular structure. The new researches in elasticity show that the mechanical behavior of rubber is fully and precisely predicted by purely mechanical principles, *providing they are not degraded by so-called "approximations"*. In respect to application to physical phenomena, this was a triumph of the principles of mechanics and of generality. Out of RIVLIN's work has grown a new branch of engineering which might be called "applied mechanics for rubber".

6. Non-linear Viscosity, and Non-linear Theories in General

In 1945 a new pathway was opened by MARCUS REINER[18] of Haifa. Originally he dealt with fluids rather than solids, but it was soon realized that a new general method in the mechanics of non-linear materials was available. Before going further I must explain the terms "linear" and "non-linear". In a linear theory, two causes applied at once produce the same effects as if first one, then the other, were applied singly. For example, if we first shear, then compress a block, in a linear theory the result is the same as if we compress and shear simultaneously (Fig. 122). In a non-linear theory, this need not be so. In fact, in a non-linear theory

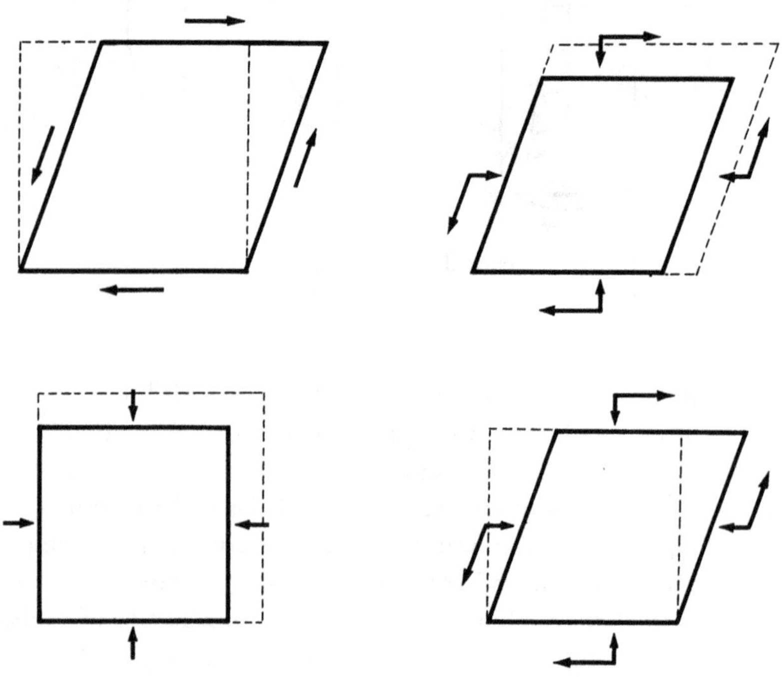

Fig. 122. Compression and shear in a linear theory

[18] M. REINER, "A mathematical theory of dilatancy," *Am. J. Math.* **67**, 350—362 (1945). Reprinted in *Rational Mechanics of Materials*, Gordon & Breach, 1965.

Cf. also W. PRAGER, "Strain hardening under combined stresses," *J. Appl. Phys.* **16**, 837—840 (1945), where the same algebraic theorem is inferred in a different context.

it is generally necessary to apply compressive forces in order for a shear without compression to be possible. The compressive force appears to do nothing whatever, but if it is not applied, the body will expand (Fig. 123). A more familiar

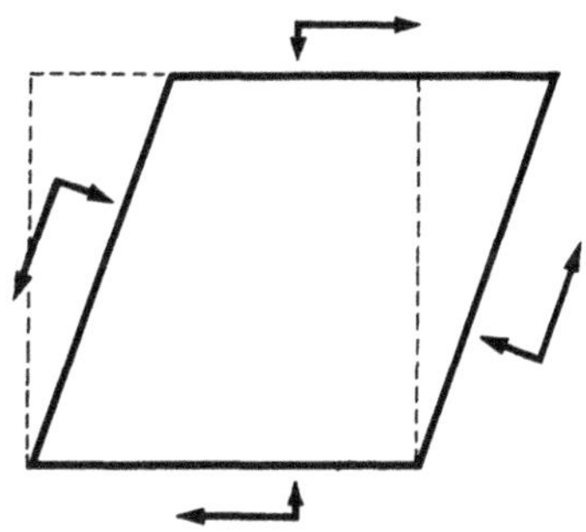

Fig. 123. Non-linear theory: Compressive forces required to effect shear without compression

effect of non-linearity is seen in the gyroscope, which usually swings over in a direction quite different from that in which you push it. Mechanics as a whole is non-linear; the special parts of mechanics which are linear may seem nearer to common sense, but this shows merely that good sense in mechanics is uncommon. We should not be resentful if materials show character instead of docile obedience.

Although mechanics is essentially non-linear, it is little exaggeration to say that for 150 years only linear mechanics and its mathematics were studied. It became standard practice after deriving the equations for a phenomenon to replace them at once by a linear so-called "approximation". It would be wrong to regard this as the original tradition of mechanics. In fact, about 100 years of development was required before it was generally recognized that linearity is synonymous with easiness. The linearizing set in about 1780 but did not gain undisputed mastery until toward the end of the last century, under the influence of RAYLEIGH. Truly, there are many physical situations where a linear theory is adequate. This does not alter the fact that in many other situations linearization is unjustified; and both the mathematics and the mechanical principles associated with non-linear phenomena are much more interesting.

While a century of linearization may have largely exhausted its capacities for explaining physical phenomena and have allowed its mathematics to be explored almost to the full, it has fostered a certain rigidity of approach which might be called "linear thinking". This has made a return to non-linear problems more difficult in fact than perhaps a few years hence it may appear to have been. In particular, it has engendered an emotion somewhere between apathy and fear in the reception accorded to non-linear work by the great majority of researchers, who today remain absorbed in conventional linear problems.

The work of Reiner in 1945 and similar work of Rivlin[19] in 1948 put into our hands a new tool for exploration of non-linear phenomena in materials. Their main result is a new theorem[20] in algebra, having broad usefulness in mechanics. It is applicable to any material, whether fluid or solid, so long as the internal forces arise solely in response to a single measure of deformation. The example above concerning simple shear is an entirely typical consequence of this new approach. Another, at bottom the same, is: *When an elastic rod is twisted a little, it will elongate in proportion to the square of the angle of twist.* Another application of the same principle is: *When a fluid is rotated about a fixed cylinder, it will climb up the cylinder, the force required to prevent such a rise being proportional to the square of the rate of rotation* (Fig. 124). The former effect had long been known experimentally. While the latter had been observed specifically only at about the same time, it was quickly recognized as familiar. In the paint industry, for example, rotary stirrers were found to have little effect because the paint agglomerated upon them. Various physical and chemical explanations were given for this effect, but the theory of non-linear fluids accounts for it easily and naturally on purely mechanical grounds. Rivlin calculated exact solutions for the configurations occurring in viscometers, and when viscometers which could measure the new effects had been constructed, experimental agreement was found.

It would be misleading to emphasize unduly the success of the non-linear theory in predicting these phenomena. To form a particular theory which will agree

[19] R. S. Rivlin, "The hydrodynamics of Non-Newtonian fluids, I," *Proc. R. Soc. London* **193**, 260—281 (1948); reprinted in *Rational Mechanics of Materials*, Gordon & Breach, 1965.

[20] *Note added for the reprinting.* As a theorem concerning polynomial functions, the result was in fact old, though little known. Rivlin's argument, unlike Reiner's, suggests a theorem on *general* functions, and such a theorem was proved by him and Ericksen in 1955; a simpler proof, due to Serrin, is given in § 12 of the second treatise cited in Footnote 15.

Fig. 124. Rotation of fluid in a cup, as reported schematically by Weissenberg in 1947 (the "special" fluid includes linear ones)

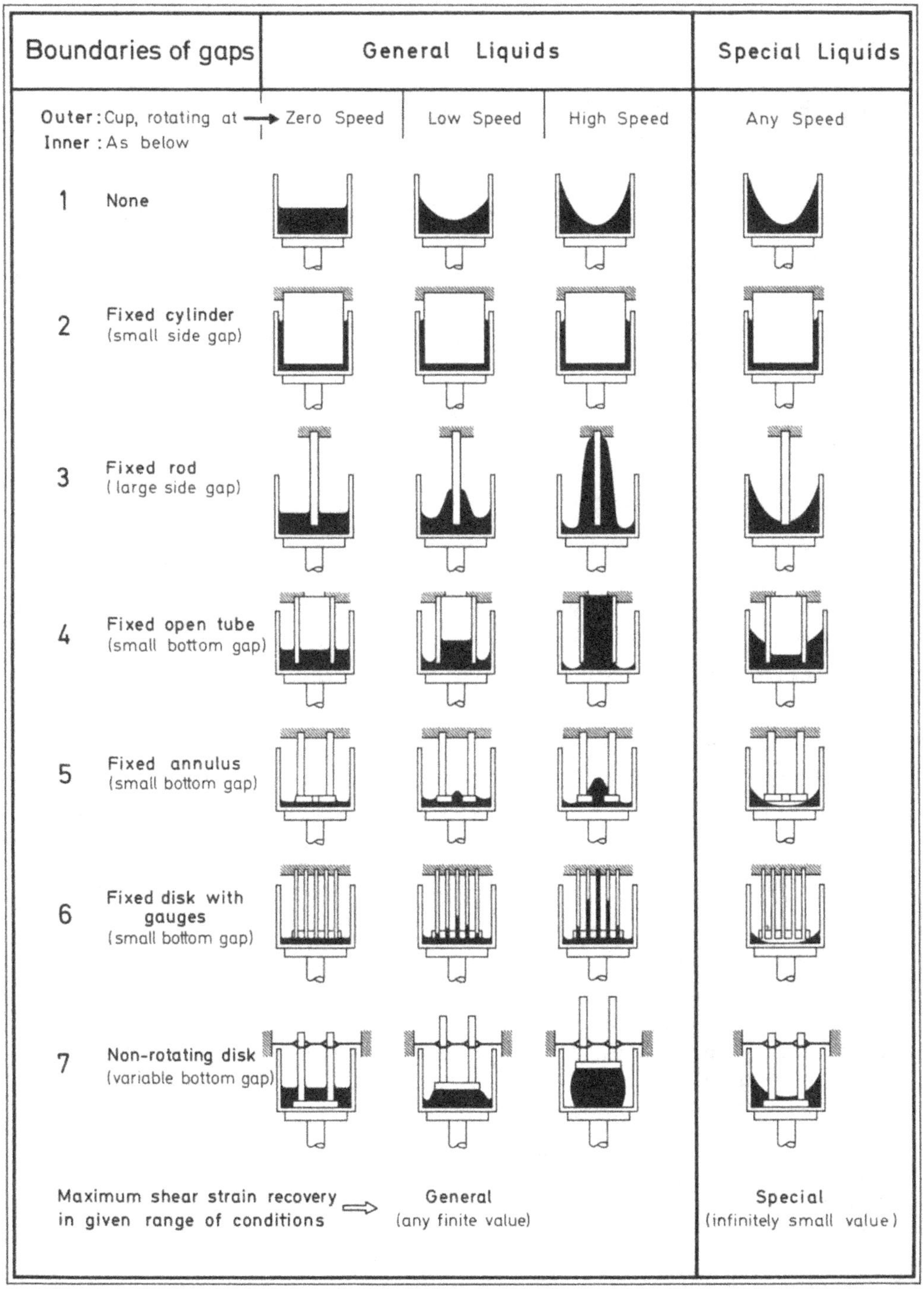
Boundaries of gaps
General Liquids
Special Liquids
Outer: Cup, rotating at
Inner: As below
Zero Speed
Low Speed
High Speed
Any Speed
1 None
2 Fixed cylinder (small side gap)
3 Fixed rod (large side gap)
4 Fixed open tube (small bottom gap)
5 Fixed annulus (small bottom gap)
6 Fixed disk with gauges (small bottom gap)
7 Non-rotating disk (variable bottom gap)
Maximum shear strain recovery in given range of conditions
General (any finite value)
Special (infinitely small value)

with any particular measurement is not nearly so difficult as might seem. Indeed, *special* theories of non-linear elasticity and non-linear fluids were constructed in some numbers during the 1930's or 1940's, and it is only the rule that such researches always include experimental data fully confirming the theory. What is new in the work I was just describing is its *independence* from experiment. Rather, the new approach gave us a unified view, a grasp upon a whole class of non-linear theories. The major result established, in respect to experimental application, is that virtually *any* reasonable non-linear theory will predict these effects, so that their experimental occurrence, while indeed showing non-linearity in real materials, does not confirm any of the numerous special theories.

You have noticed that our tone has changed. While the advances I first discussed concerned specific problems in established theories, now we are surveying the discovery of *new theories* of mechanical behavior. This, the most absorbing aspect of modern mechanics, dominates the field today.

While I have praised generality, we must have a care of its dangers. It is easy to think of a material which instead of using up work when it is deformed, gives it out. Such a material, indeed, we can think of, but it is not interesting in mechanics. This and more subtle misbehaviors we wish to exclude in the theories we create. In 1948 I showed that without surrendering the inclusiveness of the work of REINER and RIVLIN we could classify non-linear materials by dimensional analysis[21]. The results enable us to discard at one stroke many materials that might otherwise seem reasonable. In regard to elasticity, in 1952 I proposed a more specific requirement[22]: *When a body has been deformed, additional deformation requires additional stored energy.* This is not so trivial as it sounds, since what is meant by "additional deformation" is not obvious. This principle turned out to imply restrictions on the stored-energy function. In 1954 M. BAKER & JERALD ERICK-

---

[21] *Note added for the reprinting.* While this work of mine is summarized in §§ 119 A and 125 of the second treatise cited in Footnote 15, I no longer consider the results gotten by dimensional analysis to be anything but trivial. My early papers did illustrate the far-reaching effects of isotropy in cutting down what might seem superfluous generality, but here, too, the idea was not really new, and the methods I then used were neither rigorous nor general.

[22] *Note added for the reprinting.* The whole matter of *a priori* inequalities in elasticity is elaborate and unsettled, though much has been found out about it. To the names mentioned in the text should be added HADAMARD, COLEMAN, and NOLL. All results known up to 1965 are presented in §§ 51—53, 71, 87, 90, and *passim* in the second treatise cited in Footnote 15.

SEN of Washington replaced this idea by a more general one applicable to fluids as well as to solids. In 1955 I pointed out that requirements of this type are closely connected with wave propagation: A body so unreasonable as to refuse passage to waves may also do other and more obviously objectionable things. At the same time ERICKSEN & RICHARD TOUPIN of Washington established a connection between these requirements and those of uniqueness. According to one of their results, a deformed body properly receptive toward waves is also unwilling to deform further in more than one way in response to given small additional displacements upon its boundaries. This class of problems, whose objective is to distinguish liberty from license, is still under study.

7. Experiment and Theory

Here I pause to answer a heartfelt but unspoken question from the floor: Why don't we measure all these things? First, measurement in mechanics is not so easy as it might seem. Despite objections from philosophers of science, mechanics has always been expressed largely in terms of quantities themselves usually not measurable. Had there been operationalists alive in 1750, the design of a bridge in 1956 would proceed more awkwardly than it does in any engineering office today. Second, we have had more than a century of experimentation, often on a large scale and at great cost, regarding the mechanical behavior of materials. Had experiment settled these matters, I should not be talking about them here. Experimental mechanics is a recognized field which needs no defense; it has its own problems, its own methods, its own results, its successes and its failures. It may, but need not, co-operate with applied mechanics and with rational mechanics; these, in their turn, may, but need not, co-operate with it. Without *experience*, there would be no rational mechanics, but I should mislead you if I claimed that experiment, either now or 200 years ago, had greatly influenced those who study rational mechanics. In this connection experiment, like alcohol, is a stimulant to be taken with caution. To consult the oracle of a fine vintage at decent intervals exhilarates, but excess of the common stock brings stupor.

Students of rational mechanics spend much effort thinking *how materials might possibly behave*. These thoughts have not been fruitless of information on how some materials do behave. Real materials are not naive; neither are they irrational.

8. Principles of Invariance

In their behavior materials differ from women also in another respect: They act in just the same way, no matter who happens to be looking. The response of a material is independent of the observer. Now this statement becomes less trivial than it might seem when we make two further observations: First, the laws of classical mechanics themselves change most noticeably when the same body is regarded by different observers; second, to express the response of body in mathematical language we always employ the reference frame of *some* observer. The problem, then, is to find how to limit our ideal materials to those whose response is properly *invariant* with respect to change of observer.

9. Rate Theories of Material Behavior

For the old, linear theories, this problem is trivial and is solved automatically by using well known principles of tensor analysis. For the non-linear theories I have just discussed, the problem is relatively easy. But when the interior forces depend on more than one measure of deformation, or when the basic laws connect time rates, the problem becomes more difficult. For a very special case of time rates, the problem was set and solved by CAUCHY[23] a century ago; in rather more general but still limited circumstances, by ZAREMBA[24] in 1903. The nature of the general problem and a correct solution of it were first indicated in 1950 by J. G. OLDROYD[25], then working in Maidenhead. A more satisfying treatment was given in 1955 by WALTER NOLL[26], then in Bloomington. NOLL's formulation is in terms of a principle he calls *the isotropy of space*, by which the basic equations of any proposed ideal material can be tested for invariance. Similar methods, some-

---

[23] A.-L. CAUCHY, "Sur l'équilibre et le mouvement intérieur des corps considérés comme des masses continues," *Ex. Math.* **4**, 293—319 (1829) = *Œuvres* (2) **9**, 342—369.

[24] S. ZAREMBA, "Sur une forme perfectionnée de la théorie de la relaxation," *Bull. inter. Acad. Sci. Cracovie* 1903, 594—614.

[25] J. G. OLDROYD, "On the formulation of rheological equations of state," *Proc. R. Soc. London* A **200**, 523—541 (1950). Reprinted in *Rational Mechanics of Materials*, Gordon & Breach, 1965.

[26] W. NOLL, "On the continuity of the solid and fluid states," *J. Rational Mech. Anal.* **4**, 3—81 (1955). Reprinted in *Rational Mechanics of Materials*, Gordon & Breach, 1965.

what different in detail and scope, were introduced in two papers written by RIVLIN jointly with ERICKSEN and with BARBARA COTTER of Providence; also by T. Y. THOMAS of Bloomington[27].

By the use of his principle, NOLL was able to construct the first adequate theory of the *continuity* of the solid and fluid states. In this unified theory appear materials which show the responses both of solids and of fluids, the non-linear theories of pure elasticity and pure fluidity being included as extreme possibilities. The idea of such materials had been put forward a century earlier by MAXWELL[28] and had been explored with partial success by ZAREMBA, but NOLL's work is the first to combine exactness and completeness. NOLL obtained exact solutions for certain special situations and showed that his theory implied breakdown of smooth motion at high speeds. A possible theory of continuous transition from ordinary behavior to ultimate failure or breakage has thus been initiated.

In 1953 I had proposed a more modest but at the same time more definite theory of elastic behavior. My idea was, instead of relating the interior forces to the deformation, to connect the rate of change of the interior forces with the rate of change of deformation. The resulting theory, called *hypo-elasticity*[29], agrees with the old linear theory for small changes of shape but for large changes is entirely different. This new theory has turned out to be fairly general: In 1955 NOLL proved that it includes the classical theory of large elastic strain of isotropic materials, and in 1955—1956 ALBERT GREEN of Newcastle and T. Y. THOMAS proved that, apart from an assumed condition of yield, it contains all the usual theories of plastic flow, provided those theories be corrected in several respects. The reservation "apart from an assumed condition of yield" is important. For decades, theories of plasticity had employed a specially assumed condition to express plastic flow, a condition of semi-empirical nature resulting from a compromise between experiment and theory unrelated to other parts of mechanics.

---

[27] While the methods of these authors are described in the first of the treatises cited in Footnote 15, they have been superseded by the clearer and more general approach due to NOLL and made one of the main bases of the second treatise cited there.

[28] J. C. MAXWELL, "On the dynamical theory of gases," *Phil. Trans. R. Soc. London* A **157**, 49—88 (1867) = *Phil. Mag.* (4) **35**, 129—145, 187—217 (1868) = *Papers* **2**, 26—78.

[29] My own work and that of NOLL, GREEN, and THOMAS on this subject are presented and interrelated with later work in §§ 95—103 of the second treatise cited in Footnote 15, where full references to the sources may be found.

I had long felt that this was unfortunate; that in fact yield is a phenomenon which should be *predicted*, not assumed, by a proper theory; that a mechanical theory should include plastic flow not as a primary condition but as the *result* of previous circumstances leading up to elastic failure. In 1955 I showed that in certain cases hypo-elasticity furnishes a smooth and simple theory of just the kind desired. For certain hypo-elastic bodies in shear, I proved that the force required to effect the shear follows a curve of the type shown in Fig. 125.

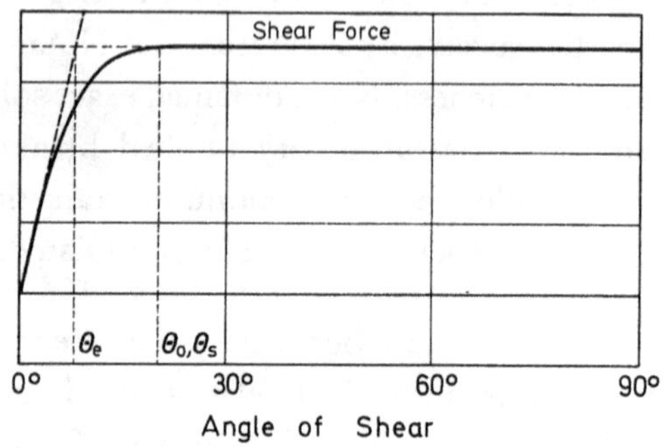

Fig. 125. Primary and secondary yield according to a particular theory of hypo-elasticity

10. Khinchin's Ergodic Theorem[30]

To conclude the list of advances, I wish to return from exploration of new theories and consider progress in an old, established theory: statistical mechanics. In statistical mechanics, apparently continuous matter is represented as a very numerous assembly of points called "molecules", which are in rapid and largely independent motion. The object is to relate quantities of experience, such as force and velocity, to *averages* over the molecular assembly. Averages of this kind are called *phase averages*. The conditions defining a particular motion of the molecules are not known. Rather, a whole class of such motions is laid down as possible, and each motion within this class is assigned a *probability* of occurrence.

---

[30] For a fuller discussion of the ergodic problem, with detailed references to the sources and further elaboration of the ideas of KHINCHIN, see my lectures of 1960, "The ergodic problem in classical statistical mechanics," pp. 21—56 of *Ergodic Theory*, Academic Press, 1962.

In introducing such averages about eighty years ago, MAXWELL remarked the difficulty of convincing oneself that such purely mathematical quantities are appropriate for comparison with physical measurements. In the particular kind of phase average he recommended, every motion admitted has the same energy, and all such motions are equally probable. He suggested that a complicated mechanical system if left to itself would eventually assume nearly every configuration consistent with its assigned energy. If we think of the motion as a curve, this curve would be so tortuous as to occupy, at one time or another, nearly every point on the surface consisting of its admissible configurations (Fig. 126). For such

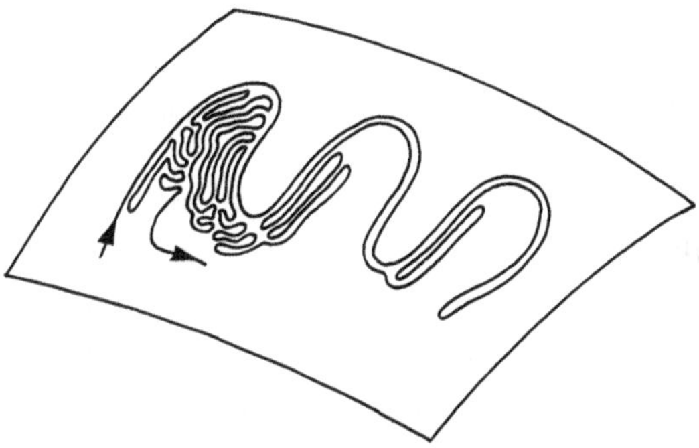

Fig. 126. Schema to suggest ergodic motion

a motion, the average behavior of a given system over a long time would equal the average over all admissible configurations at any one instant:

Time average = Phase average.

Motion of the type MAXWELL suggested was shown later to be impossible, yet the chance remained that his conclusion could be true.

The *ergodic problem*, as the problem of proving this conjecture came to be called, drew the attention of many mathematicians, and in the 1930's it was solved. The conclusion, as far as mechanics is concerned, was negative. For the time averages to be *exactly* equal to the phase averages, for *all quantities* and for *all but utterly improbable* ways of setting the system in motion, was derived a condition which was inappropriate to the mechanical problem and could not reasonably be expected to hold. While an extensive and interesting part of pure mathematics grew from this idea, no progress in mechanics resulted.

In any case, MAXWELL's type of average, which came to be called *micro-canonical*, for reasons of mathematical difficulty usually could not be evaluated. Even before MAXWELL's work, BOLTZMANN had introduced another kind of phase average later called *canonical*. No mechanical principle or idea lies behind this kind of average. Rather, its reason for existence is mathematical simplicity: With canonical averages, a patient person can get answers, and virtually all explicit statistical mechanics as applied to chemistry and physics has rested upon canonical averages from that day to this. Nowadays many scientists are willing to compare canonical averages with measurements, find them valid, and thus accept the theory based upon them as justified directly by experiment. This was not BOLTZMANN's idea at all. BOLTZMANN attempted to *prove* that as the number of molecules in the system becomes very large,

$$\textit{Micro-canonical average} \sim \textit{Canonical average}.$$

In several attempts, BOLTZMANN claimed to give analytical proofs of this conclusion, which is called *Boltzmann's law*. One of these proofs is often reproduced in courses today, but, like the others, it is unsound. Such was the insecurity of the conclusion that in his book on statistical mechanics, GIBBS did not even mention it. In the 1920's DARWIN & FOWLER succeeded in giving a satisfactory but rather difficult proof.

This was the situation when the Russian mathematician A. Y. KHINCHIN took up the subject in 1946. First he gave a new, somewhat simpler, proof of BOLTZMANN's law along lines nearer to the ideas of the classical theory of probability. If this were all, I should not be mentioning it here, but this was an essential preliminary to his solving a much deeper problem, the true ergodic problem of statistical mechanics.

In looking back on the ergodic problem, KHINCHIN realized that the mathematicians of the 1930's had asked too much. First, they had required that the time average be exactly equal to the phase average, while for mechanics it would be enough that the error be small. Second, they had required that equality hold with only utterly improbable exceptions, while it would be enough that the exceptional cases be rather improbable. Third, they had required that the equality hold for every sort of quantity that might be averaged, while in fact only a limited class of quantities are appropriate for averaging in a mechanical system. The third idea gave the real clue. A quantity that has *nearly the same value* for all possible configurations will certainly have a nearly constant average in time for a given system. It is almost trivial to remark that for such a quantity, the phase average and the time average are almost the same. But what KHINCHIN was apparently the first to realize is that BOLTZMANN's law itself implies that the quantities that are to be averaged are indeed nearly constant over most possible configurations. He then went back over his proof of BOLTZMANN's law and sharpened it so as to estimate the average error. With this estimate, he was then able to establish the ergodic principle along the lines just indicated:

Time average $\sim$ *Micro-canonical average* $\sim$ *Canonical average.*

In this sequence, proof of the second stage was shown most elegantly to yield proof of the first stage as a by-product. KHINCHIN's result, in full, may be put as follows: *The cases when the time average differs very much from the canonical average become, as the number of molecules in the system is taken larger and larger, of arbitrarily small probability.* In this statement we perceive another difference from ergodic theory, which is just as applicable to two marbles in a box as it is to a solid body with trillions of molecules: KHINCHIN's result, like BOLTZMANN's law, is a theorem appropriate only to complicated and numerous systems.

KHINCHIN's brilliant success has been little appreciated. On the one hand, most physicists have lost interest in the problem. On the other hand, mathematicians trained in ergodic theory find the case too special to be interesting. In my opinion, KHINCHIN's analysis is a masterpiece in rational mechanics, fully justifying the original view of BOLTZMANN[31]. It shows that every time a chemist calculates a canonical free energy, he is calculating in fact, for all but utterly improbable cases, the mean free energy in a system of fixed total energy left to itself for a long time.

---

[31] *Note added for the reprinting.* While my admiration for KHINCHIN's work has not diminished since 1956, I no longer think it justifies the standpoint of BOLTZMANN, because nothing could. BOLTZMANN's approach presumes tacitly that there is such a thing as *the* probability, not just an arbitrary probability. KHINCHIN's refinement brings out into the open the sets of events (trajectories) of small probability that must be neglected in order to get a specific result. Nevertheless, the arbitrariness of the probability with respect to which these events are improbable remains. The cultivators of modern probability recognize the challenge and meet it head on by stating that everybody estimates probabilities, *etc.*

11. Relation between Statistical and Continuum Theories

While the foregoing analysis refers to systems in statistical equilibrium, the statistical mechanics of deforming media furnishes even more interesting problems. Until recently the statistical mechanics of fluids, for example, was regarded as different and presumably more accurate than theories representing fluids as continuous. That this was not so, was shown in 1950 by JACK IRVING & JOHN KIRKWOOD[32] of Pasadena. They contributed the basic idea, but their proof was not entirely satisfactory, and their approach not as general as it might have been, and in 1955 a better treatment was given by NOLL[33], then in Berlin. His form of the result will now be summarized. Given any assembly of molecules, be they alike or different from one another, few or numerous, free or subject to any outside forces or forces of interaction, consider a selection of probability according to any admissible rule. Then by appropriate phase averages it is possible to define mean velocity, mean internal force, mean internal energy, mean flow of energy, in such a way that *they satisfy exactly the field equations for continuous materials.* Thus statistical mechanics and field mechanics are united. By considering a fluid or a solid as an assembly of molecules, it is impossible to derive anything in contradiction to the view of matter as continuous. Conversely, there is nothing in the continuum view of matter that is not also accessible to a molecular picture. With this beautiful unification of the two apparently opposing views of matter, I close the list of examples.

12. Apology

What I have presented shows the power and versatility of modern rational mechanics. You may expect me to say now that these achievements are only a small part of the great things that have been done, but that would be misleading. Indeed, in a longer time I would describe the work of T. Y. THOMAS on the paths of dynamical systems and on the stability of shock waves, of RINGLEB,

---

[32] J. IRVING & J. G. KIRKWOOD, "The statistical mechanical theory of transport processes, IV, The equations of hydrodynamics," *J. Chem. Phys.* **18**, 817—829 (1950).

[33] W. NOLL, "Die Herleitung der Grundgleichungen der Thermomechanik aus der statistischen Mechanik," *J. Rational Mech. Anal.* **4**, 627—646 (1955).

Tollmien, and Friedrichs on trans-sonic gas flows, of Neményi & Prim on rotational gas flows, of Hicks, Prim, and Ericksen on the stream-lines of gas flows, of Lavrentiev, Gilbarg, and Serrin on water flows past cavities, of Einstein, Bose, and Hlavatý on the new unified field theories, of E. Hopf on turbulence, of Toupin on elastic dielectrics, some further studies of vorticity, and some other things. However, the list would not be very long.

Indeed, I must conclude with an apology for rational mechanics. It is not likely to become a popular field. No prizes are awarded for rational mechanics, and it would make a poor showing in a poll of the public or of scientists in general. The monetary cost of a century of rational mechanics will not equal what is spent this year on computing machines. The work I have described was done slowly, by individuals working alone or with a single other of like tastes. The great teams that produce bombs and vaccines would not have multiplied or deepened the output here. In an age and country where numbers, cost, and statistics count, rational mechanics will never gain much notice. In itself, notice is not what we need, but nowadays he who is not noticed is unlikely to survive. In rational mechanics the financial need is on so small a scale that often it goes entirely unrecognized, in favor of more costly and more glittering projects. Whether in universities or outside them, several of the persons whom I have named here tonight lack a proper library, secretarial help, and even adequate stationery, not to mention a reasonable allowance of time for work free of teaching or other community responsibility. There is no society for rational mechanics, nor are its individualist students likely ever to be numerous enough to afford one, cooperative enough to establish one. In the National Academy of Sciences, for at least twenty-five years no member has been elected for achievement in rational mechanics.

As far as costs and numbers, there has been little change. It would be most incorrect to presume either that the "classical" mechanics we learn today as the first step in physics and engineering was produced by a co-operative effort of organized science, or that in those old fashioned days no large projects existed. Indeed, 200 years ago money was spent on science: on calculating great numerical tables, on extensive experiments for the betterment of mankind, on the design of warships, on boards and committees to organize science. But these efforts did not produce the "classical" mechanics, which was the work of a handful of men, scattered over a continent and a century: willful, uncompromising, quarrelsome, arrogant, and creative.

Thus I have come before you with a purpose. Some years hence, when tonight's initiates are deans and vice-presidents, the persons whose names I have mentioned may well be retired or dead in mind if not in body. I ask those deans and vice-presidents, remembering what I have said tonight, to give a chance to the queer and arrogant young man without a laboratory and without a computing machine who claims he can do research in mechanics just by thinking.

Bibliographical Note

This lecture was delivered at the Sigma Xi initiation at the State University of Iowa, Iowa City, May 16, 1956. A slightly edited version appeared in *Science* **127**, 729—739 (1958). I thank the American Association for the Advancement of Science for its permission to reprint the article here, but I have preferred to use my original text. The footnotes are essentially those printed in *Science*, with some more recent references and comments added. These do not reflect the almost entirely different choice of material I should make, were I to compose an article under this title today.

Index of Names

Index of Subjects

Note: For areas specifically associated with a person, *e.g.* "NEWTON's mechanics", use the Index of Names, especially the pages cited in bold type.

Universitätsdruckerei H. Stürtz AG Würzburg

www.ingramcontent.com/pod-product-compliance
Lightning Source LLC
Chambersburg PA
CBHW081133070625
27828CB00005BA/206
* 9 7 8 3 6 4 2 8 6 6 4 9 4 *